Bayesian Analysis Made Simple

An Excel GUI for WinBUGS

Chapman & Hall/CRC Biostatistics Series

Editor-in-Chief

Shein-Chung Chow, Ph.D.
Professor
Department of Biostatistics and Bioinformatics
Duke University School of Medicine
Durham, North Carolina

Series Editors

Byron Jones
Biometrical Fellow
Statistical Methodology
Integrated Information Sciences
Novartis Pharma AG
Basel, Switzerland

Jen-pei Liu
Professor
Division of Biometry
Department of Agronomy
National Taiwan University
Taipei, Taiwan

Karl E. Peace
Georgia Cancer Coalition
Distinguished Cancer Scholar
Senior Research Scientist and
Professor of Biostatistics
Jiann-Ping Hsu College of Public Health
Georgia Southern University
Statesboro, Georgia

Bruce W. Turnbull
Professor
School of Operations Research
and Industrial Engineering
Cornell University
Ithaca, New York

Chapman & Hall/CRC Biostatistics Series

Frailty Models in Survival Analysis
Andreas Wienke

Generalized Linear Models: A Bayesian Perspective
Dipak K. Dey, Sujit K. Ghosh,
and Bani K. Mallick

Handbook of Regression and Modeling: Applications for the Clinical and Pharmaceutical Industries
Daryl S. Paulson

Measures of Interobserver Agreement and Reliability, Second Edition
Mohamed M. Shoukri

Medical Biostatistics, Second Edition
A. Indrayan

Meta-Analysis in Medicine and Health Policy
Dalene Stangl and Donal A. Berry

Monte Carlo Simulation for the Pharmaceutical Industry: Concepts, Algorithms, and Case Studies
Mark Chang

Multiple Testing Problems in Pharmaceutical Statistics
Alex Dmitrienko, Ajit C. Tamhane,
and Frank Bretz

Sample Size Calculations in Clinical Research, Second Edition
Shein-Chung Chow, Jun Shao
and Hansheng Wang

Statistical Design and Analysis of Stability Studies
Shein-Chung Chow

Statistical Evaluation of Diagnostic Performance: Topics in ROC Analysis
Kelly H. Zou, Aiyi Liu, Andriy Bandos,
Lucila Ohno-Machado, and Howard Rockette

Statistical Methods for Clinical Trials
Mark X. Norleans

Statistics in Drug Research: Methodologies and Recent Developments
Shein-Chung Chow and Jun Shao

Statistics in the Pharmaceutical Industry, Third Edition
Ralph Buncher and Jia-Yeong Tsay

Translational Medicine: Strategies and Statistical Methods
Dennis Cosmatos and Shein-Chung Chow

Chapman & Hall/CRC Biostatistics Series

Bayesian Analysis Made Simple

An Excel GUI for WinBUGS

Phil Woodward

CRC Press
Taylor & Francis Group
Boca Raton   London   New York

CRC Press is an imprint of the
Taylor & Francis Group, an **informa** business

A CHAPMAN & HALL BOOK

CRC Press
Taylor & Francis Group
6000 Broken Sound Parkway NW, Suite 300
Boca Raton, FL 33487-2742

First issued in paperback 2020

Version Date: 2011916

ISBN-13: 978-0-367-57688-2 (pbk)
ISBN-13: 978-1-4398-3954-6 (hbk)

Library of Congress Cataloging-in-Publication Data

Woodward, Phillip.
  Bayesian analysis made simple : an Excel GUI for WinBUGS / Phil Woodward.
    p. cm. -- (Chapman & Hall/CRC biostatistics series ; 45)
  Includes bibliographical references and index.
  ISBN 978-1-4398-3954-6 (hardback)
  1. Bayesian statistical decision theory. 2. Microsoft Excel (Computer file) 3. WinBUGS. I. Title.

QA279.5.W655 2012
519.5'42028553--dc23   2011033774

Visit the Taylor & Francis Web site at
http://www.taylorandfrancis.com

and the CRC Press Web site at
http://www.crcpress.com

Dedicated and in memory of ...

Sydney and Brenda ... my prior

Samuel, Louise, and Benjamin ... my posterior

Susan ... my lovelihood

Contents

Case Studies

(continued)

Case	Subject Matter	Statistical Issues
7.1	Pharmacology biomarker (preclinical laboratory experiment)	Emax model, single and grouped data, MCMC failure, GFI prior elicitation tool, DIC
7.2	Pharmacology biomarker (clinical study)	Emax model, informative prior
7.3	Systematic review of replicated experiments (compound potency)	NLMM: Experiment level summaries, derivation of strongly informative prior from function of model's parameters
8.1	Observational environmental study (pollution)	Variable selection: Normal errors, covariates with mixture prior, plotting activity probabilities
8.2	Controlled laboratory experiment (assay method development)	Variable selection: Normal errors, factors with mixture prior, MCMC convergence checking
8.3	Engineering reliability testing (transmission shaft)	Variable selection: Log-normal errors, censored data, predictive distribution, MCMC failure with Gamma model
9.1	Cross-Over Design Clinical Study (pain treatment)	GLMM: Bernoulli response, cross-over design, carry-over effect
9.2	cf. 5.4	NLMM: Auto-regressive residual correlation, DIC
9.3	cf. 6.2	GLMM: Poisson transition model, DIC
9.4	Observational political study (politician approval ratings)	Time series: Normal errors, first-order auto-regressive process
10.1	cf. 3.1	t-distributed errors, prior for degrees of freedom, mixture model to accommodate outliers, modeling outliers when error distribution is non-normal
10.2	cf. 4.1 and 6.1	GLMM: Binomial-logistic model with mixture to accommodate outliers, t_4-distribution for random effects
10.3	cf. 5.4	NLMM: bivariate t_4-distribution for random coefficients
Section 11.1	cf. 4.3	Changing Weibull linear predictor from the median to the mean
Section 11.2	cf. 6.4	Altering the MCMC initial values
Section 11.3	cf. 6.1	Editing the WinBUGS code directly to estimate additional quantities of interest

Preface

Although the popularity of the Bayesian approach to statistics has been growing rapidly for many years, among those working in business and industry there are still many who think of it as somewhat esoteric, not focused on practical issues, or generally quite difficult to understand. This view may be partly due to the relatively few books that focus primarily on how to apply Bayesian methods to a wide range of common problems. I believe that the essence of the approach is not only much more relevant to the scientific problems that require statistical thinking and methods, but also much easier to understand and explain to the wider scientific community. But being convinced of the benefits of the Bayesian approach is not enough if the person charged with analyzing the data does not have the computing software tools to implement these methods. Although WinBUGS (Lunn et al. 2000) provides sufficient functionality for the vast majority of data analyses that are undertaken, there is still a steep learning curve associated with the programming language that many will not have the time or motivation to overcome. This book describes a graphical user interface (GUI) for WinBUGS, BugsXLA, the purpose of which is to make Bayesian analysis relatively simple. Since I have always been an advocate of Excel as a tool for exploratory graphical analysis of data (somewhat against the anti-Excel feelings in the statistical community generally), I created BugsXLA as an Excel add-in. Other than to calculate some simple summary statistics from the data, Excel is only used as a convenient vehicle to store the data, plus some meta-data used by BugsXLA, as well as a home for the Visual Basic program itself. Excel also has charting features that are perfectly adequate for many of the graphs needed to begin an exploratory analysis; this is another much criticized feature of Excel, unfairly in my opinion (see Walkenbach (2007) for details of how to get the most from Excel charts). Since BugsXLA resides within Excel, anyone who regularly stores, summarizes, or plots their data using Excel should be particularly interested in this book.

More generally, this book is aimed at people who wish to apply Bayesian methods, but either are not keen or do not have the time to create the WinBUGS code and ancillary files for every analysis they undertake. Even for those who would not routinely use Excel, providing they are prepared to import their data into this spreadsheet and are familiar with the Windows environment, this book provides the simplest way to get started in applying Bayesian methods. Experience with fitting statistical models using one of the major statistical packages such as SAS, R, or Genstat would be an advantage but is not essential. Since BugsXLA allows a wide range of model types to be fitted, from simple normal linear models to complex generalized linear mixed models and beyond, it is envisaged that the time and effort expended understanding the case studies in this book will soon be repaid once the program is used on one's own data. BugsXLA will be most useful to those users who want to be able to quickly apply Bayesian methods without being immediately distracted by computing or mathematical issues. In my experience, this is not only those who use statistics purely as a means to understand their own discipline better, but also professional statisticians who do not want to spend a lot of time creating WinBUGS programs and other files for every analysis they wish to undertake.

It is important to be aware that the purpose of this book does not extend to explaining in any detail the statistical theory underpinning the methods discussed. The onus is on the user to ensure their knowledge is sufficient to determine the appropriate analysis method, as well as to interpret the output. Numerous references are provided throughout the text that could be used as part of a self-study program if needed. The book is designed to act both as an instruction manual for implementing a Bayesian analysis via BugsXLA as well as a source reference for the program once the user is familiar with its features. For this reason, Chapter 2 has been kept relatively concise so that it can be used as a quick look-up. It is recommended that on the first read of the book this chapter be skipped, other than to follow the instructions on how to download the program (http://bugsxla.philwoodward.co.uk). Even if the reader is very familiar with the Bayesian approach to fitting normal linear models, it is strongly advised that Case Study 3.1 be worked through first, as this explains the frequently used features of BugsXLA in detail. In order to minimize repetition of simple tasks, knowledge of these basic features is assumed in the rest of the book.

The majority of this book is dedicated to case studies. These are used to explain how to interpret the statistical models and their parameters. The classes of models considered are broad and are applicable to a wide range of disciplines. Although BugsXLA offers a default prior distribution for every parameter, the book explains how to think about informative prior specifications in many cases. Many of the data sets used in the case studies were chosen as they conveniently lead to a discussion of various general issues, either in the application of BugsXLA or in the discussion of the Bayesian approach. It is not implied that the initial analyses and interpretations were inadequate, as it is strongly believed that in virtually all experimental work the best analyses, be they Bayesian or otherwise, are as a result of an intimate collaboration with the subject matter experts at the time. Some of the case studies describe instances in which I was personally involved, and I attempt to broaden the discussion in those cases. However, most data sets have been chosen with a focus on the main purpose of this book, that is, to show how simple it is to apply Bayesian methods using BugsXLA. Also, I have not attempted to show all that I would have done in analyzing each data set, focusing instead on the role that BugsXLA can play, as well as providing some examples of the ease with which Excel can be used to produce useful plots of the output imported back from WinBUGS. Some of the examples were also chosen to illustrate particular difficulties that sometimes arise, with suggestions on how to proceed in order to make useful inferences. Hence, the frequency in these case studies with which WinBUGS crashes, or other peculiarities occur, should not be taken as typical of what happens more generally. Unlike in these deliberately enriched problems, in the vast majority of cases that are encountered in practice the WinBUGS analysis is undertaken without any drama to resolve, other than the usual issues of convergence and adequate MCMC precision. All the data analyzed in the case studies are contained in the Excel spreadsheet, "BugsXLA Book Case Studies.xls", downloaded into the same folder as BugsXLA itself.

The title, *Bayesian Analysis Made Simple*, is not meant to underestimate the difficulties in understanding the philosophical issues, the mathematical theory, nor in the years of hard work needed to become a knowledgeable and experienced designer and analyzer of experiments, that is, a statistician (even if not by formal qualification). The title of the book refers to the relative ease with which BugsXLA allows a Bayesian analysis to be specified, and the way in which the output of the analysis is laid out for interpretation. By providing a Windows-based GUI for WinBUGS, users familiar with the Windows' environment can quickly learn how to fit complex statistical models without knowing how to code, or otherwise use, the WinBUGS program. By having a quick and easy way to implement Bayesian methods, it is hoped that users new to this approach will be motivated to read some of the Bayesian texts referenced in order to understand the theory that underpins the analyses in more detail. This approach is strongly encouraged, since the only way to ensure excellence when applying any statistical method is to have a sound understanding of the assumptions that justify the method. With this knowledge, one can determine the best ways to explore the data, and other sources of information, in order to assess the credibility of the assumptions that lay beneath the analysis undertaken.

Acknowledgments

Like most statisticians, and any others for that matter who have attended courses on statistics, my initial education in the subject was almost completely from a classical perspective. In my final undergraduate year, Bayesian inference was covered in the last three weeks before my finals; I did not attempt the set question on this topic. Hence, I graduated with the vague belief (pun intended) that Bayesian methods were just another set of techniques alongside the others taught. Also, it appeared that they required a very high level of mathematical knowledge to understand, which, once acquired, allowed you either to obtain results that were numerically identical to the classical methods or very quickly led to high-dimensional integrals that could not be computed; this was in the pre-MCMC early 1980s, just after the dinosaurs became extinct. Not surprisingly, like most other statisticians in those days, I was quite content to focus on statistical methods that could be used to fit models that contained more than a couple of parameters. Fortunately, my employer at the time, Rolls-Royce in Derby, allowed me to study part-time toward a research degree at the local college. It was there, under the mentorship of the late John Naylor, that I was introduced to what he, and my other research supervisor Adrian Smith, referred to as the Bayesian paradigm. The beauty and simplicity of this approach to statistical inference was immensely appealing. Over the next few years, I became convinced, mainly through my and my colleagues' attempts to better understand the manufacturing processes at Rolls-Royce through experimentation and observation, that the Bayesian approach to inference essentially described what we were trying to do, that is, update our beliefs about the world around us after observing new data, in light of other information already available to us. Although I left Derby for Kent in 1990, I will be forever indebted to Paul Whetton, my first boss who encouraged me to continue my studies, and John Naylor, who not only introduced me to Bayesian methods, but as an expert statistical computer programmer helped me enormously to develop a coding style that enabled me to create BugsXLA.

The spark that motivated me to develop BugsXLA was a casual conversation with Andy Grieve nearly 10 years ago, a colleague of mine at Pfizer at the time, in which we lamented on the absence of a simple way to specify a Bayesian analysis. "Wouldn't it be great if one could specify a Bayesian analysis using simple model statement commands, similar to the way classical methods are applied?," was the underlying theme of our conversation. About two years after that initial conversation, version one of BugsXLA was completed. During the last seven years additional features were added as I learnt from experience in its use and from feedback provided by numerous users who had downloaded the program from my website (http://bugsxla.philwoodward.co.uk). The current version, v5.0, is discussed in full detail in this book.

The book itself owes its existence to another casual conversation, this time with Byron Jones, another Pfizer colleague, who suggested that there might be enough people interested in BugsXLA to warrant writing a book. Although skeptical at first, I was encouraged by early positive feedback from some anonymous reviewers on a couple of draft chapters, and this persuaded me to complete the project. I must also thank David Grubbs, Marsha Pronin, and the many others at Chapman Hall who have worked to make this book a reality. John Lawson acted as a reviewer and provided valuable feedback on later drafts of the book; John also influenced an earlier version of the program itself following suggestions on additional types of variable selection models that would be useful.

I would also like to thank numerous colleagues who allowed me to base case studies on real projects with which they were involved and helped me to describe the context in sufficient detail to inform the reader without revealing any matters of confidentiality: Ieu Jones, Rich Allan, Debbie Kraus, Sharlene Phillips, David Cooke, Ed Casson, Ros Walley, Nikki Robbas, Jem Gale, and Dave Collins. Debbie Kraus and Alison McLeod both volunteered to read earlier drafts and give feedback. Ros Walley has helped me on more than one occasion, but in particular in our collaboration on the prior for multilevel variable selection factors. Prior to writing the book, feedback from many of those who downloaded BugsXLA has also been a motivation to continue with its development. Øivind Moen provided the screenshots shown in Figure E.1, explaining how he overcame the non-English-language problem discussed in Appendix E. Thank you to all the others who sent me an e-mail, however brief, informing me that you have downloaded BugsXLA. These are mainly from the United States, the United Kingdom, and Australia, but also from Germany, Canada, Finland, India, Iran, New Zealand, Spain, Afghanistan, Brazil, Egypt, Greece, Ireland, Lithuania, Malta, Mexico, Norway, Pakistan, Portugal, Saudi Arabia, and Switzerland. The major application areas of these folk are medicine and other biological sciences, various types of engineering, ecology, epidemiology, economics, and environmental monitoring, with others working in insurance, finance, marketing, teaching, agriculture, politics, fisheries, sociology, and educational research.

A big thank you to my daughter Louise who designed the BugsXLA logo that I wanted to exist for the version of the program discussed in this book. My eldest son, Sam, regularly helps me with technical matters when my computer or the Internet "misbehaves," and also redesigned my Internet page. Together with my youngest son, Ben, my children have helped me to complete the book in a reasonable time frame, if only by occasionally asking, "Have you nearly finished yet?" But the biggest thank you goes to my wife, Sue, who has often wondered where I have been all evening, and many weekends, as I spent the many hours both creating BugsXLA and then more recently writing this book. I do appreciate how lucky I am (and all the cups of tea were much appreciated also). Love, kisses, and hugs to Sue, Sam, Louise, and Ben.

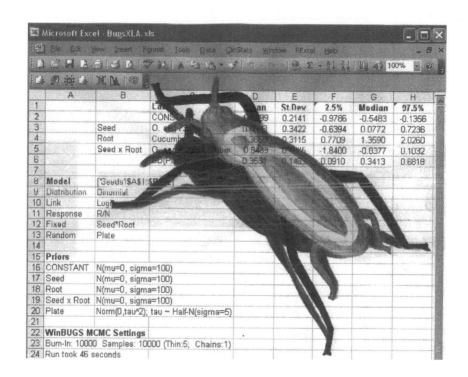

Microsoft Excel - BugsXLA.xls

File Edit View Insert Format Tools Data ClinStats Window RExcel Help

	A	B	C	D	E	F	G	H
1			La...	...an	St.Dev	2.5%	Median	97.5%
2			CONS...	...99	0.2141	-0.9786	-0.5483	-0.1356
3		Seed	O. ...	...	0.3422	-0.6394	0.0772	0.7236
4		Root	Cucumber	...58...	0.3115	0.7709	1.3590	2.0260
5		Seed x Root	O. ...	0.8483	...5	-1.8400	-0.6377	0.1032
6			SD(Pl...	0.3531	0.148...	0.0910	0.3413	0.6818
7								
8	**Model**	['Seeds'!$A$1:$B...						
9	Distribution	Binomial						
10	Link	Logit						
11	Response	R/N						
12	Fixed	Seed*Root						
13	Random	Plate						
14								
15	**Priors**							
16	CONSTANT	N(mu=0, sigma=100)						
17	Seed	N(mu=0, sigma=100)						
18	Root	N(mu=0, sigma=100)						
19	Seed x Root	N(mu=0, sigma=100)						
20	Plate	Norm(0,tau^2); tau ~ Half-N(sigma=5)						
21								
22	**WinBUGS MCMC Settings**							
23	Burn-In: 10000 Samples: 10000 (Thin:5; Chains:1)							
24	Run took 46 seconds							

1

Brief Introduction to Statistics, Bayesian Methods, and WinBUGS

Statistics is the science that provides the most logical approach to understanding the world around us via the interpretation of our observations. Clearly we need hypotheses derived from knowledge or beliefs about the underlying scientific mechanisms that make the world behave in the way that it does. But it is only by making observations that we are able to differentiate between the scientific beliefs that are close enough to the truth to be useful from those that should be rejected or refined. Statistics provides the methods by which we extract information from the observations we make on the world around us, and just as importantly, guides us when we can choose which observations to make. George Box, one of the most influential statisticians of the last 60 years, frequently discussed how statistics could and should be used as a catalyst to learning (see Box (2000) for numerous examples), and argued that such learning occurs either through "passive observation" or "active experimentation" (as explained in Box and Bisgaard (1987)). It could be argued, and it is certainly my belief, that it is in the area of experimental design that statistics has its biggest impact on the advancement of knowledge in the sciences, engineering, or any other discipline based on logical principles. The twentieth century saw a revolution in the development of methods aimed at improving the way we plan, run, and analyze experiments, starting with the pioneering work of Ronald Fisher (1925) at Rothamsted in England between the two world wars. Although my book only discusses methods of analyzing data once they have been obtained, it is essential that statistical thinking and methods be applied from the very beginning of the experimental process, even before the objectives of the study have been finalized. This is true even in those cases where experiments cannot be designed in the manner of a controlled laboratory experiment, since statistical ideas are still relevant when purely observational studies are being planned. It is unfortunate that much of the scientific research currently being undertaken and reported still does not make use of these powerful ideas first published nearly 100 years ago. A good introduction to experimental design, explaining the key principles, is provided by the classic text by Cox (1958). Another classic text that covers the most widely used types of designs in more detail is Cochran and Cox (1992). Numerous books exist that cover design for specific application areas such as Wu and Hamada (2000) or Box et al. (1978) for the more technological sciences, Fleiss (1986) or Senn (2002) for medical trials, and Rosenbaum (2009) for observational studies.

Once the experiment has been appropriately designed and run, or the study properly planned, there remains the task of extracting the information from the data collected. For most of the twentieth century, by far the most dominant approach to statistical inference was that advocated by the hugely influential statistician and geneticist, Ronald Fisher, referred to as the "classical" or "frequentist" approach. Inferences are made in classical statistics via concepts such as the p-value and its close cousin the confidence interval. A variant of this approach, invented by Jerzy Neyman and Egon Pearson to provide a mathematical solution to the problem of making decisions, involves the very formal process of defining "null" and "alternative" hypotheses. In the Neyman–Pearson approach, which has been most ardently adopted in the pharmaceutical industry, the data obtained from the study are used to assess the credibility of the null hypothesis, most typically with the hope that it can be rejected in favor of the alternative. The practical details of the classical approach to inference can be obtained from any introductory statistics text such as Clarke and Cooke (2004), or for a more mathematical introduction refer to any undergraduate text such as Mood et al. (1974). Another inferential approach that, although advocated by many of the great statisticians of the last 50 years or more, has not found much wider popularity is the "likelihood" approach. The likelihood approach is, in some ways, similar to the Bayesian approach and shares with it many of the criticisms of classical statistics. However, likelihoodists reject the routine use of prior distributions to represent one's beliefs regarding unknown quantities, and so have invented methods such as profile likelihood to circumvent their need; see Edwards (1992) for a good introduction to this approach to statistical inference. This leaves us with the inferential approach that is the basis for the methods discussed in this book, what some have referred to as the Bayesian paradigm.

1.1 Bayesian Paradigm

It is not the purpose of this section to provide a detailed explanation of the Bayesian approach to statistical inference, but more to discuss the major differences from other statistical approaches and their practical implications. It is important that anyone planning to use Bayesian methods has a good understanding of the basic principles as provided by an introductory book such as Bolstad (2007) or Lee (2004).

1.1.1 Probability

In order to properly understand the Bayesian approach, as opposed to be simply content to apply its methods, one needs to understand the Bayesian concept of probability. A very accessible brief discussion of the history of probability, which nicely compares the different statistical schools, is provided by Chapter 10 of Raiffa (1968). Although there are some Bayesians that believe probability can be defined in a logical or objective manner, I only discuss the idea of probability as a subjective measure of degree of belief as this is the predominantly held view. A subjective probability is still a number between 0 and 1, with 0 implying an event is impossible and 1 implying it is certain. More typically one is uncertain whether an event will occur or not, so its probability lies somewhere between these two extremes. To the subjective Bayesian, the probability that should be assigned to such an uncertain event is determined by the notion of a "fair bet." To understand this notion one needs to understand how a bet works. The person selling the bet, the "bookmaker," states the odds for or against an event occurring in order to encourage the person making the bet, the "punter," to do so. The odds are a ratio, expressed in terms like 2-to-1 against, which determine how much the punter will win for each unit of money staked. In this case, for each one unit staked the punter will receive two units, plus his stake, from the bookmaker if the event occurs. A fair bet is one in which the odds are set such that someone is indifferent to whether he is acting as the bookmaker or the punter, that is, he does not believe either side has an unfair advantage in the gamble. There is a direct link between odds (O) and probabilities (P):

$$O = \frac{P}{1-P} \quad \text{or} \quad P = \frac{O}{1+O}$$

where O equals the ratio X/Y for odds of X-to-Y in favor, and the ratio Y/X for odds of X-to-Y against an event. Hence, if the 2-to-1 against odds were considered fair, then this would imply that the probability associated with the event was one-third; the punter would win one-third of such bets, gaining two units, and lose two-thirds of such bets, losing one unit, and therefore would expect to neither gain nor lose from such a bet. It is intuitive that different people might determine different probabilities as being fair for the same event depending upon their knowledge of the event being considered. It is for this reason that these are known as subjective or personal probabilities. Most of the controversy in applying Bayesian methods lies in the inherent need to combine such subjective probabilities with the probabilities that emanate from the distribution of the data or via the likelihood. Some argue that these latter probabilities are more objective and hence, should not be treated in the same way as the former. To the Bayesian, since the distribution of the data is conditional on an assumed probability model and its unknown parameters, these probabilities are also subjective since the model is itself a personal choice. Even though the data themselves usually provide information with which to assess the adequacy of the assumed model, there are nearly always alternative models that the data alone cannot distinguish between, and so it is necessary to use personal judgment in choosing between them. Also, as models become more complex, such as when hierarchical models are adopted, the distinction between the assumptions underpinning the likelihood and the prior beliefs become increasingly blurred. Although probability is regarded as a personal belief, this does not mean that it is subjective in the sense of being biased, prejudiced, or unscientific. In fact, there is now a large body of research into methods that are designed to elicit judgments that are as carefully considered, evidence based, honest, and scientific as possible (see O'Hagan et al. (2006) for details). Where the more conventional definitions of probability apply, these should be used to support one's beliefs. So, for example, in the case of expressing the probability that a thrown six-sided die will show a six, unless we have good reason to doubt that the die is unbiased, it would be unreasonable to assign anything but a probability of one-sixth to this event. The advantage of the subjective definition of probability is that if we had other evidence to suggest that the die might be loaded in some manner, perhaps some prior knowledge of the dubious character of the person offering a bet on this event, we may adjust our probability accordingly.

1.1.2 Prior "Belief" Distribution

Although one might be reasonably content expressing belief concerning an observable event, such as whether it will rain tomorrow, in terms of a subjective probability, in order for the Bayesian approach to be applied more generally it is necessary to extend this to the unobservable parameters of a statistical model. To illustrate this concept, consider a very simple study designed to estimate the mean of a population whose individual values are assumed to independently follow the Normal Distribution with a known standard deviation (sd). The statistical model for the individual values has only one unknown parameter, μ, the mean that we wish to estimate. In order to undertake a Bayesian analysis, it is first necessary to specify, in advance of observing the data from the study, a probability distribution that represents one's belief about the value of μ. This "belief distribution" is referred to as the prior distribution for μ since it represents one's knowledge prior to observing the data from the study. The problem of how to justify a prior distribution is probably the most difficult aspect of the Bayesian paradigm. However, although the controversy is far from resolved, there is now much wider acceptance of the legitimacy of including prior distributions in a formal statistical analysis. Chapter 5 of Spiegelhalter et al. (2004) provides an excellent discussion of the various methods for determining priors that have gained widespread acceptance. We suppose that a Normal Distribution with a mean of 10 and sd of 2 is determined as a fair representation of one's prior belief about the value of μ. More formally we can represent the prior distribution by the formula

$$p(\mu) = (8\pi)^{-0.5} \exp\left\{ \frac{-(\mu - 10)^2}{8} \right\}$$

This implies that we believe it is quite unlikely that μ is smaller than about 6 or larger than about 14; strictly speaking we believe there is less than a 5% chance that it is more extreme than these values. It also implies that we believe the most likely value of μ is 10, with about a 50% chance that μ lies within ±1.4 of this "best estimate." It is recommended that whenever a prior distribution is specified that has the potential to be both influential and controversial, one thinks about its implications in various ways in order to assess whether it remains a credible formal representation of one's beliefs. For models that are more complex than the one being considered here, and when the parameters are difficult to interpret directly, this assessment is more easily done by simulating observable quantities using the assumed model and the parameters' prior distributions. Typically it is easier to think about the credible range of an observable quantity, such as a sample mean, rather than the parameters of a model that are always unobservable.

Often prior distributions are adopted that are not intended to represent anyone's beliefs, but to act as "vague beliefs" that allow the information in the data to dominate. Typically these distributions are flat, or approximately flat, over the range of values supported by the data. Historically, much effort has been put into developing general methods for determining such vague priors for models of arbitrary complexity. Chapter 5 of Spiegelhalter et al. (2004) and Sections 2.9 and 2.10 of Gelman et al. (2004), as well as most other general Bayesian texts, discuss these approaches under the labels of "reference," "noninformative," "locally uniform," "vague," and "flat," with Jeffreys' prior, based on his invariance principle, still receiving much coverage. Despite the vast amount of research into this topic, there is still no consensus on how vague priors can be automatically specified for any but the simplest of models. Hence, when working with any but the simplest of models, it is important to assess the sensitivity of the inferences to any priors that are intended to be vague. BugsXLA tries to specify default priors that are vague for data sets that have a reasonable amount of data with which to estimate the parameters in the model. Although some Bayesians criticize this approach, for example, Berger (2006) who refers to it as "pseudo-Bayes," most consider it reasonable in the absence of strong prior knowledge. Experience, skill, and judgment are needed to identify when more informative priors are necessary, although some diagnostic tools are provided to help with this process as discussed in later chapters.

1.1.3 Likelihood

Information in the observed data is formally represented by the likelihood, which is determined by the assumed statistical model. In classical statistics, the likelihood is defined as a function of the unknown parameters, conditional on the data. In the Bayesian paradigm, the likelihood is the probability density of the data conditional on the unknown parameters. Using standard notation, for the simple case being considered here

$$l(\mu \mid x) = p(x \mid \mu) = (2\pi\sigma^2)^{-n/2} \exp\left\{\frac{-0.5\,\Sigma(x_i - \mu)^2}{\sigma^2}\right\}$$

where
  n is the number of observed values x_i
  σ is the known population sd

1.1.4 Posterior "Belief" Distribution

Bayes Theorem provides the method by which we should logically update our beliefs concerning μ in the light of the data. This updated belief distribution is referred to as the posterior distribution for μ since it represents one's knowledge after running the study. Bayes Theorem in its simplest form as a means to calculate a conditional probability is uncontroversial:

$$p(A|B) = \frac{p(B|A)p(A)}{p(B)}$$

As indicated previously, the controversy arises when we interpret the 'A' in the formula not as an observable event with a classically defined probability of occurring, but as a set of unobservable parameters in a statistical model. The more controversial formulation is, in our specific case:

$$p(\mu \mid x) = \frac{p(x \mid \mu)p(\mu)}{p(x)}$$

That is, the posterior is proportional to the likelihood multiplied by the prior. The proportionality constant is provided by the unconditional probability of the data, also known as the "normalizing constant" as it ensures that the posterior distribution integrates to one. The normalizing constant is given by

$$p(x) = \int p(x \mid \mu)\, p(\mu)d\mu$$

More generally, the normalizing constant requires a multidimensional integral to be evaluated since the calculation involves integrating over all the parameters in the model. In our simple example, the posterior distribution can be derived analytically and it is relatively easy to show that, like the prior, the posterior is also a Normal Distribution. This is a case of what is referred to as a conjugate prior and likelihood, that is, the prior distribution and form of the likelihood are such that the posterior has the same distributional type as the prior. To complete our example, suppose the observed data were such that the sample mean and its standard error were 8 and 1, respectively,* then

---

* In this simple example, the sample mean and its standard error are sufficient statistics with which to evaluate the likelihood.

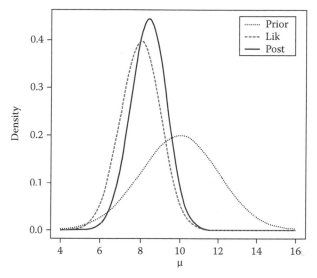

FIGURE 1.1
Plot of Normal prior, likelihood, and resultant posterior distribution.

the posterior distribution would be Normal with mean 8.4 and sd 0.89 to two decimal places. Figure 1.1 provides a graphical representation of the updating process, with the prior, likelihood and posterior all plotted together. In this case, the posterior distribution is more similar to the likelihood than the prior due to the relatively greater amount of information in the data. The amount of information is evidenced by the sharper peak and tighter distribution for the likelihood compared to the prior. Clearly, for parameters of interest at least, we want the likelihood to dominate over the prior otherwise this would indicate that the study has added little to our knowledge. It should be intuitive that as the sample size increases, the likelihood will become more sharply peaked about the point estimate best supported by the data, and hence, the posterior will become increasingly less influenced by the prior.

Having obtained the posterior distribution, one's updated knowledge of μ can be expressed in the same way as was done for the prior. So, we should now believe it quite unlikely that μ is smaller than about 6.6 or larger than about 10.2, with about a 50% chance that μ lies within ±0.6 of the "best estimate" of 8.4.

1.1.5 Inference and Decisions

It is important to understand that it is only in the Bayesian paradigm that probability statements such as those above can be made concerning the value of model parameters, such as μ, and any derived functions of these parameters. It is very likely that most users of classical statistical methods interpret their inferences in this Bayesian fashion. Although there are some circumstances where there is a legitimate Bayesian interpretation of the classical interval estimates, it is important that the user of such inferences be aware of the implied prior distribution and be prepared to defend these prior beliefs if he wishes to interpret the results in the Bayesian way.

One of the implications of the Bayesian approach is that all parameters are considered to be random variables. This contrasts with the classical approach, in which most parameters are considered to be fixed constants, and only random effects are thought of as random variables. A Bayesian still believes that each parameter has only one true value, it is just that his uncertainty causes him to think about the parameter's value in exactly the same way as he does any other uncertain outcome. Another important concept that is widely used in the Bayesian approach is the idea of exchangeability. In simple terms, if a group of parameters (or more generally random variables) are considered exchangeable, then we have no reason to partition them into more similar subgroups or order them in any other way that would infer we could distinguish between them. A more formal definition can be found in Gelman et al. (2004, pp. 121–125). An assumption of exchangeability is often used to justify a prior distribution for a group of parameters that is analogous to the classical random effects assumption, for example, $\theta_i \sim N(0, \tau^2)$. These types of parameters are discussed in Chapters 5 and 6 when mixed models are explained.

The ultimate aim of many studies is to facilitate a decision, for example, whether to progress a product to the next stage of development, where to focus effort in order to improve a process or whether to adopt a new system following a pilot run. A fully Bayesian solution to the problem of making decisions involves the specification of a utility function, and choosing the decision that maximize the expected utility. This topic is way beyond the scope of this book, but those interested can obtain a good introduction from either Lindley (1985) or Raiffa (1968).

1.1.6 Markov Chain Monte Carlo

The major practical problem with the Bayesian approach stems from the fact that the normalizing constant requires a multidimensional integral to be evaluated. Up until relatively recently, the numerical methods available were not able to cope with problems of even moderate complexity. It was only with the availability of Markov chain Monte Carlo (MCMC) algorithms in the final two decades of the twentieth century that the Bayesian approach became a feasible option for many problems. For details of the MCMC approach see Gelfand and Smith (1990) or Tierney (1994). Here we provide an explanation sufficient to allow someone to understand how to interpret the output from software that utilizes this approach. Within the context of a Bayesian analysis, MCMC is an iterative process that generates a series of values, known as a chain, for every parameter in a specified model. The theory behind the method states that providing the MCMC process is run for a sufficiently long time, the process will converge such that the values it generates will behave as if they come from the joint posterior distribution of the parameters. Hence, once the MCMC process has converged, the posterior distribution can be characterized with arbitrary precision simply by generating a sufficiently large number of simulated values from the posterior distribution and using standard summary statistics to estimate quantities of interest, for example, the posterior mean, sd, or percentiles. Since this is an iterative process in which every generation of the simulated values is dependent upon the previous generation, as defined by the Markov chain algorithm, it is necessary to provide initial values with which to start the chain. Typically, these initial values will not be close to the ultimate converged state and so it is necessary to include what is termed a "burn-in" phase during which the MCMC process should converge. Assessing convergence is a critically important part of using this method and this is discussed in Section 1.2. Only once convergence has occurred and, by definition, the burn-in phase completed, can the simulated values be used to summarize the posterior distribution, that is, all the values generated during the burn-in phase are discarded. It is very fortunate that freely available software exists that has implemented the MCMC approach to Bayesian inference. The program that BugsXLA uses as its computing engine is WinBUGS, which is now discussed.

1.2 WinBUGS

WinBUGS is an interactive Windows program for Bayesian analysis of complex statistical models using MCMC techniques. Lunn et al. (2000) is the definitive reference for WinBUGS, which is by far the most widely used Bayesian data analysis program in the world. It provides an easy to learn language that allows a wide variety of models to be specified. Its popularity is probably due to the user not needing to know very much about the MCMC methods that underpin it, and that it is free to download from the BUGS website: http://www.mrc-bsu.cam.ac.uk/bugs.

Although the main purpose of this book is to show how to use BugsXLA, eliminating the need to learn the WinBUGS software and its idiosyncrasies, WinBUGS still needs to be downloaded in order for BugsXLA to work. Also, it is very strongly recommended that the user knows how to use WinBUGS to assess convergence of the MCMC run, since BugsXLA does not provide any separate functionality for this purpose. In order to gain this understanding, at least a basic knowledge of WinBUGS is required. Only the most advanced users need to learn the WinBUGS programming language, so others should view this section purely as a beginner's guide to assessing MCMC convergence (Section 1.2.5 onward).

As briefly explained in Section 1.1, in the Bayesian approach inferences are made by conditioning on the data to obtain a posterior distribution for the parameters in the model. The MCMC approach is used to marginalize over the posterior distribution in order to obtain inferences on the main quantities of interest. WinBUGS contains a programming language that allows an arbitrarily complex statistical model to be defined. This code is then compiled into a form suitable for efficient computation, and a sampler operates on this structure to generate appropriate values of the unknown quantities in order to feed the MCMC algorithms. Instead of calculating exact or approximate estimates of the necessary numerical integrals, MCMC generates a stream of simulated values for each quantity of interest. WinBUGS also has tools to monitor the stream for convergence and summarize the accumulated samples.

1.2.1 Programming Language

The best way to understand how WinBUGS can be used to undertake a Bayesian analysis is to work through a relatively simple example. Here we show how WinBUGS could be used to fit an analysis of covariance Normal linear model to a toy data set. After starting WinBUGS, a new program is created by first selecting 'File: New' from the main menu bar. The code is typed into the new window that appears, and can be saved by selecting 'File: Save As…'. By default, files are saved in a WinBUGS specific file type with the '.odc' suffix. WinBUGS also recognizes plain text files, which is the file type that BugsXLA uses to interact with WinBUGS.

```
MODEL - START                                               1
{                                                           2
  ### PRIORS (all vague)                                    3
  alpha ~ dnorm( 0, 1.0E-6 )                                4
  G.eff[1] <- 0                                             5
  for (i in 2:N.grp){                                       6
    G.eff[i] ~ dnorm( 0, 1.0E-6 )                           7
  }                                                         8
  beta ~ dnorm( 0, 1.0E-6 )                                 9
  prec ~ dgamma( 1.0E-3, 1.0E-3 )                          10
  var <- 1/prec                                            11
  sigma <- sqrt(var)                                       12

  ### LIKELIHOOD                                           13
  for (k in 1:N.obs)                                       14
  {                                                        15
    Y[k] ~ dnorm( mu[k], prec )                            16
    mu[k] <- alpha + G.eff[ G[k] ] + beta*X[k]             17
  }                                                        18

  ### Estimate pair wise treatment contrasts              19
  for (i in 1:N.grp-1){                                    20
    for (j in i+1:N.grp){                                  21
       G.con[i, j] <- G.eff[i] - G.eff[j]                  22
    }                                                      23
  }                                                        24
}                                                          25
MODEL - END                                                26
```

FIGURE 1.2
WinBUGS code for simple ANCOVA model.

The WinBUGS code is shown in Figure 1.2; the line numbers on the right-hand side are not part of the code but are included to facilitate the explanation that follows. Lines 3, 13, and 19 do not perform any function other than to comment on the program, since text that follows the hash symbol (#) is ignored by WinBUGS. Lines 1 and 26 are also not mandatory, but are useful to delineate the code from any other text that one might want to include in the file. The code defining the model is enclosed between two curly brackets, see lines 2 and 25.

1.2.2 Data

The data being analyzed here consists of

Y[]   the response data
X[]   the continuous covariate
G[]   the level indicator for the grouping factor

which are in lines 16 and 17 of the code. Data, as well as other constants used to define the model, can be read into WinBUGS in two ways. Either separate files can be created that contain the data, or separate data lists can be created in the same file as the code. It is generally good practice to keep data and programs separate and so we only show the former method here. Figure 1.3 shows the two types of data format that WinBUGS recognizes; rectangular format on the left and list format on the right. It is usually convenient to store

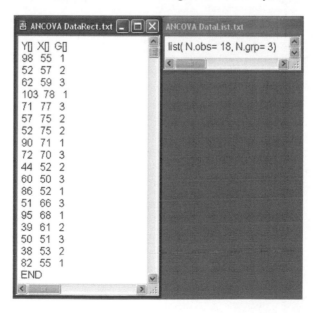

FIGURE 1.3
Data files used in ANCOVA example.

all the response data as well as the covariates in rectangular format, while the other constants used to define the model have to be stored in list format.

Note that the last line of the rectangular file contains the word END, and that the first row of this file contains the node names followed by a pair of square brackets, '[' and ']'. The data in the list file define the number of observations 'N.obs' and the number of levels of the grouping factor 'N.grp'; lines 6, 14, 20, and 21 show how these are used in the code. These data files can also be created from within WinBUGS, although in this case they were created using a simple text editor and saved with the '.txt' extension. Files already created can be read into WinBUGS by selecting, 'File: Open…', and then locating the file using the standard Microsoft Windows form that appears.

1.2.3 Variables: Deterministic and Stochastic Nodes

The model being fitted, as defined by lines 14–18 in the code, can be represented algebraically:

$$y_{ij} = \alpha + \gamma_i + \beta x_{ij} + e_{ij} \quad i = 1, 2, 3 \quad j = 1, 2, \ldots, 6$$

where
   y_{ij} is the response on the jth experimental unit in the ith group
   α is a constant in the model
   γ_i represents the effect of the ith level of the grouping factor
   β is the regression coefficient associated with the simple linear effect of the covariate
   x_{ij} is the covariate measured on the jth experimental unit in the ith group
   e_{ij} is the residual random error associated with the jth experimental unit in the ith group
   $e_{ij} \sim N(0, \sigma^2)$

In the WinBUGS code, the parameters are represented:

```
alpha = α
G.eff[i] = γᵢ
beta = β
prec = σ⁻²
```

WinBUGS parameterizes the Normal Distribution using the mean and precision, which is defined as the inverse of the variance. Note that in the WinBUGS code, rather than make the response variable a two-dimensional array as represented in the model formula above, a simpler vector representation is coded (line 16). To ensure that the appropriate grouping factor effect is associated with each observation, the level indicator G[] is used as a nested index for the parameter G.eff[] (line 17).

All the variables in a WinBUGS program are referred to as nodes, which can be either stochastic or deterministic. All the unknown parameters are stochastic, and their prior distributions are defined using the tilde (~) relationship symbol, which means "is distributed as." So, for example, line 4 defines the prior for the parameter alpha as a Normal Distribution with mean zero and precision one-millionth. This prior for alpha is intended to be very vague, approximating a uniform over the range of values supported by the likelihood. The effects of the grouping factor, G.Eff[], are parameterized with the first level set equal to zero, using notation explained later, (line 5), and giving all the other levels the same vague prior as that for alpha (lines 6–8). This prior is the Bayesian equivalent of the "fixed effects" assumption in classical statistics (see Chapter 3 for further discussion). The regression coefficient, beta, is also given a vague prior (line 9). The prior for the precision parameter, prec, is a much dispersed Gamma Distribution (line 10). Although this prior is not intuitively vague, it is now the standard default vague prior for this parameter and is approximately equivalent to a uniform distribution for log(sigma). As well as the parameters, the response data, Y, are also defined as stochastic nodes (line 16).

Deterministic nodes are defined using the arrow (<-) relationship symbol, which means "is to be replaced by." Note that this is in fact two symbols, the less than sign and the hyphen, placed together to look like an arrow. So, for example, lines 11 and 12 define the residual variance, variance, and sd, sigma, via the stochastic parameter, prec. The WinBUGS programming language allows repetitive structures to be succinctly described using for loops (lines 6–8 and 14–18). These for loops can be nested to run over more than one dimension of an array as shown in lines 20–24 where pair wise treatment contrasts are defined. Note that these loops constitute a declarative description of the model and should not be interpreted in a procedural way, that is, they are simply interpreted as a replacement for the same lines of code repeated numerous times with only the index of the looping variable changed each time. Although not done in this simple example, it is good practice to center, and sometimes also to scale, covariates when including them in a WinBUGS model in order to improve convergence as discussed later. BugsXLA does this automatically as shown in Appendix C.

1.2.4 Compiling and Initializing the Markov Chain

The WinBUGS program, together with the data, must be compiled before it can be run. This is done via the model specification tool that is obtained by selecting 'Model: Specification…' from the menu bar. Figure 1.4 shows all three of the tools that WinBUGS provides to run the analysis, with the top one being this 'Specification Tool'. After first ensuring that the mouse cursor is somewhere within the code that is to be compiled, click on the 'check model' button on the 'Specification Tool' form. If the code has been correctly entered as shown in Figure 1.2 (excluding the line numbers), then the message 'model is syntactically correct' will appear at the bottom left of the WinBUGS window. The two data files shown in Figure 1.3 must now be

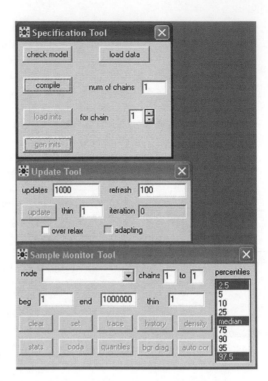

FIGURE 1.4
WinBUGS tools used to compile, run, and monitor a specified analysis.

loaded. The data files must have been opened via the menu bar as explained earlier so that they now reside inside the WinBUGS program. Click on one of the data files so that it is active, and then click on the 'load data' button on the 'Specification Tool' form. The message 'data loaded' will appear at the bottom left of the WinBUGS window. Repeat this for the other data file. WinBUGS allows multiple chains to be run, which is useful for assessing convergence of the MCMC process. In this simple example, we will only create a single chain, and so click on the 'compile' button, which will display 'model compiled' in the usual place for such messages. WinBUGS requires the MCMC process to be initiated with starting values, that is, each of the parameters in the model that are stochastic nodes require an initial value to be set. In many cases, WinBUGS can generate these initial values itself from the specified prior distributions. However, for numerical reasons, it is not able to do this for the precision parameter in this model, and so another list formatted data file, virtually identical to the one shown on the right-hand side of Figure 1.3, is required that contains the line:

```
list(prec=1)
```

After first ensuring this file is active by clicking on it, click on the 'load inits' button on the 'Specification Tool' form. The message 'this chain contains uninitialized variables' appears, informing the user that not all the stochastic nodes have had initial values set. Since the rest can be generated by WinBUGS, click on the 'gen inits' button, which brings up the message 'initial values generated, model initialized'. The model has now been compiled and is ready to run.

1.2.5 Updating the Chain and Assessing Convergence

WinBUGS refers to running the analysis as "updating," to reflect the fact that each iteration of the MCMC process produces a new simulation that is updated from the previous. The 'Update Tool', the second form shown in Figure 1.4, is obtained by selecting 'Model: Update...' from the menu bar. As discussed in Section 1.1, before we can reliably use the output from an MCMC run we must be confident that the process has converged, and that only samples generated after the burn-in phase are used. Checking for convergence is discussed later, but for this particular example an experienced user would recognize it as sufficiently simple that a small burn-in phase will be adequate. By default the number of updates is set to 1000 on the 'Update Tool' and this does not need to be changed. Click on the 'update' button and as well as the 'iteration' counter turning over very rapidly, the message 'model is updating' will appear fleetingly before changing to 'updates took 0s' (unless your computer is incredibly slow). In order to make inferences we need to monitor the nodes of interest in the model, and this is done via the 'Sample Monitor Tool', the last form shown in Figure 1.4, obtained by selecting 'Inference: Samples...' from the menu bar. The name of the node to be monitored is entered into the 'node' text box and the 'set' button clicked. Repeat this for each of the nodes listed here: alpha, G.eff, beta, sigma, and G.con. By clicking on the down arrow head to the right of the node text box, it is possible to check which nodes have been set for monitoring; this is also a convenient way to select specific nodes for summary after more samples have been generated. Click on the 'update' button on the 'Update Tool' to generate another 1000 samples. We now have 1000 simulated values from the posterior distributions of the nodes set. Rather than summarize the nodes individually, WinBUGS provides a short cut by which all nodes can be summarized together: type an asterisk '*' in the 'node' text box and click on the 'history' button. All the samples are plotted in the order generated. Figure 1.5 shows the plot for the covariate coefficient parameter beta.

The history plot provides a simple way to visually assess convergence. If all the values appear to be drawn from a single distribution, with no evidence of the chain drifting away from a previous location, then the plot is consistent with a converged Markov chain. Figure 1.5 is an example of the simplest case to assess: a parameter with a symmetrical posterior distribution, not heavy tailed and little autocorrelation in the simulations. The lack of autocorrelation can be seen more clearly via the plot obtained by clicking the 'auto cor' button on the 'Sample Monitor Tool' as shown in Figure 1.6.

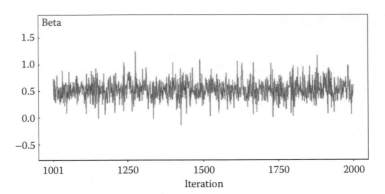

FIGURE 1.5
History plot for beta parameter in ANCOVA example.

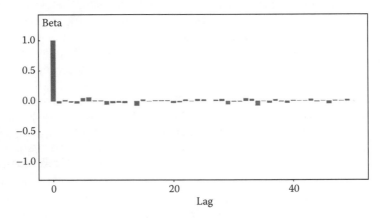

FIGURE 1.6
Autocorrelation plot for beta parameter in ANCOVA example.

For most models of at least moderate complexity, autocorrelation will exist for some of the parameters even after the MCMC process has converged. Sometimes, it is possible to reparameterize the model to reduce this correlation, for example, by centering and scaling covariates, but more generally this is a skill that comes with experience of developing models that is beyond the scope of this book to explain. Autocorrelation does not invalidate using the simulated values to summarize the posterior distribution, but it does make assessing convergence more difficult, as well as leading to the need for longer runs of the process to obtain the same precision in the summaries. A measure of this precision is provided by WinBUGS in the form of the 'MCMC error', obtained with the other summary statistics on clicking the 'stats' button on the 'Sample Monitor Tool'. Table 1.1 shows the results provided for all the parameters monitored in this ANCOVA example.

TABLE 1.1

Node Statistics for the ANCOVA Example

Node	Mean	Sd	MC Error	2.5%	Median	97.5%	Start	Sample
G.con[1,2]	44.87	4.13	0.1485	36.62	44.77	53.04	1001	1000
G.con[1,3]	30.81	4.118	0.1346	22.15	30.88	38.53	1001	1000
G.con[2,3]	−14.06	4.114	0.1046	−22.43	−14.11	−5.265	1001	1000
G.eff[2]	−44.87	4.13	0.1485	−52.95	−44.77	−36.49	1001	1000
G.eff[3]	−30.81	4.118	0.1346	−38.42	−30.85	−22.14	1001	1000
alpha	59.36	11.67	0.3346	35.06	59.66	82.18	1001	1000
beta	0.5217	0.1784	0.005097	0.168	0.5148	0.9109	1001	1000
sigma	6.901	1.461	0.05271	4.745	6.674	10.49	1001	1000

The MC error is an estimate of the Monte Carlo standard error of the mean, which takes account of the auto correlation in the samples. Intuitively, providing this is small relative to the estimate of the posterior sd, there is little to be gained from obtaining a larger sample, providing one is convinced that the MCMC process has converged. A reasonable rule of thumb is to aim for an MC error of less than 5% of sd, with an MC error of up to 10% being fine for parameters that are not directly of interest. It is important to be aware that this MC error applies only to the mean and not the percentiles of the distribution, which will have bigger Monte Carlo standard errors. The final simple method that should be used to help assess convergence is to inspect the kernel density plot obtained by clicking the 'density' button on the 'Sample Monitor Tool.'

Although the density shown in Figure 1.7 for the beta parameter appears rough, due to the relatively small number of simulations with which to build up a complete distribution, the distribution already looks reasonably close to a Normal or t-distribution as one would expect for this model. Unless

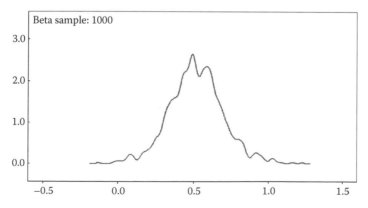

FIGURE 1.7

Kernel density plot for beta parameter in ANCOVA example.

one had reason to expect otherwise, posterior distributions are typically uni-modal and approximately symmetric; a common exception being variance components with relatively few effective degrees of freedom that tend to be skewed to the right. The smaller the sample size of the data set, or effective sample size for hierarchical parameters, the more likely it is that parameters will have posterior distributions that deviate from these typical shapes making it more difficult to spot convergence issues with this plot. Another important approach for assessing convergence is to simply inspect the summarized posterior distribution and consider whether the credible range is credible to you. This latter approach may be difficult for more complex models where it may be harder to obtain intuition for sensible values of the parameters. Also, it is possible that an incredible posterior distribution could be due to a poorly specified model that has converged, rather than an adequate model that has not. As with many other aspects of statistical model building, these skills can only be developed with experience.

1.2.6 Interpreting the Output

Once we are convinced that the MCMC process has converged, and that we have sufficient precision with which to make inferences, it is then important to assess whether the specified model is a good fit to the data. Any model checking functions, such as residuals, have to be explicitly coded up in WinBUGS. Chapter 10 of Ntzoufras (2009) and Section 9.4 of Spiegelhalter et al. (1996) discuss this aspect of the model-building process in detail. Although there are some specific Bayesian approaches to model checking, the general approach is essentially the same as in classical statistics. In fact, some have argued, for example, Box (1983), that the model-checking process is essentially frequentist in nature. That is, we are assuming a known model and then assessing whether predictions made conditional on this model being true are credible. Model checking via WinBUGS will not be discussed in detail here, but we come back to this important topic when discussing some of the case studies in the latter chapters of the book.

Table 1.1 shows summaries of the posterior distributions for each of the parameters as well as the pair wise contrasts between the treatment means. As discussed in Section 1.1.6, the summaries are simply the sample mean, sd, and percentiles of the simulated values. These can be interpreted in a similar manner to how most non-statisticians interpret classical point estimates, standard errors, and confidence intervals. However, it is only in the Bayesian paradigm that the probabilities associated with these summaries relate directly to the credible values of the parameters. So, for example, providing the prior distribution appropriately represents one's knowledge prior to observing the data, and conditional on the model being an adequate representation of the underlying real-world process that generated the data, it is legitimate to state that there is a 95% probability that the value of beta lies between 0.17 and 0.91. An immediate conclusion is that the covariate explains a "statistically significant," at least, amount of the variation in the response, since zero is not a credible value for beta. Other than the way in which inferences can be interpreted, and the ease with which other quantities of interest can be added to the code, the rest of the statistical analysis and reporting process is very similar in nature to that when using classical statistical methods.

1.2.7 Identifying and Dealing with Poor Mixing

As mentioned above, assessing convergence of the MCMC process is more difficult when the simulated values exhibit strong autocorrelation. Figure 1.8 shows a history plot of a parameter from a different model and data set that clearly shows such correlation. Figure 1.9 shows the estimated autocorrelation function.

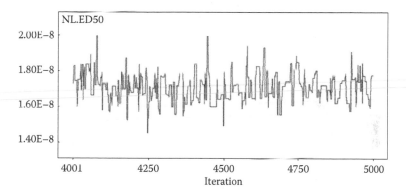

FIGURE 1.8
History plot of parameter exhibiting clear autocorrelation.

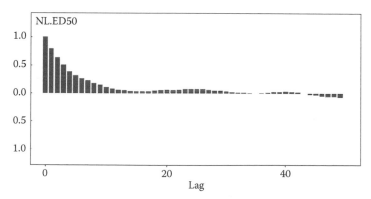

FIGURE 1.9
Autocorrelation plot of parameter exhibiting clear autocorrelation.

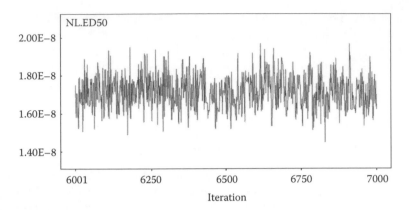

FIGURE 1.10
History plot of parameter with one-fifth thinning.

Although Figure 1.8 shows that consecutive simulations have a tendency to drift either upward or downward, a facet of the autocorrelation, when considered over a sufficiently long run, the chain appears to be fluctuating around the same location. The autocorrelation makes it more difficult to spot any underlying more gradual drift of the whole process. A technique that helps in these circumstances is to thin the chain of simulated values by only recording every Kth value. An informed choice for K can be made by inspection of the autocorrelation plot. It will still be necessary to generate a larger number of simulations than would be the case if little autocorrelation was present, but thinning makes it easier to assess convergence. Figures 1.10 and 1.11 shows the history and autocorrelation plots for this same parameter when one-fifth thinning is applied.

Thinning is specified by entering the thinning rate in the text box labeled 'thin' on the 'Sample Monitor Tool'. It is clear from Figures 1.10 and 1.11 that

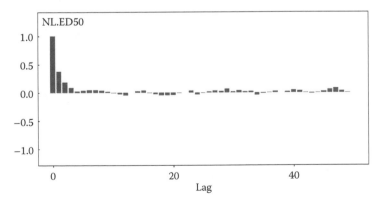

FIGURE 1.11
Autocorrelation plot of parameter with one-fifth thinning.

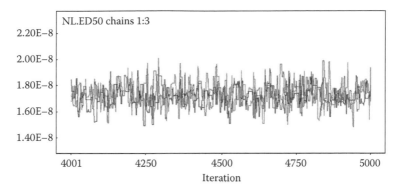

FIGURE 1.12
History plot of parameter with three parallel chains.

thinning has reduced the autocorrelation, making it easier to see that the MCMC process appears to have converged.

For the more complex models or data sets, it is recommended that convergence be checked by running multiple chains, each with different initial values for the stochastic nodes, with the aim being to demonstrate that each of the chains converge to the same distribution. Three chains should be sufficient, which will require three separate list formatted data files to be created containing the initial values for each chain. These initial values should be diverse in order to provide confidence that the end state of the MCMC process is not sensitive to the starting state. The number of chains is specified on the 'Specification Tool' prior to compiling the model. The initial values are then loaded in turn. The rest of the analysis and summarizing process is the same as the case when only a single chain is being run. Figure 1.12 shows a history plot when three chains are generated.

The simulated values from each chain are overlaid, with each chain being represented by a different color. It is clear from Figure 1.12 that the chains are fluctuating around the same location, with what appears to be the same degree of variation. This pattern is consistent with what we would expect from chains that have converged to the same distribution. A more formal assessment of the convergence of multiple chains is provided by the Brooks–Gelman–Rubin plot obtained by clicking on the 'bgr diag' button on the 'Sample Monitor Tool'. Gelman and Rubin (1992) and Brooks and Gelman (1998) provide details of the method; here we simply explain how to interpret the plot.

Figure 1.13 shows an example of the Brooks–Gelman–Rubin diagnostic plot. The plot is derived from an analysis of the multiple chains comparing the variation within each chain with the variation that exists between the chains, analogous to the one-way analysis of variance. The blue and green lines in the plot represent the variation within and between the chains respectively. The measures of variation are scaled so that once the chains have converged these two lines should converge to the same value. The red line

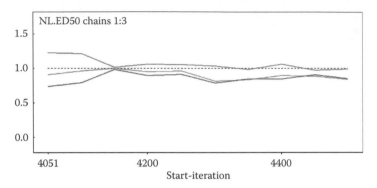

FIGURE 1.13
Brooks–Gelman–Rubin plot of parameter with three parallel chains.

is the ratio of the between to the within chain variation, and so should converge to one, which is shown on the plot as a horizontal dashed line. It follows that prior to the chains converging the red line will be significantly greater than one, reflecting the fact that the chains have different locations. Figure 1.13 is indicative of an MCMC process that has converged, since from about iteration 4150 onward the conditions stated above are approximately true.

As will be shown in later chapters, BugsXLA can be used to generate multiple chains with a diverse set of initial values. However, it is up to the user to assess convergence using the tools from within WinBUGS and the techniques discussed above. WinBUGS also provides a scripting language that allows the whole process described in this section, from checking the model and loading the data through to generating the simulations and saving the results, to be run in batch mode, that is, without having to manually click on the menu items and buttons. BugsXLA works by creating all the files needed by WinBUGS—the code, data, and initial value files, as well as the script to run the analysis—and then calling WinBUGS in batch mode. Once WinBUGS completes the specified analysis, BugsXLA reads the output files created and imports the results back into Excel.

1.2.8 BugsXLA's Role

The WinBUGS package contains many more features than have been illustrated in this section, and anyone wishing to become proficient in its use will need to become familiar with the manual that is built into the program ('Help: User Manual' from the menu bar, or Spiegelhalter et al. (2003)), as well as working through some of the examples ('Help: Examples Vol I and II'). The book by Ntzoufras (2009) is also recommended, as it provides details on every aspect of the WinBUGS package with worked examples of simpler models than those packaged with WinBUGS. Although WinBUGS provides an incredibly powerful statistical programming language, there is inevitably a learning curve barrier associated with any new software that requires sufficient motivation to overcome. The purpose of BugsXLA is to provide a tool that makes it easy for anyone involved with data analysis to try Bayesian methods. By allowing models to be specified in a similar manner to that adopted by most of the main stream statistical packages such as SAS, S-PLUS, R, and Genstat, it removes one of the major barriers to those people who are curious but uncommitted to the Bayesian approach. Also, by integrating BugsXLA within the Microsoft spreadsheet application, Excel, it opens up the power of the Bayesian approach to statistics to the vast number of scientists, engineers and other technologists who routinely use Excel to store and understand their data. BugsXLA should also be a valuable tool for the applied statistician as it can save many hours coding and debugging programs, which is notoriously difficult for the novice WinBUGS user due to the extremely unhelpful error trap messages. By removing the need to know how to code the models, import the data, and export the results, BugsXLA allows the user to focus on the more important issues:

- Is the model appropriate?
- What priors can I justify?
- What inferences am I trying to make?

The purpose of this book is to explain how BugsXLA can be used to facilitate these aims.

1.3 Why Bother Using Bayesian Methods?

Before discussing the benefits of taking a Bayesian approach to statistical inference, it is important to be aware that there are still some who are fundamentally opposed to the approach on mainly philosophical grounds. They reject the definition of probability as a measure of degree of belief as outlined in Section 1.1.1, and will only admit the use of Bayes theorem when the probabilities have a classical or frequentist interpretation. For the majority of the twentieth century, due in no small part to the huge influence of Fisher, the dominant statistician of his time and an uncompromising opponent and vocal critic of Bayesian inference, virtually all statistical work was frequentist in nature. The case for Bayesianism was also greatly hampered by the computational issues that severely limited its practical applications. Since the advent of fast and affordable computers as well as the development of MCMC algorithms in the latter part of the twentieth century, the use of Bayesian methods in academia has increased at an exponential rate. The philosophical debate will never go away, but even among academics, it is the practicality of being able to solve real problems that has greater influence. It is somewhat ironic that today one of the advantages of the Bayesian approach is the ease with which it can fit many of the more complex models widely used. For example, whereas there are still unresolved issues in how to make inferences in mixed effects models using purely likelihood based methods, the Bayesian approach is conceptually straightforward (typically involving integrating out the nuisance parameters) and often simple to implement using freely available MCMC based software. As well as making inferences on the model's parameters directly, it is now computationally trivial to make inferences on arbitrarily complex functions of these parameters, or determine predictive distributions for future quantities of interest. The statistical journals are now awash with examples of inferences made from models fitted to complex data sets that simply would not be possible using non-Bayesian methods.

Another practical advantage of the Bayesian approach is the natural way it combines information obtained from the most recent data with that already available. It does this by allowing the previously available information to be formally represented in the prior distribution. Contrary to the view of some who see the need for a prior as a weakness, it should be seen as one of the strengths of the approach. Used appropriately, the prior can show the sensitivity of our inferences to our initial beliefs. If strong prior beliefs can be justified, typically based on historical evidence, there are now few who object to the use of Bayesian methods to incorporate this additional information. Even when moderately informative priors can be justified there are cases, such as some nonlinear or complex mixed models, where this additional information is enough to overcome numerical convergence problems that would occur with likelihood based methods. In those cases where the data set is large there is little controversy since the information in the likelihood will dominate that in the prior.

As I indicated in the Preface, what ultimately convinced me to adopt the Bayesian approach to inference is that it provides the simplest and most intuitively reasonable formal description of the process by which one's beliefs should be logically updated in the light of new information. It does this by recognizing probability as a measure of degree of belief that allows us, via Bayes theorem, to quantify how new data should logically change our level of certainty in any particular scientific hypothesis. No other inferential approach can do this. For example, the Bayesian 90% credible interval for a quantity of interest can be interpreted as the range of values within which the quantity lies with probability 0.9, or odds of 9-to-1 in favor. No such interpretation can be made for the frequentist's 90% confidence interval. Although not discussed in this book, by describing the uncertainty in quantities of interest using probability the Bayesian approach provides, once again, the simplest and most intuitively reasonable way to tackle decision making problems. Books such as Lindley (1985) or Raiffa (1968) explain how such a remarkably simple theorem can be used to guide decision makers in any situation where the options can be listed and their consequences quantified.

2

BugsXLA Overview and Reference Manual

BugsXLA is an Excel Add-In that provides a graphical user interface for the Bayesian software package WinBUGS; Woodward (2005) describes an earlier version of the program discussed in this book. It allows the user to fit a wide class of generalized linear mixed models, as well as other more complex models, without the user needing to know how to program or otherwise interact with the WinBUGS program. BugsXLA has been designed to be easily learnt by anyone familiar with using Microsoft Windows style interfaces, and the specification of statistical models via one of the common statistical packages such as SAS, S-PLUS, R, or Genstat. The main purpose of this chapter is to explain how to obtain this freely available software, and then describe each of the program's features. This chapter should be seen as a reference manual where specific details can be looked up when they are needed, rather than guidance to be read through in the order it is written. It is recommended that once the software has been loaded, the user learns how to use it by working through relevant examples in later chapters, referring back to this chapter when more details on a specific feature are required. In particular, Case Study 3.1 illustrates most of the frequently used features of BugsXLA and should be followed carefully before moving on to the more complex analyses later in the book.

2.1 Downloading and Installing BugsXLA

BugsXLA can be downloaded from http://bugsxla.philwoodward.co.uk. This book refers to Version 5.0, so if a more recent version is available, it is advisable to be aware of the new features and bug fixes that have been implemented; a link to a document that describes these changes will exist on the web page. After following the instructions to download BugsXLA (v5.0) and saving the self-extracting zip file to a convenient location on your PC, follow these instructions to install BugsXLA as an Excel add-in. It is necessary to have Windows 98 or higher (tested on 98, 2000, XP, Vista, and Windows 7), Microsoft (MS) Excel (tested on versions 2000–2007; will not work on 97), MS Notepad, and WinBUGS v1.4.x already loaded onto your PC. If you want to take advantage of the R scripts that BugsXLA can create, you will also need to have the R package loaded.

It is important that Excel be closed before starting.

1. Run the self-extracting zip file; by default this puts the files in 'C:\Program Files\WinBugsXLA50'
2. Open MS Excel
3. Make Excel recognize the BugsXLA add-in

   EXCEL 2000–2003

   > Select Tools: Add-Ins

   EXCEL 2007

   > Select MS Office Button: Excel Options: Add-Ins

   > (listed in left hand column)

   > At the bottom of the form ensure Manage: Excel Add-Ins is selected, and click 'Go ...'

4. Browse in the folder to which the zipped files have been extracted (see step 1)
5. Select the file BugsXLA50.xla and make sure that BugsXLA has been checked in the Add-Ins dialog box before clicking 'OK'
6. If you are warned that either WinBUGS or Notepad is not in 'specified location', click 'Yes' and alter the folder locations so that these programs can be found
7. A toolbar titled 'BugsXLA50' should appear that can be moved to your favored location on the Excel frame. In Excel 2007, the toolbar will appear on the Add-In tab
8. If at anytime you have problems running the program, try a clean reinstall by first removing all the old components:

   a. Open Excel and remove the BugsXLA toolbar:

      EXCEL 2000–2003

      > Tools: Customize

      > Select 'BugsXLA' from the list of Toolbars

      > Delete (confirm OK)

      EXCEL 2007

      > Right mouse click on any BugsXLA icon and select

      > 'delete custom toolbar'

   b. Remove the BugsXLA add-in

      EXCEL 2000–2003

      > Tools: Add-Ins

      EXCEL 2007

      > Select MS Office Button: Excel Options: Add-Ins
      (listed in left hand column)

> At the bottom of the form ensure Manage:
> Excel Add-Ins is selected, and click 'Go ...'
> Uncheck 'BugsXLA' from the list of Add-Ins
> OK

   c. Close Excel and manually delete the WinBUGSXLA folder created in step 1

If you have a previous version of BugsXLA loaded, the new version should automatically uninstall this previous version. However, it may be safer to manually uninstall the old version first by opening the 'BugsXLA Options' form and clicking on the 'Unload BugsXLA' button and confirming 'OK'. You may also wish to delete the WinBUGSXLA folder associated with this previous version before starting the new installation.

Note: BugsXLA will only function correctly if the English version of Excel is set. If you have loaded and tried to use the program under a different language, you will need to set it to the English version, and then on the 'BugsXLA Options' form click on the 'Clear Old Analysis' button.

As well as the Excel add-in file, BugsXLA50.xla, 14 other files are unzipped to the folder created in step 1 above:

BugsXLA Changes from v4.0.txt

As the name suggests, this file gives a brief description of changes that have been made since the previous version.

BugsXLA Egs1.xls
BugsXLA Egs2.xls
BugsXLA Egs3.xls

These three files contain example data sets with output produced by Version 5.0 of BugsXLA. In most cases, they also contain results from alternative analyses with which to compare.

BugsXLA Book Case Studies.xls

This contains all the data sets used in this book.

BugsXLA ReadMe.txt

This provides a very brief description of each option available on the BugsXLA toolbar.

BugsXLA Zip content.txt

This lists all the files that are downloaded.

WinBUGS Script Template.txt

This provides a template that facilitates the creation of a WinBUGS script. It is the default template used by the 'Edit WinBUGS Script' utility described later.

Six small bitmaps (file extensions.bmp) containing the new icons used on the toolbar.

2.2 BugsXLA Toolbar

All of the functionality of BugsXLA is accessible via the toolbar created when the add-in is attached to Excel. The toolbar displayed will depend upon whether the version of Excel on your computer is before or after 2002 (Figures 2.1 and 2.2). The four left-most icons provide utilities that facilitate the interaction between Excel and WinBUGS:

Export to WinBUGS
Edit WinBUGS Script
Run WinBUGS (Script)
Import from WinBUGS

More information about these is provided in Section 2.13. The main focus of this book is the functionality provided by BugsXLA for the fitting of complex statistical models. The final three icons provide this functionality:

Bayesian Model
Post Plots
BugsXLA Options

The first of these icons is the place to start when specifying the model to be fitted. It brings up the model specification form, and leads the user through all the steps needed to undertake a Bayesian analysis. The

FIGURE 2.1
BugsXLA50 v5.0 toolbar (Excel 2002 or later).

FIGURE 2.2
BugsXLA50 v5.0 toolbar (pre Excel 2002).

'Post Plots' icon provides a tool for plotting imported samples from the inferred posterior distributions. Many of the default settings associated with BugsXLA can be altered via the 'BugsXLA Options' icon. The rest of this chapter provides more details on the Windows forms that make up BugsXLA, and should be seen as a reference rather than a way to learn how to use the program. As stated previously, it is recommended that the user learns how to use BugsXLA by working through relevant examples in later chapters, referring back to this chapter when more details on a specific feature are required.

2.3 Bayesian Model Specification

The first form that appears after clicking on the 'Bayesian Model' icon is the main model specification form (Figure 2.3).

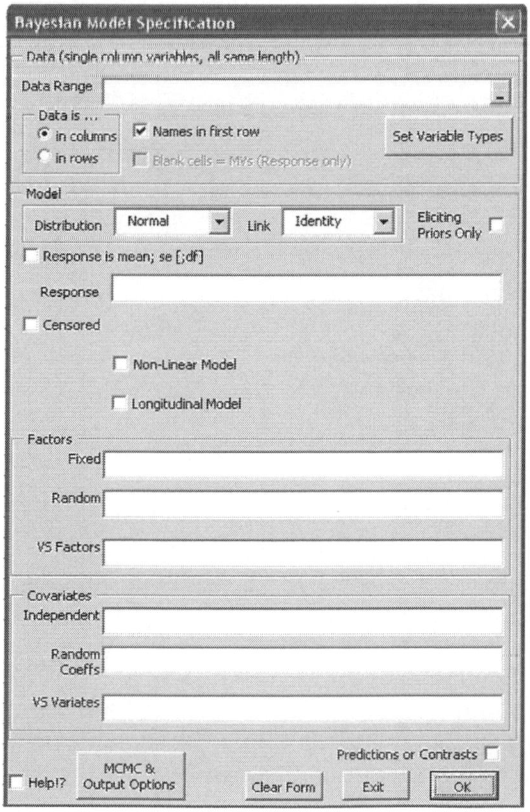

FIGURE 2.3
Bayesian model specification form.

This is where the data are read into the program, and the likelihood part of the model is specified. As with most forms, there is a '**Help!?**' check box in the bottom left corner. By checking this box, all parts of the form with help text available are highlighted in yellow. To obtain help on a particular aspect of the form, move the mouse over the highlighted area and a dialog box will appear with some information to assist the user.

If the form already contains entries that are not relevant to the current analysis, then this, as well as all the other forms, can be cleared by clicking on the '**Clear Form**' button. Note that this button will also reset the options on the other forms to their defaults.

2.3.1 Variable Names and Types

The first thing to do on this form is tell the program where the data are stored on the spreadsheet. The '**Data Range**' is the range of cells on the Excel sheet that contain all the data. The first row of this range must contain the names of the variables. Variable names must be no longer than 32 characters, beginning with a letter and consisting of alphanumeric, '_' or '.' characters. If any spaces are found in a variable's name these will be automatically replaced by '_', and the user warned that this has been done.

Although the program allows data to be stored in rows or columns, it is strongly recommended that columns be used, and this book assumes this is the case. To the right of the 'Data Range' label is a ref edit box in which the range of cells containing all the data needs to be specified. The easiest way to specify this range is to simply click in the empty box, and then use any standard Microsoft Windows approach to select the rectangular range of cells, for example, drag the mouse over the range of cells while holding down the left mouse button. Note that the first row of cells containing the variable names must also be included in this range.

By default, all columns of data are assumed to be **variates**, that is, quantitative continuous or count data. If any of the variables need to be treated as **factors**, that is, categorical data, or are **censored** variates, then this needs to be specified by clicking on the '**Set Variable Types**' button (refer to Section 2.4).

2.3.2 Error Distributions and Link Functions

Once the location of the data has been specified and any factors or censors defined, the likelihood part of the model must be set. Like the generalized linear model, this consists of three parts: the linear/nonlinear predictor, the error distribution, and the link function. The **distribution** and **link** are set via drop down boxes, with the following error distributions available:

Normal, t-Distribution, Log-Normal, Gamma, Weibull, Poisson, Binomial (or Bernoulli), Multinomial (counts), Categorical (individual responses), Ordinal (counts), or Ordered Categorical (individual responses).

Depending upon the distribution, some of these link functions can be used:

Identity: $\mu = \beta X$
Log: $\log(\mu) = \beta X$
Logit: $\log(\pi/(1 - \pi)) = \beta X$
GenLogit: $\log(\pi^{(1/m)}/(1 - \pi^{(1/m)})) = \beta X$
Probit: $\pi = \Pr\{N(0,1) < \beta X\}$
CLogLog: $\log(-\log(1 - \pi)) = \beta X$

Note that log is to base e, that is, $\log_e$, throughout this book unless stated otherwise. The linear predictor is denoted by βX here. When the Log-Normal or Weibull is specified then μ is the median. Refer to Prentice (1976) for details of the generalized logit (GenLogit) link.

2.3.3 Response Variables

In the text box labeled '**Response**', the name of the variable whose variation is being modeled is entered. Typically, the response must be a variate, but exceptions are discussed below. When a Binomial Distribution is specified then both the numerator (R), for example, number survived, as well as the denominator (N), for example, number treated, is specified, with the variable names separated by a forward slash, '/', for example, Survived/Treated. Section 4.1 shows how to model a binomial response.

Bernoulli observations can be modeled by specifying a Binomial Distribution and entering a Response of the form:

AliveDead/1

where AliveDead is a factor, the first level of which is assumed to be the event of interest. Case Study 9.1 shows how to model a Bernoulli response.

When individual multinomial observations, as opposed to counts, are modeled then the response must be a factor. When Multinomial or Ordinal count data are modeled, the response should consist of a list of variates, separated by commas that provide the counts for each category, for example, Red, Green, Blue, Black, Other. Case Studies 4.6 and 4.7 show how to model ordinal and multinomial responses, respectively.

If the response data are censored then the '**Censored**' check box should be selected. On checking this box, another text box appears in which the name of the censoring variable should be entered. Censoring is handled in essentially the same way as in WinBUGS. The main response variable contains all the uncensored data, with blank cells, or 'NA', in the rows where data are censored. A separate data column is needed to specify the censored values. BugsXLA refers to this other data column as a variable of type '**Censor**'. This data column consists of blank cells, or 'NA', in the rows that are not censored, and a special entry type to define the censored values:

LE <value>: less than or equal to value specified
GE <value>: greater than or equal to value specified
IN <val1> <val2>: in the inclusive interval [val1, val2]; val1 < val2

Note that there must be a space between the initial operator (LE, GE or IN) and the first numeric value. In the case when the censored values are either all left censored or all right censored, the 'Set Variable Types' utility provides a quick way to add the prefix. The censoring variable must be defined as type 'censor' via the 'Set Variable Types' utility (refer to Section 2.4). Case Study 4.5 shows how to model censored data.

When the error distribution is Normal, it is possible to analyze data that has been collected in the form of a mean, standard error and, optionally, an associated degrees of freedom parameter. The **'Response is mean; se [;df]'** check box must be selected. A column of data should exist for each, and the response should be specified as a semicolon delimited list of their names in the order: mean; se [;df]. If the degrees of freedom are not specified then the standard errors are assumed to be known values rather than estimates. This type of data is typically associated with a **meta-analysis** of a series of published studies that have only reported summary statistics. More generally, it can be used to undertake a **systematic review** of any series of studies or experiments run to obtain information on a common quantity of interest. Case Study 5.3 shows how to model such summarized responses.

When the error distribution is Poisson, it is possible to define an **'Offset'**. The Poisson model is modified to take account of a fixed contribution to the linear effects for each unit, supplied by a variate referred to as the offset. The log of this variate is added to the linear predictor with a known coefficient of one. Hence, all the values must be positive. Case Study 4.4 shows how to model a Poisson response containing an offset.

2.3.4 Model Statements

The linear predictor consists of factors and covariates. In the classical terminology, factors can be considered 'Fixed' or 'Random' effects, and the covariates can be considered 'Independent' or 'Random Coefficients'. The 'Variable Selection' (VS) factors and covariates will be discussed later. Models are defined by using model-statement operators in a similar fashion to statistical packages such as SAS, S-PLUS, R, or Genstat. The following are legal operators for defining the model terms involving factors:

+   addition of another term
−   subtraction of a term
:   interaction between two terms
/   nesting of second term inside first, for example, A/B = A + A:B

*   addition of second term plus interactions with first, for example, A*B = A + B + A:B

@   only expand terms to order specified, for example, A*B*C@2 = A*B*C − A:B:C

Brackets, () or [], can be used to affect order of operator action. Brackets can also be used to shorten complex model definitions, for example, (A + B)*C is equivalent to A + B + C + A:C + B:C.

When categorical models are specified, the categorical factor must not be included in the model specification. For unordered categorical data, all terms are implicitly assumed to have separate effects for each category, that is, all the effects are implicitly defined as interactions with the response factor.

All the variables included in model statements in the 'Factors' part of the form must be defined as factors as explained in Section 2.4.

2.3.5 Continuous Covariates

The **independent covariates** model statement defines that part of the model consisting of variables assumed to have a simple linear relationship with the response. Cross-product terms can be defined in the model by the use of the ':' operator. The '*' operator has a similar use to that for factors, for example, A*B is equivalent to A + B + A.B, that is, both 1st order terms plus their cross-product. The '/' operator is not permitted. Factor by variate interactions can be specified; this being the usual way to fit separate regression coefficients for each level of a factor. It is not possible to specify different priors for these factors by variate regression coefficients (refer to Section 2.7). Note that if separate intercepts are also required then the factor should also be included as a Fixed term in the factor's model statement. Power terms can be specified, for example, A^2 is the quadratic term for variable A. The '@' operator is useful when polynomial models are being specified, for example, (X1*X2*X3*X4)@2 + (X1+X2+X3+X4)^2 would fit all terms up to the second order.

If one wants to model a covariate with a separate coefficient for each level of a grouping factor, but for these separate coefficient parameters to be "similar," and "exchangeable," rather than completely independent, then this should be specified in the '**Random Coeffs**' model statement. Random coefficients can be fitted by specifying a 'factor/variate' term. This specifies both the intercepts and the slopes, which will be given a bivariate Normal prior. Hence, the grouping factor should not be included in any factor model statement. Note that although it is strongly recommended that terms be specified using the '/' operator (variate within factor), ':' is recognized but will still lead to both the intercepts and slopes being given a bivariate prior. The current version only permits each factor to be modeled jointly with one other variate. Case Study 5.4 illustrates the fitting of random coefficients.

2.3.6 Variable Selection Terms

The VS factors and covariates are specific to Bayesian models as they define a mixture prior for these parameters that provide a hybrid approach between estimation and "significance" testing. For more details on Bayesian VS, refer to Section 8.5 of Wu and Hamada (2000), Chipman et al. (1997), or Woodward and Walley (2009). VS factors and variates are given a prior that includes a spike associated with no effect. Main effects and two factor interactions can be included. If any parents of the interaction terms are absent from the model statement, it is essential that these are included as either Fixed or Random terms. Bayesian VS parameters are discussed further in Section 2.7. Chapter 8 is dedicated to models that include VS terms.

2.3.7 Emax Function

As well as the linear predictor, a non-linear component can also be added to the model. By selecting the '**Non-Linear Model**' check box, another text box appears in which the non-linear model part can be specified. BugsXLA v5.0 only offers the **Emax** (also known as the sigmoidal logistic curve) non-linear model. An Emax model is specified by entering a statement of the form:

$$\text{Emax}\left\{[\text{Group}/]\,X\right\}$$

where Group is an optional grouping factor, and X is the covariate in the Emax function. Curly brackets, '{}', must surround the variables' names. When Group is specified, separate parameters can be estimated for each level of the grouping factor. In BugsXLA the Emax model is parameterized:

$$\text{Emax}\,X^{\text{Hill}}\,/\left(X^{\text{Hill}}+\text{ED}_{50}^{\text{Hill}}\right)$$

or (alternative equivalent parameterization)

$$\text{Emax}\,/\left(1+10^{-\text{Hill}(\log X\,-\,\log \text{ED50})}\right)$$

where logX and logED50 are $\log_{10}$ of X and ED_{50}, respectively.

   Note that the E0 parameter is absorbed into the model constant, or can be estimated separately for each group by including 'Group' as a fixed or random effect in the factors' model statement. Any of the three parameters, Emax, ED_{50}, or Hill, can be fixed to pre-specified constant values via their prior distributions (refer to Section 2.7). Chapter 7 is dedicated to the fitting of Emax models.

2.3.8 Serial Correlation and Measurement Error

BugsXLA allows the assumption of independent residual errors to be altered through the specification of a "**longitudinal**" component to the error structure. BugsXLA v5.0 only allows two types of longitudinal error structures to be specified here: a first-order autocorrelation structure ('exponential correlation model') with or without an additional independent measurement error component on each observation. These are specified by entering a statement of the form:

$$AR1\{Unit \, [/Time]\}$$

or

$$AR1e\{Unit \, [/Time]\}$$

where Time is an optional variate that denotes the relative time of each observation within each Unit, and Unit is a factor that defines the groups within which the errors are correlated. Curly brackets, '{}', must surround the variables' names. A single time-series can be specified by typing '1' instead of Unit. If the observations are equally spaced and listed in time order then the time variate is not required. There are speed advantages in avoiding the need to specify the time variate, but if this is essential then MCMC performance is better if integer values are used and a metric chosen that avoids 'large' time intervals between successive observations. If non-integer 'time deltas' exist then the within unit correlation is constrained to be positive. Measurement error models should only be considered when the data and/or prior are sufficiently informative to estimate this variance parameter. Chapter 9 is dedicated to the fitting of models with serial correlation in the residuals.

To apply a compound symmetry correlation structure (uniform correlation model) to the within unit errors, do not use this field but fit the Unit factor as a random effect. Note that this forces the within unit errors to be positively correlated. If covariates have been obtained to hopefully explain some of the within unit variation, then it is recommended that random coefficient terms are used to model their effects. It is often possible to simplify the explicit correlation structure needed for the errors by including such random effects.

If the error distribution is Poisson, then only the AR1{Unit [/Time]} model statement is permitted. This fits the Poisson transition model of Zeger and Qaqish (1988), which is analogous to a first-order autocorrelation structure. Case Study 9.3 discusses this approach in more detail.

2.3.9 Saving Informative Priors

It is possible, and if one planned to use informative priors probably quite desirable, to use BugsXLA to record these priors in advance of obtaining the experimental data. In this case, one should select the '**Eliciting Priors Only**' check box. You will need to specify the model as usual, and enter the name of the response. Providing you do not run BugsXLA again on this Workbook before entering the response data, your specified prior will be saved. You will also be prompted to save the prior to a new Excel sheet. If this option is selected then, after specifying the priors, instead of WinBUGS being run, the prior will be saved to the spreadsheet. Once the data has been obtained, this should be entered into the spreadsheet and the program run without any changes to the previously saved prior.

The final option on the form is the **'Predictions or Contrasts'** check box. Select this box if you wish to obtain predictive distributions for average or individual responses given fixed, or 'random', values for the predictor variables. It is also possible to specify contrasts to be estimated via this option (refer to Section 2.6).

The **'MCMC & Output Options'** button brings up another form that allows the user to control the WinBUGS run from within BugsXLA (refer to Section 2.5).

2.3.10 Loading a Previously Specified Model

Figure 2.4 shows an Excel sheet with output created after a BugsXLA analysis (see Case Study 3.1 for details of the analysis that led to this output). If the active cell on the worksheet contains the word 'Model' when clicking on the 'Bayesian Model' icon, the 'Model Specification' form is loaded with the entries saved from this previous analysis. In the case of Figure 2.4, cell A14 is active, which would lead to the form loading as shown in Figure 3.1. More generally, if 'Model' is not in the active cell then the form will load the last model specified in the current workbook. If no previous analysis has been run, or the 'Clear Old Analysis' button has been clicked on the 'Options for BugsXLA' form (see Figure 2.29) then the 'Model Specification' form will be blank as shown in Figure 2.3.

	A	B	C	D	E	F	G	H	I	J	K
1			Label	Mean	St.Dev.	2.5%	Median	97.5%		**WinBUGS Name**	
2			CONSTANT	-0.0394	0.0458	-0.1291	-0.0398	0.0520		Beta0	
3			Intercept at 0	0.0448	0.0637	-0.0784	0.0448	0.1693		alpha	
4		TRT	A	0.0000	0.0000					X.Eff[1,1]	
5		TRT	B	0.0922	0.0653	-0.0380	0.0923	0.2207		X.Eff[1,2]	
6		TRT	C	0.1490	0.0586	0.0342	0.1490	0.2638		X.Eff[1,3]	
7		TRT	D	0.0222	0.0645	-0.1060	0.0222	0.1468		X.Eff[1,4]	
8		TRT	E	0.0543	0.0639	-0.0693	0.0543	0.1818		X.Eff[1,5]	
9			FEV1_BASE	-0.0658	0.0418	-0.1470	-0.0661	0.0175		V.Coeff[1,1]	
10			SD(residual)	0.1835	0.0141	0.1587	0.1827	0.2137		sigma	
11	Note: CONSTANT & Factor effects are determined at the mean of the covariate(s).										
12	Interpret these cautiously when Factor x Covariate terms have been fitted.										
13											
14	**Model**	['clin study'!$A$1:$D$96]									
15	Distribution	Normal									
16	Link	Identity									
17	Response	FEV1_CFB									
18	Fixed	TRT									
19	Covariates	FEV1_BASE									
20											

FIGURE 2.4
Output sheet with cell containing 'Model' selected so that previous model specification will be loaded into BugsXLA's Model Specification form.

2.4 Set Variable Types

If any of the variables in the data need to be treated as **factors**, that is, categorical data, or are **censored** variates, then this needs to be specified by clicking on the 'Set Variable Types' button on the main model specification form. This loads a utility that facilitates the defining of the type of each variable. Variables can be of type:

Variate: Real numbers used to define a quantitative variable. All variables are assumed to be variates by default, and so there is no need to use this utility unless some need to be changed.

Factor: Text or numbers used to define a categorical variable. Be warned that any commas found in a factor's level label will be automatically replaced by '_'.

Censor: Special text plus real numbers to define censoring.

2.4.1 Defining Factors

On clicking the 'Set Variable Types' button the '**Create Factors**' form is loaded (Figure 2.5). Initially, all the variables in the data range will be in the 'Variates' list box. Highlight all of the variables that need to be treated as factors, using any standard Microsoft Windows approach, and click on the button with the left pointing arrow marks to move them to the 'Factors' list box. It is now necessary to define the level labels for each of the factors. To do this, highlight all of the variables in the 'Factors' list box and click on the '**Edit Factor Levels**' button. This will bring up the 'Edit Factor Levels' form.

Each of the factors highlighted in the 'Factors' list box on the 'Create Factors' form will be loaded in turn into this form. The name of the factor having its levels defined is shown above the list box in Figure 2.6. By default, it will load the list box on this form with all the unique entries in the

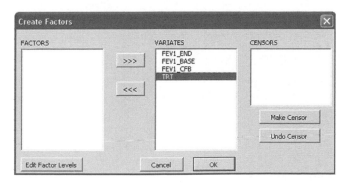

FIGURE 2.5
Create factors form.

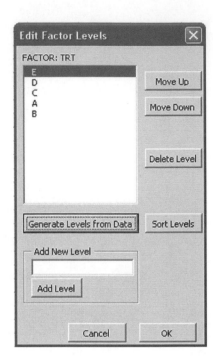

FIGURE 2.6
Edit factor levels form.

column of data for this factor. If any of these level labels originally contained commas, a message will be displayed warning the user that these will have been replaced by the underscore ('_') character; this is to avoid problems that would occur if a comma-separated format data-file was created for use in the R script facility (refer to Section 2.5). Unless the user wants to change the order of the factor levels, using the 'Move Up', 'Move Down', or 'Sort Levels' buttons, the 'OK' button should be clicked to accept these changes and load up the next factor for editing. The main reason for changing the order of a factor's levels is to specify which one should be used as the reference level for a fixed effect, Bernoulli or Categorical response; these issues are discussed when these types of model are exemplified in later chapters.

2.4.2 Defining Censors

Not only are factors defined using this utility, but also **censoring variables** are defined here. Refer to the explanation of how BugsXLA handles censored variables given in Section 2.3. If the censored values are all left censored, that is, less than or equal to a value, or all right censored, that is, greater than or equal to a value, then it is simplest to just enter the censored values into the appropriate cells on the spreadsheet in the column for the censoring variable. The 'Set Variable Types' utility can then be used to prefix all these values with either 'LE' or 'GE' as appropriate. This is done on the '**Create Factors**' form (see Figure 2.5) by highlighting the variable that needs to be defined as a Censor, clicking on the button '**Make Censor**', and then selecting the appropriate option. If the data are censored in more than one way, or there is interval censoring, then the prefixes have to be manually entered. Note that it is still necessary to define the variable as a censor, either by creating a comment with the text 'CENSOR' in the cell containing the variable's name, or by using the utility described above and choosing the '**Mixed censoring (manual edit)**' option after clicking on the 'Make Censor' button. Case Study 4.5 provides an example of how censored data are defined in BugsXLA.

Note that the variable types only have to be set up once, as the information provided is stored in comments on the spreadsheet in the cells containing the variables' names. However, if you subsequently edit the data, it may be necessary to rerun the 'Set Variable Types' utility to ensure the factors are defined properly, for example, if you make additional, or remove completely, factor levels. In this case, the '**Generate Levels from Data**' button should be clicked on the 'Edit Factor Levels' form in order to refresh the list of levels for this factor (see Figure 2.6).

2.5 MCMC & Output Options

Clicking on the 'MCMC & Output Options' button at the bottom of the main Model Specification form brings up the following form (shown after the 'Use Preset Values' check box has been selected) (Figure 2.7).

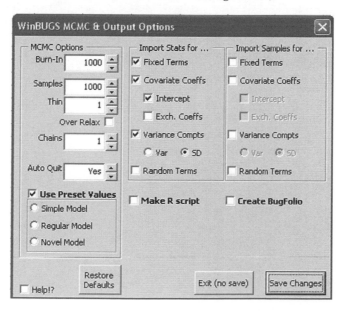

FIGURE 2.7
WinBUGS MCMC & output options form.

It is here where both the options for controlling the WinBUGS MCMC run, as well as the summary statistics and simulated sample values to be imported, are set. It is also possible to request that all the files be saved that are created by BugsXLA to define the WinBUGS analysis. Chapter 11 discusses how one might use the files saved to undertake an analysis beyond the current capabilities of BugsXLA. R scripts can also be requested to facilitate further analysis using the R environment. Appendix D provides more details on the R scripts that BugsXLA can create. Refer to Chapter 1 for a brief introduction to WinBUGS and an overview of the MCMC approach.

2.5.1 WinBUGS MCMC Options

The following **MCMC Options** can be set:

2.5.1.1 Burn-In

The number of initial MCMC samples to be discarded. Convergence is assumed past this point. See Section 1.2 for information on checking convergence.

2.5.1.2 Samples

The number of MCMC samples to be generated from the posterior distribution. Together with the degree of thinning, and whether over-relaxation is applied, this will determine how precisely the posterior distribution's parameters are estimated. Note that when multiple chains are specified, this number of samples is generated for each chain. As a rule of thumb, providing the burn-in is long enough to achieve convergence, an MC Error less than 5% of the posterior standard deviation of quantities of interest is sufficient precision for most inferences.

2.5.1.3 Thin

If set to K, then only every Kth sample is saved. Use when high autocorrelation is present to reduce the number of samples required to give good coverage of the posterior distribution.

2.5.1.4 Over Relax

This will lead to an over-relaxed form of MCMC as detailed by Neal (1998). It is another approach to obtaining samples with low autocorrelation (see also 'Thin'). Like thinning it also increases the run time per iteration. Over-relaxation is not always effective.

2.5.1.5 Chains

The number of separate chains to be generated. Select more than one to check convergence (Auto Quit will be turned off). Use Brooks–Gelman–Rubin diagnostic check (bgr diag) in WinBUGS to assess convergence. It is strongly recommended to do this when fitting a novel model or the data's structure is relatively complex.

2.5.1.6 Auto Quit

Change to 'No' if want to stay in WinBUGS after MCMC run. This is necessary when wanting to check convergence using WinBUGS.

The default settings of these values are not expected to be appropriate for many analyses. Some alternative **MCMC preset values** have been provided as a guide to values that might give reasonable parameter estimates.

2.5.1.7 Simple Model and Data Set

An orthogonal or nearly orthogonal design. All the covariates in the model should be only weakly correlated with other independent variables. All variance estimates have four or more effective degrees of freedom.

2.5.1.8 Regular Model and Data Set

A model or data set that is more complex than that suggested for the 'simple settings', but you are still confident will converge relatively quickly. I would recommend that the data set is assessed using 'novel settings' and the WinBUGS diagnostics before making inferences and decisions based on this analysis.

2.5.1.9 Novel Model or Complex Data Set

Use these settings when you wish to work within WinBUGS to assess convergence and mixing before finally determining the appropriate MCMC settings. This is recommended whenever a novel model has been developed, usually by building up slowly from simpler models that converge and mix reasonably well.

Mixing can often be greatly improved if informative priors are provided for poorly estimated parameters, particularly for non-linear terms and variance components. These need not be strong priors, just sufficiently precise to effectively rule out incredibly extreme values. You may also need to review the initial values automatically generated by BugsXLA, but these should be reasonable in most cases as they are designed not to be too extreme relative to the priors set (but see Section 11.2 for an exception). Appendix B provides more information on how BugsXLA determines initial values for the MCMC chains.

2.5.2 Posterior Summaries and Samples

The '**Import Stats for…**' section on the form allows the user to specify which parameter types will have posterior distribution summary statistics imported back into Excel. Select the check boxes against those terms whose posterior distribution summary statistics you wish to import. The following terms, if in the model, are imported by default:

Constant, Weibull r, generalized logit power parameter, Categorical model intercepts and ordinal cut points, t df parameter and the non-linear model Emax, ED50, and Hill parameters that are not 'group specific'.

Any non-linear 'Fixed effect' parameters are imported with the Fixed Terms, any 'Random effect' parameters are imported with the Random Terms and the 'Random effects' variance components are imported with the Variance Compts. The AR1 'longitudinal model' parameter is imported with the Variance Compts.

Similarly, the types of parameters for which the simulated sample values are to be imported back into Excel, can be altered using the check boxes under '**Import Samples for...**'. Note that this will result in columns of length specified by 'Samples' being imported into Excel. These can be used to assess the shape of the full posterior distribution, or to derive posterior distribution samples of functions of these terms. The 'Post Plots' utility on the BugsXLA toolbar can be used to produce a histogram of imported samples with the prior distribution overlaid if it can be inferred from the specified model (refer to Section 2.11).

WARNING: if the model contains numerous variables of the type you select to import, and there are many thousands of samples being simulated, Excel can be incredibly slow to import all the samples (and in extreme cases can simply hang). It is recommended that you first ensure you have imported and saved the summary statistics output from your analysis before rerunning and specifying samples to be imported. Also, consider using thinning and importing only 1000 samples, if this will give sufficient precision (refer to Section 2.10 for instructions on how to thin the samples at the import stage). If you are familiar with the R package, you might find it more convenient to let BugsXLA make an R Script and use this to import the samples into R via BRugs instead.

2.5.3 Bugfolios

If the user is interested in using the files created by BugsXLA, or simply in viewing them, then they can be saved by selecting the '**Create Bugfolio**' check box. If this is not selected, then the automatically generated WinBUGS code, script, data, initial values, and 'log' files will not be saved unless specifically requested at the importing stage or the default save option set on the 'BugsXLA Options' form. Usually these files are not of interest to the user and they are deleted after the results have been imported back into Excel. However, if the user wishes to manually edit the code generated by BugsXLA and reanalyze the same data, perhaps to fit a more complex model than the program currently offers, then this box should be selected so that these files will be saved in a sub-folder of the active directory (name entered by user in the text box visible when this option selected). A BugFolio is automatically created if R Scripts are requested. Note that if the code is manually edited, it may be necessary to edit the script, data, or initial values files in order for WinBUGS to run without crashing. An advantage of explicitly specifying a BugFolio name here, rather than relying on the default, is that the edit and run script utilities will link directly to the files created (refer to Section 2.13). Refer to Chapter 11 for some examples of how to work with these saved files.

2.5.4 R Scripts

Users familiar with the R programming language may also be interested in obtaining the R scripts that BugsXLA can provide. These are obtained by selecting the '**Make R script**' check box. This will lead to the creation of an R Script that, via the BRugs package, can use the files created by BugsXLA. This enables the Bayesian analysis specified to be run from within the R environment. The WinBUGS analysis is often significantly faster when run via BRugs. Hence, for very complex models, it is worth considering a strategy that uses BugsXLA to create the code and ancillary files, but with nominal Burn-In and Sample numbers of 1000, say. Then edit the R Script created by BugsXLA to run a more appropriately sized MCMC analysis. Note that the results of the BRugs analysis will not be identical to that obtained via BugsXLA, since a different set of MCMC samples is generated. More importantly, the author has on some occasions found that BRugs has failed to converge even though running the same analysis in WinBUGS directly converged very quickly. Another R 'EDA' script is also created that facilitates other 'non-Bayesian' exploratory data analyses, including in some cases a classical likelihood based method of analysis. These R scripts are saved in a sub-folder of the active directory (name entered by user in the text box visible when this option selected). See Appendix D for more details on these R scripts.

The default settings for this form can be restored by clicking on the '**Restore Defaults**' button. These defaults can be altered by the user on the 'BugsXLA Options' form, accessible via the main BugsXLA toolbar (refer to Section 2.12).

2.6 Predictions and Contrasts

If the 'Predictions or Contrasts' box is checked on the main Model Specification form, then the following form appears after clicking on the 'OK' button (Figure 2.8). The 'Predictions Range' refers to a range on the Excel sheet that has already been populated with entries to define the predictions and contrasts required. The first row in this range must contain the

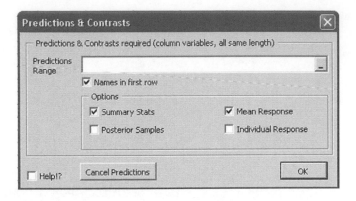

FIGURE 2.8
Predictions and contrasts form.

names of the variables whose settings are being specified. Each additional
row defines the settings with the following entries being allowed when specifying predictions:

1. Blank cells, or '-', are interpreted as

   The zero constrained level for fixed effects

   The mean of the distribution for random effects

   The mean value for covariates (cross-product/power terms are fixed
   at the cross-product/power of the covariates' means)

   Note that variables not specified at all are treated the same as a blank
   entry.

2. Specific level of a factor, or the value of a covariate.

3. * gives predictions for every level of this factor (not valid if factor in
   covariate or non-linear model only).

4. ~ gives a prediction for a new (unobserved or future) level of this
   factor (only allowed for random effects, and cannot be used in non-
   linear model). When a contrast has been set, different new levels are
   used on each side.

Contrasts are specified as follows:

$$C(<1>,<2>)$$

where <1> and <2> are the two factor levels or variate values that define the
contrast. No entry in one side is interpreted as defined for blank cells above.
The '*' character can be used on one or both sides to specify all levels for this
factor. Examples of valid contrast specifications:

$$C(1,3) \text{ or } C(,*) \text{ or } C(*,*)$$

When the link is different from the identity, the contrasts are back-transformed to the original metric before reporting. Hence, when the link is log the contrast is a ratio, and for logit (only binomial or ordered categorical allowed) it is an odds ratio.

To illustrate the various ways that predictions and contrasts can be specified, consider a model for a response, one observation per subject, that includes a fixed effects factor **treatment** with five levels labeled A–E, a fixed effects factor **sex** with two levels labeled male and female, a covariate **age**, and a random effects factor **study** with numerous levels, one of which is labeled A100. Columns 2–5 in the table below represent a range of cells that would be a valid Predictions Range for BugsXLA; the first column is included simply to aid the explanation that follows.

	treatment	sex	age	study
1	A	male	30	A100
2	*			
3	*	*		
4	A	female	20	~
5	c(A,B)			
6			c(50,20)	
7	c(A,*)	*		
8		c(female,male)	c(20,30)	
9	c(*,*)			

The header row contains the names of the variables in the model. The nine rows beneath the header instruct BugsXLA to estimate:

1. The predicted response assuming treatment A is applied to a male subject of age 30 in study A100. Note that although study is defined as a random effect, it is still possible to make inferences about a specific level of this factor.

2. The predicted responses for each of the five levels of treatment, each assuming that the sex of the subject is equal to the constrained to zero level of this factor (the first level by default), the subject's age is equal to the mean observed in this study, and the study effect is zero (its population mean). The '*' symbol simply saves the effort of having to specify five separate rows, one for each level of the factor.

3. The predicted responses for each combination of the five levels of treatment and two levels of sex, with the same assumptions as in row 2 for variables age and study. This statement generates 10 estimates.

4. The predicted response from a future, or unobserved, study that is considered exchangeable with those in this study, with treatment A applied to a female subject of age 20.

5. The contrast between the mean responses of treatments A and B ("A – B"). If no interaction terms involving the factor treatment are included in the model, then the entries in the other columns will cancel out unless these are also contrast statements (see explanation of rows 7 and 8 below).

6. The difference between the mean responses of subjects of ages 50 and 20.

7. The contrasts between the mean responses of treatments A and each of the other four levels of treatment in turn, estimated for both levels of sex. If a treatment-by-sex interaction term is in the model, then the estimated treatment contrasts will depend upon the sex of the subjects. This statement generates eight contrast estimates.

8. The contrast between the mean responses of 20 year old females and 30 year old males. This statement generates one contrast estimate.

9. The contrasts between all possible pair-wise combinations of treatments. This statement generates 10 contrast estimates.

It is only necessary to create a column in the Predictions Range for each variable that you wish explicitly to specify a value in the predictions or contrasts required. The absent variables will have implicit values set as if a blank cell was entered in a column set up for this variable. VS factors are excluded from the prediction equation when absent or blank cells specified. When a grouping factor is set for the non-linear model, blank cells for this factor can be interpreted differently for each parameter depending upon their prior. For this reason, the header that is displayed for such predictions denotes a "blank setting" by '(0)', and the user must be aware of the priors for each of the non-linear parameters when interpreting the prediction.

The zero constrained level for fixed effects is discussed in Sections 2.7 and 2.12, as well as in Chapter 3 when fixed effect factors in the Normal Linear Model are explained.

The contrast statement gives the difference, on the link scale, between the responses at the two settings, that is, first minus second setting. Whenever possible, this contrast is back-transformed to the natural metric. As illustrated in row 8, it is possible to specify a contrast for more than one variable in the same row of the Predictions Range. The contrast will still be the difference of the responses at the two settings, but each variable with a contrast specified will have a different value on each side of the contrast. If a contrast is specified for a variable that is also included in any interaction terms in the model, then the contrast will be affected by the setting for any variables with which this variable interacts. If the '~' symbol is entered for a random factor in a line in which a contrast is specified for another factor, then a different new level for the random factor is assumed for each side of the contrast.

Just as on the 'MCMC & Options' form (refer to Section 2.5), it is possible to request both summary statistics as well as individual simulated sample values. In many cases, it is also possible to request both the mean response and the response predicted for a future individual experimental unit. Strictly speaking, only the latter quantities are truly 'predictions', but this term is used more loosely here to be consistent with much other mainstream statistical software. Individual observations are not reported for any contrasts that are specified. Note that when a non-linear link function is used (e.g., logit) and a random effect has been set to its mean value, the prediction is not of the mean response averaged over the population of random effects. In this case the prediction should be thought of as the mean response for a 'typical' random effect.

If you decide not to request any predictions or contrasts, click on the button labeled '**Cancel Predictions**'. You will also need to 'Cancel Predictions' if you find an error at this stage in the cell entries defining the predictions. After the form closes, you will need to exit from each of the remaining forms in turn until you are completely out of BugsXLA. The cell entries can then be edited and BugsXLA rerun.

2.7 Prior Distributions

After clicking the 'OK' button on the initial Model Specification form, or the Predictions & Contrasts form if specified, the Prior Distributions form appears.

2.7.1 Independent Factors and Model Constant

This form contains a separate page for each parameter type in the model. The 'Ind. Factors' page allows the user to alter the priors for the independent (**fixed effects**) factors as well as the model constant term (Figure 2.9). BugsXLA provides default prior distributions for all parameters in the model; these defaults can be seen and altered on the BugsXLA Options form accessible from the BugsXLA toolbar (refer to Section 2.12). The default specification on this page is essentially equivalent to the classical approach of fitting 'fixed effects' to these terms. Corner constraints are imposed by effectively fixing either the first or last level of each factor to zero (refer to Section 2.12 for details of how to alter the corner constraints used). Although this is not strictly necessary in theory for a Bayesian analysis, it greatly improves the performance of the MCMC method.

Each of the fixed effect terms in the model can have their prior distribution altered by first clicking on the term in the list box. Note that it is only possible to alter the prior standard deviation associated with a term's effect contrast; this contrast is the difference between the effect of any unconstrained level of the factor and the constrained level. Fixed effects either have their first

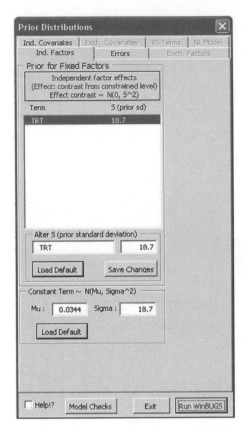

FIGURE 2.9
Prior distributions form: Independent factors page.

level set to zero (as in Genstat, R, or S-PLUS) or the last level (as in SAS); the 'BugsXLA Options' form shows the current setting, which can also be changed there (refer to Section 2.12). It is necessary to click on the '**Save Changes**' button in order for the prior standard deviation to be updated; the current prior can be seen next to the term's name under the 'S (prior sd)' column in the list box. The default setting for these standard deviations is 100 times the standard deviation of the response. This should be a close approximation to a flat prior over the range supported by the likelihood, which should give inferences similar to those obtained by maximum likelihood (ML) methods (providing similarly weak priors are used for the other parameters). Note that the prior mean cannot be altered from zero. When a link function other than the Identity is used, then the standard deviation of the response is calculated on the 'link transformed Y' (with adjustments to avoid infinite values). Similarly, log(response) is used for the Log-Normal model, when the linear predictor is on the log-scale. In these cases, care should be taken when specifying informative priors, as the parameters are specified on the link scale. When probabilities are being modeled then the prior standard deviation is set to 100.

The Normal prior for the **model constant** has a mean equal to the mean of the response and a standard deviation approximately equal to 100 times the standard deviation of the response. This should be a close approximation to a flat prior over the range supported by the likelihood, with the same caveats as above. The same adjustments for the link function as discussed above are made. The constant term is effectively the predicted mean response for the constrained levels of the fixed effects. The constant term can be set to zero by giving it a N(0, 0) prior.

2.7.2 Residual Variance, Outliers, and Degrees of Freedom

The 'Errors' page allows the user to alter the prior for the lowest level **residual error** term. It also provides the option to model outliers explicitly by using a mixture distribution for the individual observations; this distribution being a combination of the model specified and a variance inflated distribution with the same mean (Figure ? 10).

When the error distribution is Normal the errors' prior is determined by that for σ^2, the residual variance. The more natural parameter to work

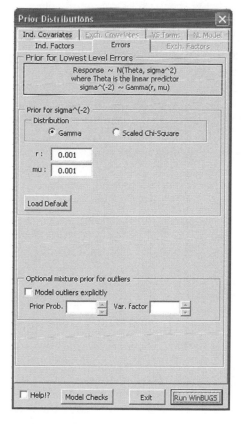

FIGURE 2.10
Prior distributions form: Normal errors page.

with in the Bayesian paradigm is the precision, σ^{-2}, the inverse of the variance. The Gamma Distribution is the conventional prior for this parameter (see, e.g., Gelman et al. (2004)). BugsXLA allows the prior for the precision parameter to be Gamma or the scaled Chi-Square, which is just a special case of the Gamma. The preset default prior is a Gamma Distribution with very small parameter values; this will lead to inferences very similar to those obtained using ML methods providing similar weak priors are used elsewhere. A scaled Chi-Square Distribution can be used to set an informative prior. One possibility is to use the conventional estimate of variance (s^2) and its degrees of freedom (df) from a previous study; this would be appropriate if there was little prior information before this previous study, and the residuals are believed to be exchangeable with those from the current study. This latter belief will rarely be held very strongly, so it is sometimes considered more reasonable to reduce the degrees of freedom associated with the estimate, say to four, so that the prior is not allowed to heavily influence the final inferences. Refer to Chapter 5 of Spiegelhalter et al. (2004) for more sophisticated ways to determine an informative prior.

BugsXLA provides a method for modeling outliers explicitly that is based on an approach first discussed by Box and Tiao (1968). By selecting the '**Model outliers explicitly**' check box, a model is fitted that will be robust to extreme observations. Each observation is assumed to come from either the distribution defined by the model or a variance inflated distribution. The latter distribution models the process that is responsible for the outliers. The user can alter the prior probability of an observation being an outlier. The user can also alter the factor by which the variance of the Normal Distribution associated with outliers is inflated. The mean of this distribution is unchanged. When **Binomial** data are being modeled, outliers can still be modeled explicitly. In this case, the outliers are assumed to come from a Binomial Distribution whose probability parameter has a Uniform Distribution between 0 and 1. For **Poisson** data the outliers are assumed to come from a Poisson Distribution whose mean is still defined by the model, but has an additional Normal error with variance equal to the inflation factor times the Poisson mean. Chapter 10 provides examples of fitting these models robust to outliers.

If the error has a **t-Distribution**, then the df parameter also needs a prior distribution; the scale parameter is treated in the same way as the variance for a Normal model (Figure 2.11).

The df parameter is given an Inverse-Uniform Distribution. The user can specify an informative prior by changing the bounds of this distribution. Alternatively, the df parameter can be fixed constant rather than trying to estimate it from the data. A common choice is four, providing an alternative outlier robust model to the mixture prior described above. The preset default prior for the df parameter is a very vague Inverse-Uniform Distribution bounded below by 0, and above by 0.5. This ensures that the df parameter cannot be smaller than 2. Note that if we had put a uniform prior on df itself this would give too much weight to large df values, which is equivalent to

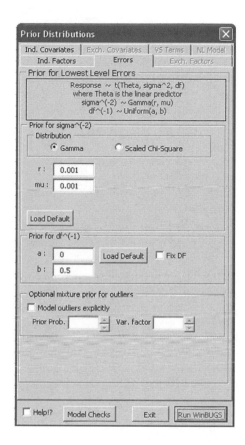

FIGURE 2.11
Prior distributions form: t-Distribution errors page.

a very strong prior belief that the model's errors are effectively Normal distributed. The t-Distribution should only be used when there is a reasonably strong belief that the Normal errors model is likely to be inadequate. If the user wanted to fix the df parameter, then this is done via the '**Fix DF**' check box; it is necessary to click on the '**Update Prior**' button to alter the saved distribution. Case Study 10.1 provides an example of a model with a t error distribution.

2.7.3 Serial Correlation and Measurement Error

If a **longitudinal error** structure is specified then the prior distributions for the additional parameters are specified on this page also. The form below shows the error prior distribution page when the within unit errors have been given a first-order autocorrelation structure (exponential correlation model) with an additional independent 'measurement error' component on each observation (Figure 2.12).

A Beta Distribution scaled to the range [−1, +1] is used as the prior for the first-order autocorrelation parameter phi. This parameter can be

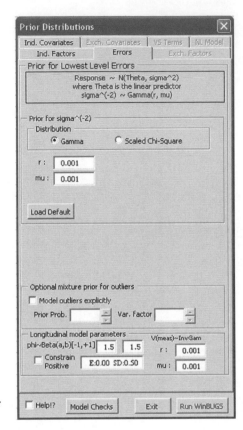

FIGURE 2.12
Prior distributions form: Longitudinal error parameters.

constrained to be positive, in which case it will be given a standard Beta Distribution prior. If a time variate is specified with non-integer time differences between successive observations, then phi is automatically constrained to be positive. The parameters of the prior can be altered, but must lie in the range [1, 99], and will be rounded to two significant figures. The mean (E) and standard deviation (SD) of the prior is shown to guide the choice. Note that the Beta (1, 1) is a Uniform Distribution across the whole range. The preset default of Beta(1.5, 1.5) was chosen as it is quite 'vague' while giving relatively low prior probability to extreme positive or negative values for phi; this is thought to be generally appropriate, and also has been found to help improve mixing for some problems when compared to the uniform. The variance parameter defining the measurement error is given an inverse-gamma prior. The preset default has scale (r) and shape (μ) parameters both equal to 0.001. Case Study 9.2 fits an auto-regressive correlation structure to the residuals.

When a longitudinal error structure is specified for a Poisson model, the log-linear transition model introduced by Zeger and Qaqish (1988) is fitted. As part of this model, it is necessary to specify a constant, d, that is used to 'restart' the 'autoregressive' Poisson process when a zero value is observed. Case Study 9.3 illustrates this Poisson transition model.

2.7.4 Distribution Shape Parameters

When the error distribution is Gamma, the linear predictor provides a log-link for the mean, which equals the ratio of the Gamma parameters: r/μ (Figure 2.13). The scale parameter, r, is itself given a Gamma prior distribution. The residual coefficient of variation (CV) is directly related to this parameter: $r = 1/CV^2$. Since on the log-scale the CV is in some ways analogous to the standard deviation on the natural metric, the implied prior for the CV here is analogous to the usual inverse-gamma for the Normal residual variance.

When the error distribution is Weibull, the linear predictor provides a log-link for the median (not the mean), with the following relationship to the Weibull's parameters: median = $(\ln(2)/\theta)^{1/r}$ (see Appendix A for parameterization used for Weibull Distribution) (Figure 2.14). The preset default prior for the shape parameter, r, is a very dispersed Exponential Distribution, defined as a special case of the Gamma Distribution to allow more flexibility

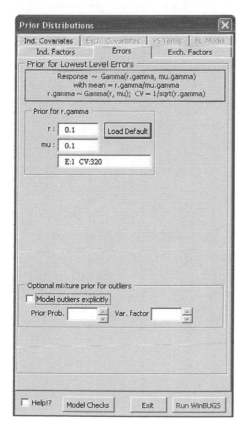

FIGURE 2.13
Prior distributions form: Gamma errors page.

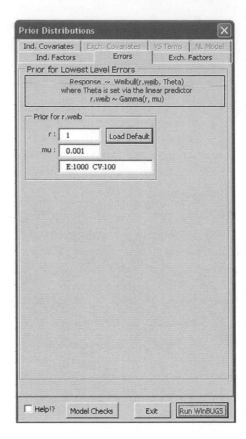

FIGURE 2.14
Prior distributions form: Weibull errors page.

in setting an alternative prior. This will favor smaller values, but decreases very slowly on the positive real line. The mean (E) and coefficient of variation (CV%) of this prior updates as the parameters are altered.

When the generalized logit (**GenLogit**) link is used, then a prior must be specified for the 'power parameter'. This is given a very dispersed Gamma prior distribution with a mean of one; a Gamma(0.25, 0.25). The mean (E) and coefficient of variation (CV%) of this prior updates as the parameters are altered.

2.7.5 Random Effects (Factors)

The 'Exch. Factors' page allows the user to alter the priors for the exchangeable (**random effects**) factors (Figure 2.15). The default specification is essentially equivalent to the classical approach of fitting 'random effects' to these terms. The distribution of the random effects can either be a Normal (the standard formulation of the model) or a t-Distribution with four degrees of freedom, with the scale parameter ("standard deviation") being denoted by tau. There are five types of prior distribution available for tau:

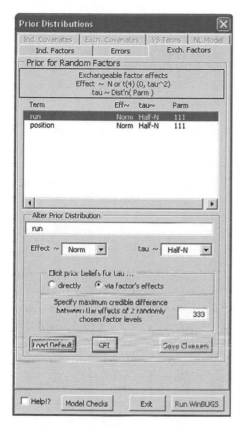

FIGURE 2.15
Prior distributions form: Exchangeable factors page.

Half-Normal (Half-N)

This has been recommended by Spiegelhalter et al. (2004).

Half-Cauchy (Half-C)

This has been recommended by Gelman (2006). Note that some problems have been noticed when using the Half-Cauchy with the Probit or CLogLog link.

Uniform, bounded below at 0

This has been suggested by various people, including Spiegelhalter and Gelman.

Gamma: note this is for tau^{-2}, not tau

This allows the 'naive' default of Gamma(0.001, 0.001) to be used.

Scaled Chi-Square: note this is also for tau^{-2}

This allows a more informative prior as discussed for the residual variance term.

The prior for the variance component parameter can either be defined directly, or via the implications for the size of the factor's effect. If done directly this implies you will specify the maximum credible value for tau;

this being interpreted as the approximate 95%ile of its prior distribution for Half-N and Half-C Distributions (2× and 12× scale parameter, respectively), and the upper limit for the Uniform Distribution. If done via the factor's effects, this implies you will specify the maximum credible difference between the effects of two randomly chosen levels for this term; this being interpreted for Half-N and Half-C Distributions as the approximate 95%ile of the prior distribution for the difference between these effects. This is approximated by 3× (Half-N), 14× (Half-C), and 2× (Uniform) the scale parameter of this prior when the effects have a Normal Distribution, and by 4× (Half-N), 18× (Half-C), and 2.5× (Uniform) this scale parameter when the effects have a t_4 Distribution. These factors relating the scale parameter to the ~95%ile of the distribution of the difference between the effects of two randomly chosen levels were calculated via simulation.

The preset default setting for the parameter defining the prior for tau is five times the standard deviation of the response. This should be a close approximation to a flat prior over the range supported by the likelihood, which should give inferences similar to those obtained by ML methods, providing similarly weak priors are used for the other parameters. Note that the Half-Normal and Half-Cauchy Distributions for informative priors gives more weight to smaller values, effectively shrinking the posterior estimates of variance toward zero. When a link function other than the Identity is used, then the standard deviation of the response is calculated on the 'link transformed Y' (with adjustments to avoid infinite values). In these cases, care should be taken when specifying informative priors, as the parameters are specified on the link scale. When probabilities are being modeled then the prior parameter is set to five. You must click on 'Save Changes' after altering the prior distribution.

When the Half-Normal prior is selected, the **'GFI'** (Graphical Feedback Interface) button can be clicked to facilitate elicitation of an informative prior. This enables the user to see the prior distribution inferred by the value elicited. The graph is dynamically updated to facilitate the elicitation process. Refer to Section 2.8 for more information.

2.7.6 Independent Covariates

The 'Ind. Covariates' page allows the user to alter the priors for the **independent covariates** (Figure 2.16). The default specification is essentially equivalent to the classical approach of fitting fixed effect linear regression terms. The default prior is a Normal Distribution. The Half-Normal Distributions allow the user to specify that the coefficient must be positive (HN+) or negative (HN−). Note that the Half-Normal Distribution for informative priors gives more weight to smaller values, effectively shrinking the posterior estimate of the coefficient toward zero. The default setting of the prior standard deviation (S) is 100 times the ratio of the standard deviations of the response and the covariate. This should be a close approximation to a flat prior over

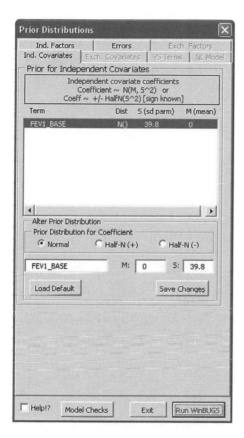

FIGURE 2.16
Prior distributions form: Independent covariates page.

the range supported by the likelihood, which should give inferences similar to those obtained by ML methods providing similarly weak priors are used for the other parameters. Note that when a full Normal Distribution is specified, the prior mean can be altered from the default of zero. When a Half-Normal is specified the 'S' parameter is not the standard deviation, but the Half-Normal's scale parameter. When a link function other than the Identity is used, then the standard deviation of the response is calculated on the 'link transformed Y' (with adjustments to avoid infinite values). In these cases, care should be taken when specifying informative priors, as the parameters are specified on the link scale. When probabilities are being modeled then the default value for S is 100 divide by the standard deviation of the covariate. You must click on 'Save Changes' after altering the prior distribution.

2.7.7 Random Coefficients

The 'Exch. Covariates' page allows the user to alter the priors for the exchangeable covariates (**random coefficients**) (Figure 2.17). The model

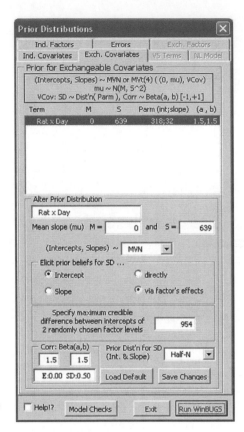

FIGURE 2.17
Prior distributions form: Exchangeable covariates page.

specified is essentially equivalent to the classical approach of fitting 'random coefficients' to these terms, although here the intercepts and slopes are given a Bivariate Normal prior distribution. The mean intercept is fixed at zero since the model constant term will include any offset. The preset default for the mean slope is a Normal Distribution with zero mean and standard deviation equal to 100 times the ratio of the standard deviations of the response and the covariate. The prior variance–covariance of the intercepts and slopes is set via their standard deviations and correlation. The preset default prior distribution for the standard deviations is a Half-Normal with scale parameter of five times the standard deviation of the response (intercept) or five times the ratio of the standard deviations of the response and the covariate (slope). A Beta Distribution with default parameters (1.5, 1.5), scaled over the range [−1, +1], is used for the correlation. These priors should be a close approximation to a flat prior over the range supported by the likelihood. When a link function other than the Identity is used, then the standard deviation is calculated on the 'link transformed Y' (with adjustments to avoid infinite values). In these cases, care should be taken when specifying informative priors, as the parameters are specified on the link scale. When probabilities are being modeled, the prior standard deviation for the mean slope is set to 100 divide by the standard deviation of the covariate, and the standard deviation scale parameters are set to five (intercept) and five divide by the standard deviation of the covariate (slope).

The priors for the variance component parameters can either be defined directly, or via the implications for the size of the factor's effect. Three types of prior distribution are available: Half-Normal, Half-Cauchy, and Uniform. Refer to the 'Exchangeable Factors' page above for further information on these options. You must click on 'Save Changes' after altering the prior distribution.

A Beta Distribution scaled to the range [−1, +1] is the preset default prior for the correlation. The parameters can be altered, but must lie in the range [1, 99], and will be rounded to 2 significant figures. The mean (E) and standard deviation (SD) of the prior is shown to guide the choice. Note that the Beta(1, 1) is a Uniform Distribution across the whole range, [−1, +1]. A default of Beta(1.5, 1.5) was chosen as it is quite 'vague' while giving relatively low prior probability to extreme positive or negative correlation; this is thought to be generally appropriate, and also has been found to help improve mixing for some problems when compared to the uniform.

It is important to note that BugsXLA automatically centers covariates, and the 'intercepts' are therefore the mean responses at the mean of all the covariates. Centering the covariates affects the interpretation of the correlation parameter as well as the intercept parameters, and so the user should be aware of this both when determining priors and when interpreting the parameter estimates. Case Study 5.4 provides an example of a model that includes random coefficients.

2.7.8 Variable Selection Terms

The 'VS Terms' page allows the user to alter the priors for the **VS** factors and covariates (Figure 2.18). A mixture prior for the effects (factors) or coefficients (variates) is specified. With probability pi each term is active, that is, included in the model, otherwise it is excluded by setting its value to zero. Active terms are given a Normal prior with mean zero and variance as described below. Factors are fitted using sum-to-zero constrained dummy variables with Multivariate Normal priors inferred from the prior specified on the factor effects. A Gibbs VS approach is adopted similar in essence to that first proposed by Kuo and Mallick (1998).

The preset default prior for pi is a Beta Distribution with parameter settings that represent a reasonably strong belief that pi is close to 0.25: a Beta(30, 90). The prior probabilities of activity for the interaction terms determine the type of heredity principle that is applied (see Chipman et al. (1997) for more on heredity principles). Prior probabilities of activity for terms whose parents are all active (p_{11}), whose parents are all inactive (p_{00}), or who has some

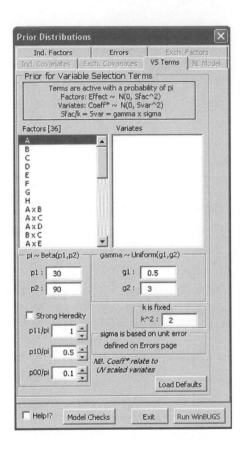

FIGURE 2.18
Prior distributions form: VS terms page.

but not all parents active (p_{10}) can be set as a multiple of pi. Strong heredity is forced if both p_{10} and p_{00} are zero. Strict weak heredity is allowed if only p_{00} is zero, while relaxed weak heredity is allowed if all three are nonzero. Note that pi $\geq p_{11} \geq p_{10} \geq p_{00} \geq 0$. Unless you have a good reason to do otherwise, it is recommended that strong heredity be forced. Irrespective of the heredity principle adopted, all terms are active with probability independent of the activity of other terms of the same order.

By specifying a prior standard deviation for the effect size that is a multiple of sigma, a measure of experimental unit variability on the link scale, makes it easier to choose default settings; this approach for Normal models was originally suggested by Box and Meyer (1993). The value of sigma depends upon the error distribution.

2.7.8.1 Normal Model

Sigma is the residual standard deviation parameter, with prior as specified on the 'Errors' page. The log-scale sigma parameter is used for the Log-Normal model.

2.7.8.2 Gamma Model

Sigma is set equal to the CV with prior inferred from that set for the scale parameter on the 'Errors' page. This was chosen based on the following argument. Gamma data can often be approximated by the Log-Normal. Sigma is the log-scale residual standard deviation for Log-Normal models. Sigma is typically a good approximation to the CV for Log-Normal models.

2.7.8.3 Weibull Model

Sigma is given the default value of 0.4, chosen based on the following argument. A log-link is used for Weibull models. The Weibull Distribution can often be approximated by the Log-Normal. Sigma is the log-scale residual standard deviation for Log-Normal models. Sigma is often a good approximation to the CV for the Log-Normal. The CV of the Weibull is determined by its shape parameter, r, which is commonly in the range two to four. Refer to Appendix A and Table 2.1 for relationship between the CV and r.

2.7.8.4 Poisson Model

Sigma is set to the square-root of the inverse geometric mean of the data, chosen based on the following argument. A log-link is used for Poisson models. Sigma is the log-scale residual standard deviation for Log-Normal models. The variance of a log-Poisson variate is roughly the inverse of the Poisson mean. The geometric mean is used as it is more robust to extreme values than the arithmetic mean.

2.7.8.5 Binomial Model

Sigma is set to $2/n^{1/2}$, based on the following argument. Sigma is the residual standard deviation for Normal models. Simulation of $\text{link}([r + 0.25]/[n + 0.5])$ has shown that, for a wide range of π, its standard deviation is approximately $2/n^{1/2}$. The numerator varies between about 1.3 and 3.3 within the probability range 0.1–0.9, sample size 10–100, and depending upon the link function. The default value for sigma has been set by substituting the mean sample size.

Note that only in the case of Normal models is there a body of evidence to support the prior defined above. The others should be seen as speculative suggestions that may work well in some circumstances. It is strongly advised

TABLE 2.1

Relationship between Weibull Shape
Parameter, r, and the Distribution's CV

r	1	1.5	2	3	4	5
CV	1	0.68	0.52	0.36	0.28	0.23

that if VS terms are fitted in Non-Normal models, that the sensitivity of the conclusions to the default BugsXLA prior be thoroughly investigated.

The gamma multiplier for sigma is given by default a Uniform prior over a range that has proven to work well for various published examples: U(0.5, 3). Note that the standard weak prior for regular regression coefficients cannot be used as this always results in the null model being selected. The constant k represents the ratio of the prior standard deviations for factor effects and variate coefficients. The preset default value of $2^{½}$ was chosen so that the prior for two-level factors is identical to that set if the factor's levels are set to ±1 and the term fitted as a covariate. Chapter 8 is dedicated to the discussion of models that include VS terms.

2.7.9 Emax Function Parameters

The 'NL Model' page allows the user to alter the priors for the **non-linear model** parameters. If no grouping factor is specified then the page will differ from that shown below; only the location (M or LM) and dispersion (S or LS) of the prior distributions for the Emax, ED50, and Hill parameters would need to be specified (Figure 2.19).

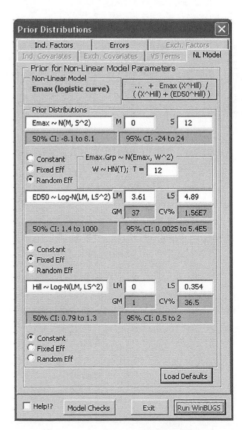

FIGURE 2.19
Prior distributions form: Non-linear parameters page.

The parameterization of the Emax model is given in Section 2.3. Note that although the preset default priors are intended to be only weakly informative, the Bayesian analysis may give estimates quite different to those obtained by ML methods, particularly when the data do not provide much information with which to estimate some of the parameters. By default, the 'missing'E0 parameter is included in the model constant whose prior is specified on the 'Independent Factors' page. Alternatively, the grouping factor specified in the Emax model could be entered as either a fixed or random effect in the model, allowing the E0 parameter to vary between groups. In this case, the prior is specified on the 'Independent Factors' or 'Exchangeable Factors' page, respectively, with the model constant estimate still including information relevant to the E0 parameter.

The Emax parameter can be assumed to be the same for all groups (constant), independent for each group (fixed), or exchangeable across the groups (random). When Emax is assumed constant or fixed, the prior is a Normal Distribution, $N(M, S^2)$, with parameters set here. The exchangeable assumption is implemented by giving each Emax a Normal prior distribution with shared hyper parameters: a mean having the Normal Distribution described above, $N(M, S^2)$, and a standard deviation having a Half-Normal Distribution with scale parameter, T, set here. To aid setting the parameters of the prior distribution for the 'population mean' Emax, implied 50% and 95% credible intervals are shown.

The ED_{50} and the Hill parameters can also be assumed to be constant, fixed or random. When they are assumed constant or fixed, the prior is a Log-Normal Distribution, $Log-N(LM, LS^2)$, with parameters set here. When treated as random, each parameter is given a Log-Normal prior distribution with shared hyper parameters: a median having the Log-Normal Distribution described above, $Log-N(LM, LS^2)$, and a scale parameter having a Half-Normal Distribution with its scale parameter, T, set here. Although the Hill parameter is constrained to be positive, a negative slope will be modeled by a negative Emax parameter. As for the Emax parameter, implied 50% and 95% credible intervals are shown to aid setting the prior distributions for the 'population medians' of the ED_{50} and Hill.

The T parameters that define the prior distributions for the non-linear random effects are given default values that attempt to make the prior vague relative to the information in the likelihood. However, if the data are such that there is very little information with which to estimate a parameter, then these may still be influential. Non-linear models can be difficult to fit, and so it is particularly important to both assess MCMC convergence (see Chapter 1) as well as the sensitivity of the inferences to the priors in these cases. The posterior plotting tool discussed in Section 2.11 is a useful aid for assessing sensitivity to the prior.

The default prior distributions for the population means of the grouped parameters are the same as those provided for the parameters when there is no grouping factor. The Emax parameter is given a Normal prior with a mean of zero and standard deviation equal to 10 times the standard deviation of the response. When a link function other than the identity is used, then the standard deviation is calculated on the 'link transformed Y' (with adjustments to avoid infinite values). Similarly, log(Y) is used for the Log-Normal model, when the linear predictor is on the log-scale. In these cases, care should be taken when specifying informative priors, as the parameters are specified on the link scale. When probabilities are being modeled, the prior standard deviation is set equal to 10. The ED_{50} parameter is given a Log-Normal prior with a median (geometric mean) equal to the geometric mean of the smallest and largest nonzero X values (the covariate with a Emax relationship to the response), and a scale parameter such that the ED_{50} is believed to lie between these two values with 50% probability. The Hill parameter is given a Log-Normal prior with a median of one and a scale parameter such that the Hill is believed to lie in the interval [0.50, 2] with 95% probability. Note also that for the Log-Normal priors, the geometric mean (GM) and coefficient of variation (CV%) are also shown. The standard deviation on the log-scale (LS) is not allowed to be larger than five, which may make the default prior for ED_{50} tighter than stated above. Chapter 7 is dedicated to discussing Emax models.

The button at the bottom of the form labeled '**Model Checks**' brings up another form that allows additional functions of the data and parameters to be calculated and imported back into Excel (see Section 2.12.2.10).

Clicking on the button '**Run WinBUGS**' tells BugsXLA to create all the files needed by WinBUGS, and then run it via a script. A new instance of WinBUGS will appear and the analysis specified undertaken. During this time it will not be possible to use Excel. Once the analysis is complete, WinBUGS will be closed down (unless Auto Quit was set to 'No' in the MCMC & Output Options) and the 'Import Results' form will appear (see Section 2.10)

2.8 Graphical Feedback Interface

As mentioned in Section 2.7, when the Half-Normal prior is selected for the hierarchical standard deviation parameter for factors with random effects, the 'GFI' (Graphical Feedback Interface) button can be clicked to facilitate elicitation of an informative prior (see Figure 2.15). Clicking this button brings up the form shown in Figure 2.20.

The top part of this form, in the frame labeled '**Prior Elicitation Controller**', shows the parameter whose prior is being elicited, as well as the link scale on which its effect is acting. Here, the user can choose to elicit the value of this parameter directly, or via the maximum credible difference between two randomly chosen factor levels (refer to the discussion in Section 2.7 for a more precise definition of these approaches). It is important that the link scale is considered when using this tool. An identity link implies that the effects are on the same scale as the response. A log-link implies that the effects are

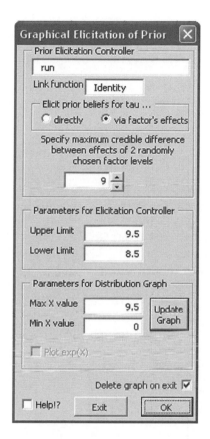

FIGURE 2.20
Graphical feedback interface form.

on the log-scale, with differences being log(ratios). A logit-link implies that the effects are on the log(odds) scale, with differences being log(odds-ratios). For other link functions, the scale is more difficult to interpret. The spinner control allows the user to alter the prior distribution for the parameter, with the graph showing this prior updating automatically.

The frame labeled '**Parameters for Elicitation Controller**' allows the user to alter the range over which the spinner control described above operates. Typically this needs to be altered from the default in order to allow finer control over the elicited prior distribution. The frame labeled '**Parameters for Distribution Graph**' allows the user to alter the range over which the prior distribution is plotted. The exponential of the parameter being elicited can be plotted when the link is log or logit. The '**Show grid**' option (only available when the exponential is being plotted) adds vertical lines to assist interpretation of the log-scale axis. The '**Update Graph**' button must be clicked for changes to occur to the plotted graph. In order to prevent the graph from being deleted after exiting from this form it is necessary to deselect the '**Delete graph on exit**' check box. If it is not deleted, then the sheet will be given a name of the form 'Prior{#}'. For the elicited prior to be applied to the model, the 'OK' button must be clicked; if the form is closed using the 'Exit' button then no changes are made. Case Study 7.1 provides an example of how the GFI can be used to help elicit an informative prior distribution.

2.9 Model Checks

As mentioned in Section 2.7, at the bottom of the Prior Distributions form is the 'Model Checks' button. When clicked the form shown in Figure 2.21 appears. The top section, labeled '**Model Comparison**', allows the user to request summaries that could be useful in assessing the overall goodness of fit of the model. As well as a summary of the posterior distribution of the deviance function, it is possible to import the individual samples generated for this node. DIC refers to the Deviance Information Criterion (refer to Spiegelhalter et al. (2002)). Note that the DIC is not defined for every model that can be fit using BugsXLA, and will simply be absent from the displayed results in those cases.

Checking the box labeled '**Checking Functions**' instructs BugsXLA to return a range of checking functions calculated for each observation. These include the raw and scaled residuals, the fitted values and the probability of getting a more extreme observation. Note that for the Gamma Distribution scaled residuals are a Bayesian version of deviance residuals:

$$\text{sign}(r_i)\left(2\left\{-\log(y_i/\mu_i) + (r_i/\mu_i)\right\}\right)^{\frac{1}{2}} \quad \text{where } r_i = y_i - \mu_i.$$

Scaled residuals are not available for Weibull errors. Also, when probabilities are being modeled (Bernoulli, Ordered Cat, and Categorical), this option

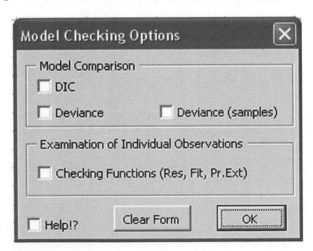

FIGURE 2.21
Model-checking options form.

only returns the posterior predictive probability associated with each observation, that is, the probability of obtaining the response given the values of its associated explanatory variables. When outliers are being explicitly modeled, as specified on the 'Errors' page of the Prior Distributions form (see Section 2.7), the probability of being an outlier is also shown.

As well as tabular output of these values, graphs are also created within Excel. An example is shown in Figure 2.22. There are two columns, the left

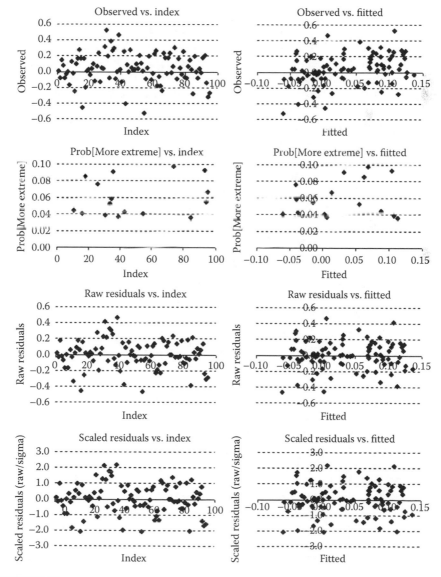

FIGURE 2.22
Model-checking plots.

hand column has the observation index (i.e., order listed in data set) on the horizontal axis, and the right hand column has the fitted (posterior mean) value on the horizontal axis. The top row shows the observed responses. The third and fourth rows show the raw and scaled (posterior mean) residuals, respectively. The second row is sometimes referred to as "Bayesian p-values." Section 9.3 "Model criticism and selection" of Spiegelhalter et al. (1996) describes how WinBUGS can be used to calculate the chance of getting a more extreme observation. This involves generating a replicate data set from the specified model, and counting how many times the replicate is more extreme than the observation (see also Appendix C). The graph shown here only plots those observations with such probabilities less than 10%. On the tabular output (not shown) "suspect" residuals and Bayesian p-values are highlighted in red; the thresholds for these signals can be set on the 'BugsXLA Options' form accessible from the BugsXLA toolbar (see Section 2.12). Note that the Bayesian p-values BugsXLA provides are, in the terminology of Marshall and Spiegelhalter (2007), full-data posterior predictive p-values. For random effects models they are likely to be very conservative since the observed data will often strongly influence these parameters, causing the replicated data to agree too well with the actual data. Hence, like the residuals, which are conditional on the random effects, they are really only of use in assessing the conditional likelihood rather than any aspect of the random effects prior distribution.

When outliers are explicitly modeled (see Section 2.7), then a graph of these "outlier probabilities" is also produced.

2.10 Import Results

After WinBUGS has completed the analysis specified via BugsXLA, the form shown in Figure 2.23 appears. Here the user decides precisely which variables and summaries to import, and where they should be written in the current Workbook. By double-clicking on a specific variable name, it can be toggled IN or OUT of the importing procedure. Note that it is only possible to alter the node statistics that are imported when accessing this form via the 'Import from WinBUGS' icon (refer to Section 2.13). When importing DIC statistics, the '**Response Node**' name should only be changed from 'Y' if you are importing from your own code; enter the name of the main response variable.

It is possible to prevent any node samples from being imported by clicking on the '**Cancel importing of samples**' check box; if these were originally requested by mistake, then it is best to cancel as the Excel file can grow very large if many samples are imported. An alternative way to keep down the size of the Excel file is to thin the samples here. The advantage of doing this here, as opposed to on the 'MCMC Output & Options' form, is that all the samples are still used when calculating the node statistics,

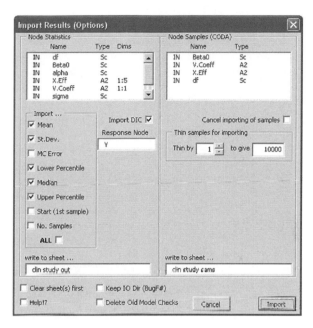

FIGURE 2.23
Import results form.

ensuring no loss of precision. If samples are imported then there is a posterior plotting tool available to visualize the estimated posterior distributions (refer to Section 2.11).

All the files created by BugsXLA to manage the input and output with WinBUGS can be retained by checking the '**Keep IO Dir**' box; these will be stored in the sub-directory 'BugF#', # being the next vacant integer. NOTE: this must remain checked if R scripts are required.

If you select the option '**Delete Old Model Checks**' then all sheets labeled 'Mdl Chks(#)' will be deleted prior to creating the new Model Checks sheet. If you renamed any of the old model checking sheets created, then this option will not delete them. Selecting the check box '**Clear sheet(s) first**' will automatically overwrite any worksheets with the same names as those specified; if this is not selected then the user is prompted to confirm before any sheets are overwritten.

2.11 Posterior Plots

This tool is accessible via the BugsXLA toolbar as shown in Figure 2.24. Clicking on the 'Post Plots' icon, highlighted in Figure 2.24, while a cell is selected in the column of samples one wishes to plot, brings up the form shown in Figure 2.25.

FIGURE 2.24
Location of posterior plots icon.

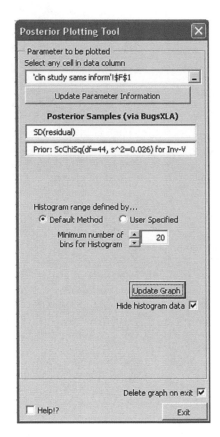

FIGURE 2.25
Initial posterior plotting tool form.

    This tool will produce a histogram of the samples imported from WinBUGS. The column of values to be plotted can be changed by first clicking inside the ref edit box labeled '**Select any cell in data column**', and then selecting any cell in the column of data you wish to plot before clicking on the '**Update Parameter Information**' button. If the prior distribution can be inferred from the specified model, then this will also be plotted on the graph. Note that it uses information stored in a comment in the first cell of the column, so do not edit or delete this comment if you wish to use this tool. Note also that

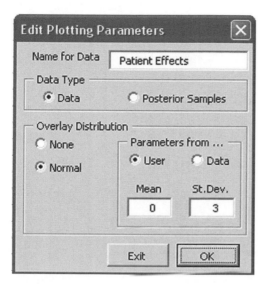

FIGURE 2.26
Edit plotting parameters form.

when Prior Distributions bounded at zero are used, then the bin centered on zero has its counts doubled to make the distribution better visually.

This tool can also be used to plot a histogram of any column of numeric data. The first cell must contain a label header for the data column, and this must either be in the first row of the worksheet or have a blank cell immediately above it. Optionally, a distribution can also be specified to overlay the histogram. The program will recognize that the column of values are not samples imported from WinBUGS after running an analysis via BugsXLA, and will display a button labeled '**Edit Plot Parms**' on the form. The user must click this button and complete the '**Edit Plotting Parameters**' form (see Figure 2.26) in order to prepare the column of data for plotting; this information will be stored in the comment in the header cell so that this plotting tool can reuse it in future. If the data being plotted were samples from the posterior distribution of a quantity with a known Normal prior distribution, by selecting '**Posterior Samples**' on the form, this prior could be specified and overlaid. If the data were not from a posterior distribution, or one wanted to overlay another Normal distribution for some reason, then the 'Data' option should be selected. In the latter case, it is also possible to overlay the best fitting Normal Distribution (matching moments) by selecting the 'Parameters from … Data' option.

After clicking on the '**Update Graph**' button shown in Figure 2.25 the specified plot is produced. If the '**User Specified**' option is then chosen, and the '**Probabilities**' check box selected, additional options on the form will be available as shown in Figure 2.27. The default histogram produced by BugsXLA can be manually edited via this plotting tool. The number of bins for the histogram and x-axis scale can be altered. The decimal places for the

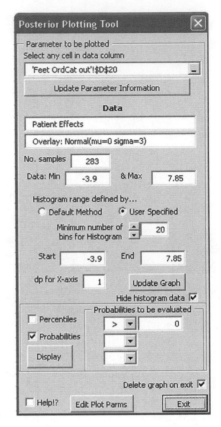

FIGURE 2.27
Posterior plotting tool form with graphing options.

x-axis scale can also be changed, up to a maximum of five. Entering six in this field will instruct the program to use scientific notation. Click on the 'Update Graph' button again to show the effects of any changes made.

More refined percentiles of the posterior distribution than can be routinely imported back from WinBUGS can be obtained. Simply select the **'Percentiles'** check box before clicking on the **'Display'** button. It is also possible to specify up to three 'tail area' probability calculations. Select the **'Probabilities'** check box, define the tail areas of interest and then click on the 'Display' button.

2.12 BugsXLA Options

Most of the default settings can be altered to suit the user's preferences. The BugsXLA Options form is accessible via the main BugsXLA toolbar as shown in Figure 2.28. Clicking on the 'BugsXLA Options' icon highlighted

FIGURE 2.28
Location of BugsXLA Options icon.

in Figure 2.28 brings up the form shown in Figure 2.29. This page allows the user to alter various settings that are relevant to most aspects of the BugsXLA program.

WinBUGS.exe directory
The location of the WinBUGS executable file. This should not change unless the software is reinstalled into a different folder on your computer's hard drive.

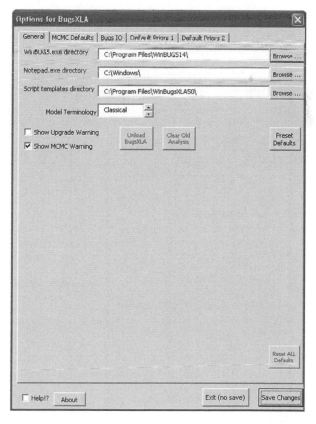

FIGURE 2.29
BugsXLA Options form: General page.

Notepad.exe directory
The location of the Notepad executable file. This should not change unless Windows is reinstalled to another directory. If Notepad is not in the specified location, then try:
C:\WINNT
C:\Windows
C:\Windows\System32

Script templates directory
The location of any WinBUGS script templates you may have created. These can then be used as a starting point when creating new scripts. If any templates are bundled with BugsXLA, they will be in the directory to which all the files were unzipped. Script templates can be useful when using BugsXLA's 'Edit WinBUGS Script' feature (refer to Section 2.13).

Model terminology
Determines how the model terms are labeled. Bayesian: uses terms such as Independent and Exchangeable. Classical: uses terms such as Fixed and Random.

Show upgrade warning
Determines whether any upgrade warning is shown each time BugsXLA is run.

Show MCMC warning
Determines whether the MCMC warning is displayed with all imported results: "BugsXLA is provided without any warranty of any kind, either expressed or implied. The user is responsible for any consequences arising from use of this program. See also the disclaimer provided with the WinBUGS package."

Unload BugsXLA
Unloads the BugsXLA Add-In and deletes the toolbar. This is sometimes necessary when BugsXLA is upgraded or one wishes to prevent BugsXLA from always loading on starting Excel. You will need to follow the process for reattaching add-ins to Excel in order to reload BugsXLA (refer to Section 2.1).

Clear old analysis
Deletes the hidden sheet that stores the most recent analysis settings. This is only necessary if you obtain error messages that suggest this sheet has been corrupted.

Preset defaults
Loads the preset defaults for the options shown on the same page as the 'Preset Defaults' button. You will still need to save the changes in order for them to become effective.

Reset ALL defaults
Resets the defaults for all the options on every page. Be warned, on clicking this button the preset defaults are automatically saved, so any changes

will be lost. It is recommended you only use this option if an error message appears every time you load the Options form. More generally, it is safer to use the 'Preset Defaults' button on each page separately, as you will then still have the option to 'Exit (no save)' if you want to restore the changed settings.

2.12.1 Default MCMC Settings

On the 'MCMC Defaults' page (Figure 2.30), you can define your own preferred settings for the four 'levels' of model complexity. The basic settings are intended for test runs rather than a proper analysis. The simple and regular model settings are intended to be generally appropriate for models and data sets that you are confident will converge relatively quickly. It is strongly recommended that unless you are extremely confident that the model and data will not cause difficulties of MCMC mixing or convergence, that you take advantage of the diagnostics built into WinBUGS to assess these issues before making inferences or decisions based on the output (see discussion in Section 1.2).

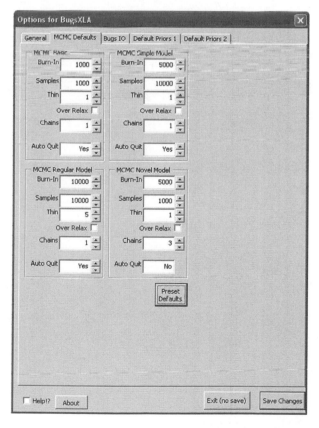

FIGURE 2.30
BugsXLA Options form: MCMC defaults page.

The novel model settings are intended for these situations; these will run multiple chains and leave you inside WinBUGS from where you should determine the appropriate MCMC settings by interactively working with your model and data.

2.12.2 WinBUGS Input and Output

On the 'Bugs IO' page (Figure 2.31), it is possible to alter how BugsXLA interacts with WinBUGS. The first two options on the form are only relevant when using the more general utilities for importing results from WinBUGS (refer to Section 2.13).

2.12.2.1 WinBUGS Node Stats

When importing node summary statistics following a WinBUGS analysis, the filename specified here will be suggested for import if found in the folder

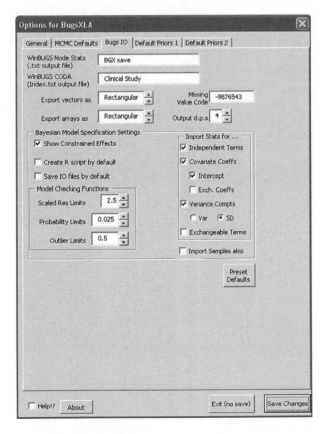

FIGURE 2.31
BugsXLA Options form: Bugs IO page.

searched. To be valid, the filename must contain both 'node' and 'stats'. Note that this is automatically updated to the most recent file used to import results, even when this follows a BugsXLA specified analysis; BugsXLA uses the default name 'BGX save'.

2.12.2.2 WinBUGS CODA

When importing CODA samples following a WinBUGS analysis, the filename specified here will be suggested for import if found in the folder searched. Note that this is automatically updated to the most recent file used to import samples, even when this follows a BugsXLA specified analysis; BugsXLA uses the default name 'BGX CODA'. An 'Index.txt' suffix, with no space between the name stem and 'Index', is assumed for the index file.

The next two options below are only relevant when using the more general utilities for exporting data from Excel into a format that WinBUGS recognizes (refer to Section 2.13).

2.12.2.3 Export Vectors as

Determines whether columns of data exported from Excel are written in rectangular or list format.

2.12.2.4 Export Arrays as

Determines whether any data array exported from Excel is written in rectangular or list format.

The next two options below are relevant to all parts of BugsXLA.

2.12.2.5 Missing Value Code

Determines the number used internally to represent a missing value. Clearly this should not be set to a value that might be a true response. Values within 10,000 of zero are not allowed.

2.12.2.6 Output Decimal Places

Determines the number of decimal places used when displaying the results of the analysis. Note that it is currently not possible to alter the output precision (i.e., significant figures) displayed by WinBUGS, with the default as of version 1.4.3 being four. So the value set here should be recognized as the maximum possible number of decimal places. If your results have a string of trailing zeroes, then it is very likely that you have specified more decimal places than WinBUGS provided. Note also that BugsXLA will ignore the decimal place setting when displaying results with very small or large absolute values.

The following discuss those settings that are only relevant when using BugsXLA via its Bayesian model specification form.

2.12.2.7 Show Constrained Effects

If this option is chosen then BugsXLA will display the terms constrained to zero on importing results back from WinBUGS. Otherwise these terms will not be shown and the user should be aware of how the effects were parameterized. See also the 'Zero Constraint Variance' setting on the 'Default Priors 1' page.

2.12.2.8 Create R Script by Default

This determines whether R scripts are created by default for each analysis specified. These scripts use the files created by BugsXLA, enabling the Bayesian analysis specified to be run from within the R environment, as well as other exploratory analyses. Refer to Appendix D for more information.

2.12.2.9 Save IO Files by Default

This determines whether the automatically generated WinBUGS code, script, data, initial values, and log files are saved by default after each analysis is run. Refer also to the BugFolio on 'MCMC & Output Options' form (Section 2.5) for more control on this feature.

2.12.2.10 Model Checking Functions

If scaled residuals or probabilities are imported as part of the model checking options, then any that are found to be more extreme than indicated by these limits will be automatically highlighted on importing back into Excel. The outlier limits are for providing a warning when outliers are being explicitly modeled; any observation with an outlier probability greater than the value set will be counted as a potential outlier.

2.12.2.11 Import Stats for ...

Check those terms whose posterior distribution summary statistics you wish to import by default for all analyses. The non-linear model parameters are included with the Independent terms. If you wish to import CODA samples by default, then check the '**Import Samples also**' option. Since the importing of samples into Excel can be quite time consuming, it is not recommended that samples be imported by default.

2.12.3 Default Prior Distributions

On the 'Default Priors' 1 and 2 pages (see Figures 2.32 and 2.33), it is possible to alter the parameters, and sometimes the distribution, of most of the default priors.

2.12.3.1 *Error Distributions*

The '**Residual Variance Gamma Parms**' define the default Gamma Distribution for the residual variance. This value is also taken to be the smallest value that can be set for any Gamma Distribution parameter when adopted as a prior. The parameters for the t-distribution, Gamma, Weibull, Longitudinal, generalized logit, and explicit outlier models are explained in Section 2.7.

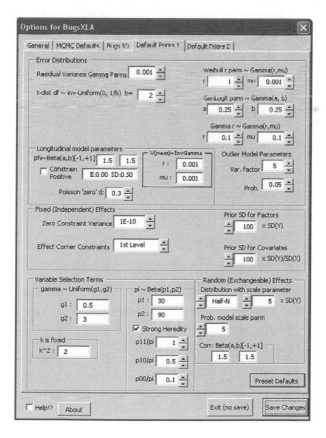

FIGURE 2.32
BugsXLA Options form: Default Priors 1 page.

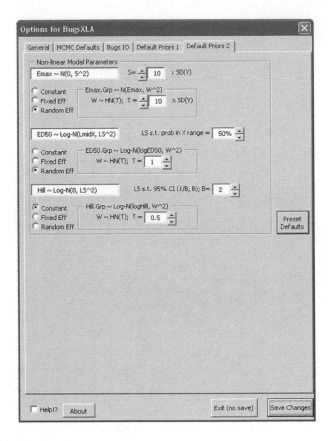

FIGURE 2.33
BugsXLA Options form: Default Priors 2 page.

2.12.3.2 Zero Constraint Variance

This is a technical issue that determines how BugsXLA instructs WinBUGS to treat effects that have been constrained to zero. Although not necessary in theory, in practice it greatly helps MCMC convergence if fixed effects have one of their levels constrained to zero. For convenience in the automatic generation of the code, this is effectively achieved by assigning a Normal prior distribution with zero mean and an extremely small variance. The value of this prior variance is shown here. Although it appears that Normal linear models will work with this set as low as 10^{-99}, problems have been observed with log-linear Poisson models when this has been set smaller than 10^{-10}. You should only consider the need to change this if you prefer the 'zero constrained effects' to be closer to zero than their current estimates (make this smaller), or WinBUGS crashes for no good reason (make this larger)—although WinBUGS may of course be crashing for other reasons. Note that this value also determines the smallest value to which any nonnegative prior parameter can be set when the identity link is used, for example, this is the assumed variance when a value of zero is entered for any Normal prior distribution.

2.12.3.3 Effect Corner Constraints

Corner constraints are applied to fixed effects terms to aid MCMC performance. Either the first level of each main effect (R or Genstat default) or the last level (SAS default) can effectively be constrained to zero. See 'Zero Constraint Variance' for details of how this is actually applied.

2.12.3.4 Prior SD for (Independent) Factors

The prior standard deviations of the model constant and fixed effect terms are given a default value that is some large multiple of the standard deviation of the response.

2.12.3.5 Prior SD for (Independent) Covariates

The prior standard deviations of the independent regression coefficients are given a default value that is some large multiple of the ratio of the standard deviation of the response and that of the covariate. The mean regression coefficient for exchangeable covariate terms is given the same prior standard deviation as set here.

When a link function other than the identity is used, then the standard deviation of the response is calculated on the 'link transformed metric' (with adjustments to avoid infinite values). When probabilities are being modeled, the 'SD(Y)' term is set equal to one in these formulae.

2.12.3.6 Random (Exchangeable) Effects

The user can define which prior distribution is used by default for the standard deviation of random effects; this includes factors and exchangeable regression coefficients. If the Half-Cauchy or Half-Normal is chosen, then the default scale parameter also needs to be specified, while if the Uniform is chosen then the upper bound has to be specified (zero lower bound is assumed). The default parameter is expressed as a multiple of the standard deviation of the response variable, or an appropriate transformation if the identity link is not used in the model. A separate scale parameter can be set for 'Probability' models (Binomial and Multinomial models).

A Beta Distribution scaled to the range [−1, +1] is used as the prior for the correlation between the intercepts and slopes in exchangeable regression coefficient models. The parameters must lie in the range [1, 99], and will be rounded to two significant figures. Note that the Beta(1, 1) is a uniform distribution across the whole range [−1, +1]. The preset default of Beta(1.5, 1.5) was chosen as it is quite 'vague' while giving relatively low prior probability to extreme positive or negative correlation; this is thought to be generally appropriate, and also has been found to help improve mixing for some problems when compared to the uniform.

The meaning of the parameters for the 'VS Terms' are explained in Section 2.7.

The 'Default Priors 2' page allows the user to change the defaults for the non-linear Emax model parameters.

2.12.3.7 Emax Parameter

When no grouping factor is specified in the non-linear model, the Emax parameter is given a Normal prior distribution with a mean of zero and standard deviation equal to some multiple of the standard deviation of the response. When a grouping factor is specified, the Emax parameter can be assumed to be the same for all groups (Constant), independent for each group (Fixed), or exchangeable across the groups (Random). When Emax is assumed Constant or Fixed, the Normal Distribution described above is taken as the default prior. The exchangeable assumption is implemented by giving each Emax a Normal prior distribution with shared 'hyper parameters': a mean having the Normal Distribution described above as the default prior, and a standard deviation having a Half-Normal distribution with scale parameter equal to some multiple of the standard deviation of the response.

2.12.3.8 ED50 Parameter

When no grouping factor is specified in the non-linear model, the ED50 parameter is given a Log-Normal prior distribution with a median equal to the geometric mean of the smallest and largest nonzero X values (the covariate in the Emax model), and a scale parameter such that the ED50 is believed to lie between these two values with probability specified here. When a grouping factor is specified, the ED50 parameter can be assumed to be the same for all groups (Constant), independent for each group (Fixed), or exchangeable across the groups (Random). When ED50 is assumed Constant or Fixed, the Log-Normal Distribution described above is taken as the default prior. The exchangeable assumption is implemented by giving each ED50 a Log-Normal prior distribution with shared 'hyper parameters': a median having the Log-Normal Distribution described above as the default prior, and a scale parameter having a Half-Normal distribution with its scale parameter specified here.

2.12.3.9 Hill Parameter

When no grouping factor is specified in the non-linear model, the Hill parameter is given a Log-Normal prior distribution with a median of one and a scale parameter such that the Hill is believed to be in the interval specified here with 95% probability. When a grouping factor is specified, the Hill parameter can be assumed to be the same for all groups (Constant), independent for each group (Fixed), or exchangeable across the groups (Random). When Hill is assumed Constant or Fixed, the Log-Normal Distribution described above is taken as the default prior. The exchangeable assumption is implemented by giving each Hill a Log-Normal prior distribution with shared 'hyper parameters': a median having the Log-Normal Distribution described above as the default prior, and a scale parameter having a Half-Normal distribution with its scale parameter specified here.

Note that the E0 parameter is included in the model constant, and so has its prior set there. When a grouping factor is specified, the E0 can be modeled as independent for each group or exchangeable across the groups by including the grouping factor as a term in the fixed effects or random effects model, respectively.

2.13 WinBUGS Utilities

Although the main function of BugsXLA is to facilitate the easy fitting of Bayesian models via WinBUGS, the main toolbar provides some other utilities for interacting with WinBUGS from within Excel.

2.13.1 Exporting Excel Data in WinBUGS Format

Figure 2.34 shows the 'Export to WinBUGS' icon. Clicking on this icon, highlighted in Figure 2.34, brings up the form shown on the right in Figure 2.35. This form can be used to export scalars, vectors, or a two-dimensional array from within Excel to a text file that is readable by WinBUGS. The example shown in Figure 2.35 is trivially small and unrealistic in order to make the illustration of how the ranges are specified simple to follow. Scalars are stored in list format, while vectors and arrays can be stored in either list or rectangular format. Missing value codes can be generated for blank cells if required. 'NA' is automatically recognized as a missing value. Any vector defined as a factor (see Section 2.4) will have its levels saved rather than its values. An Export Log is created that is useful if you plan to use the 'Edit Script' utility to automatically add commands for loading this data. Once the data ranges have been specified for exporting, clicking on the **'Export'** button brings up the form shown in Figure 2.36.

FIGURE 2.34
Location of export to WinBUGS icon.

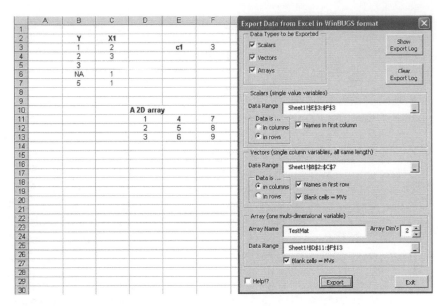

FIGURE 2.35
Export data to WinBUGS form.

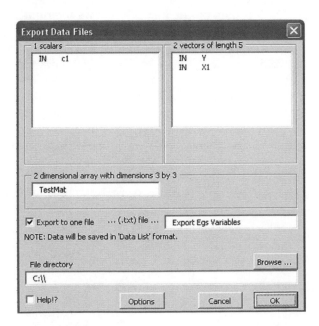

FIGURE 2.36
Export data files form.

This form allows the user to precisely define which variables to export, and the names of the files to which they should be written. By double-clicking on a specific variable name, it can be toggled IN or OUT of the exporting procedure. By default, the files are saved in the same folder as the active Excel workbook. It is possible to browse to find an alternative location. Note that if '**Export to one file**' is checked, then all data is saved in list format. If the 'Export to one file' is unchecked, then all the scalars are saved in one file, all the vectors in another file and the array in another different file. Only by exporting to separate files can vectors or arrays be saved in rectangular format.

2.13.2 Editing Scripts and Other Text Files

Clicking on the 'Edit WinBUGS Script' icon highlighted in Figure 2.37 brings up the form shown in Figure 2.38. This form can be used to create new or edit old script files. If editing an old script file, then by default the current Excel workbook's directory is searched, unless the most recent BugsXLA run was used to create a BugFolio (see Section 2.5). All files in the same directory with the text 'script' in their name will be loaded into the drop down list box. If creating a new script file (see Figure 2.39), then a template can be used; by

FIGURE 2.37
Location of edit WinBUGS script icon.

FIGURE 2.38
Edit WinBUGS (old) script form.

FIGURE 2.39
Edit WinBUGS (new) script form.

default the '**Script templates directory**' (see Section 2.12) is searched. If data have been previously exported using the 'Export to WinBUGS' utility, then the script commands for loading this data can be automatically generated by selecting the '**Add data in Export Log**' check box. Note that this utility loads the specified script file into Notepad and leaves the user to edit and save it from within that package. Hence, it can be used to edit any text file, including the WinBUGS code itself or any other of the text files created by BugsXLA.

2.13.3 Running a WinBUGS Script

Clicking on the 'Run WinBUGS (Script)' icon highlighted in Figure 2.40 brings up the form shown in Figure 2.41. This form can be used to run WinBUGS either directly or via a script file. By default the current Excel workbook's directory is searched, unless the most recent BugsXLA run was used to create a BugFolio. All files in the same directory with the text 'script' in their name will be loaded into the drop down list box. Note that once WinBUGS has completed the analysis specified in the script file, it will not automatically import the data back into Excel. This needs to be done manually using the 'Import from WinBUGS' utility.

FIGURE 2.40
Location of run WinBUGS (script) icon.

FIGURE 2.41
Run WinBUGS (script) form.

2.13.4 Importing WinBUGS Output into Excel

Clicking on the 'Import from WinBUGS' icon highlighted in Figure 2.42 brings up the form shown in Figure 2.43. This form can be used to import node statistics or samples (CODA output) from WinBUGS. By default the current Excel workbook's directory is searched, unless the most recent BugsXLA

FIGURE 2.42
Location of import from WinBUGS icon.

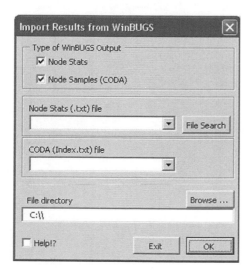

FIGURE 2.43
Import from WinBUGS form.

run was used to create a BugFolio. All files in the same directory with the text 'node' and 'stats' or 'BGX' and 'save' in their name will be loaded into the drop down list box for Node Stats, while files whose name end 'Index. txt' are loaded into the CODA list box. Note that only files with this suffix are recognized as CODA index files, and only the associated CODA data files with the suffix '1.txt' can be imported.

3

Normal Linear Models

Normal linear models (NLMs) are the workhorse of statistical analysis, consisting of the analysis of variance and regression techniques that are routinely applied to thousands of problems every year. For more details on the NLM, albeit with a focus on the classical approach to fitting, refer to a standard statistical text such as Clarke and Cooke (2004), Draper and Smith (1998), or Faraway (2004). Box and Tiao (1973) provides a thorough coverage of the Bayesian analysis of NLMs focusing exclusively on vague priors. A more modern Bayesian approach is given in Chapter 14 of Gelman et al (2004). To illustrate the essential features of an NLM, we consider a simple analysis of covariance (ANCOVA) model that consists of one grouping factor and one covariate. A common situation where this model might be appropriate is an experiment where some characteristic of interest is measured before (baseline) and after (response) some treatment is applied to the experimental units. Suppose there are three different treatments being compared and the measured characteristic is continuous, then the ANCOVA model could be represented algebraically:

$$y_{ij} = \mu + \alpha_i + \beta x_{ij} + e_{ij} \quad i = 1, 2, 3 \quad j = 1, 2, \dots, n_i$$

where
  y_{ij} is the response on the jth experimental unit receiving the ith treatment
  μ is a constant whose meaning depends upon how the treatment grouping factor is parameterized, as discussed later
  α_i represents the effects of the treatments (the grouping factor)
  β is the regression coefficient associated with the simple linear effect of the baseline measurement (the covariate)
  x_{ij} is the baseline measurement on the jth experimental unit receiving the ith treatment
  e_{ij} is the residual random error associated with the jth experimental unit receiving the ith treatment
  n_i is the number of experimental units receiving the ith treatment

This is a Normal Model if the random errors are assumed to be Normally distributed:

$$e_{ij} \sim N(0, \sigma^2)$$

It is a Linear Model since the formula given above implies that the mean response is a weighted linear combination of the unknown location parameters μ, α_i, and β. Note that the statistical definition of a linear model is not the same as the more intuitive interpretation, in which the linearity refers to the relationship between the response and a covariate. In statistics, models that include curvilinear relationships between the response and a covariate, such as polynomials, are still regarded as linear, providing the mean can be expressed as a linear function of the unknown parameters.

It is well known, and easily seen, that the parameters μ, α_1, α_2, and α_3 are not uniquely identified as the model is currently specified, and so additional assumptions or constraints are necessary in order to define parameters that are interpretable. In a standard NLM, all the location parameters are, in the classical terminology, fixed constants. Only the error terms, e_{ij}, are considered to be random variables. The NLM is referred to as a type of fixed effects model. Random effects models are discussed in later chapters, and this distinction influences how the parameters defining the grouping factor's effects are made interpretable. In the Bayesian approach to statistical inference, the terms 'fixed effects' and 'random effects' are strictly speaking inappropriate, since all parameters in the model are thought of as random variables, as discussed in Chapter 1. It is convenient to continue using these terms since there are Bayesian equivalents, with effects being defined as fixed or random depending upon the nature of their prior distribution. A Bayesian considers a factor as having fixed effects if the parameters defining them are assumed, a priori, to be completely independent, implying that knowledge of the true value of any of them tells us nothing about the true values of any of the others. Chapter 5 contrasts this assumption with that underlying a random effects model. If a factor is defined as fixed in BugsXLA, then, as in the classical approach, constraints are applied to the effects, α_i, to define a reduced set of free parameters:

$$\alpha_i^* = \alpha_i - \alpha_1, \quad i = 2, \ldots, \text{number of factor levels}$$

with α_1^* being constrained to equal zero. Constraining the first level of each factor to zero is also the default used by the statistical packages R and Genstat. It is possible to adopt the default used by SAS, constraining the last level, by changing the 'Effect Corner Constraints' option on the 'Default Priors 1' page of the 'BugsXLA Options' form (see Section 2.12).

The free parameters, representing contrasts with the effects of the first level of the factor, are given the prior:

$$\alpha_i^* \sim N(0, K)$$

where K is a fixed value with a default chosen by BugsXLA to represent little information relative to the variability in the response. Note that although the free parameters, α_i^*, are given independent priors, dependency is still induced in the prior for the unconstrained parameters, α_i. It could be argued that the prior for the α_i^* should be multivariate, taking account of the correlation due to these parameters all being contrasts to the first level of the factor. However, this has not been typically done in the literature, and providing K is relatively large should not matter much in practice. Theoretically, it is not necessary to apply any constraints to the fixed effects in a Bayesian analysis providing the prior distribution is proper (Lindley and Smith, 1972). In practice, MCMC sampling of the posterior would face severe problems of slow convergence due to the contours of constant likelihood combining with the weak priors to allow the chain of simulated values to wander over an enormous region.

As illustrated in the ANCOVA model, as well as factors, NLMs can also contain covariates. If a covariate is specified in BugsXLA, then its coefficient, β, is given the prior:

$$\beta \sim N(0, K)$$

where K is a fixed value with a default chosen by BugsXLA to represent little information relative to the ratio of the variability in the response to that in the covariate. If the sign of this coefficient is known with certainty, then a Half-Normal prior distribution can be specified to constrain the coefficient to be either positive or negative. BugsXLA automatically centers and scales each covariate using its mean and standard deviation, as this can greatly improve the performance of the MCMC algorithm. Note that the effect of scaling is reversed before the regression coefficient estimates are summarized, so this process is invisible to the user. However, the centering of the covariates, along with the corner constraints on the factors, means that the model constant parameter, μ, should be interpreted as the mean response when each factor is set at their first level and each covariate equals the mean observed in this experiment. The model constant is given the prior:

$$\mu \sim N(M, K)$$

where
  K is a fixed value with a default chosen by BugsXLA to represent little information relative to the variability in the response
  M is the mean of all the data

The purpose of centering the prior for the constant on the mean of the data is simply to ensure that the likelihood and the prior for this parameter are not in conflict; the default value of K is large enough to ensure that the prior is effectively flat over the likelihood. Refer to Section 2.7 for the precise details of the default priors for the location parameters.

The residual variance parameter, σ^2, is given the now standard non-informative Inverse-Gamma prior:

$$\sigma^{-2} \sim \text{Gamma}(r, \mu)$$

with the default settings of $r = \mu = 0.001$. The default priors chosen by BugsXLA for all the parameters in the NLM are intended to provide a good approximation to the prior density for the parameters that is uniform on $(\mu, \alpha_i^*, \beta$ and $\log \sigma)$. This is widely accepted as a sensible vague prior density for the parameters of a standard NLM (see, e.g., Box and Tiao (1973) or Gelman et al. (2004)).

Although the values of r and μ in the prior for σ^2 can be altered by the user, there may be some cases where an informative prior can be more easily specified via the scaled chi-square distribution:

$$\sigma^{-2} \sim \text{ScChiSq}(df, s^2)$$

The scaled chi-square distribution is a special case of the gamma, with $r = df/2$ and $\mu = (df\ s^2)/2$. One way of justifying an informative prior would be to use the conventional estimate of variance (s^2) and its degrees of freedom (df) from a previous study; this would be appropriate if there was little prior information before this previous study, and the residuals are believed to be exchangeable with those from the current study. This latter belief will rarely be held very strongly, so it is sometimes considered more reasonable to reduce the df associated with the estimate, so that the prior is not allowed to heavily influence the final inferences. For more sophisticated ways to determine an informative prior, refer to Chapter 5 of Spiegelhalter et al. (2004).

We now show how to fit a simple Normal linear model to data obtained from a clinical study. This case study will not only be used for this purpose, but also to illustrate the most commonly used features of BugsXLA. It is strongly recommended that the reader work through this example, since future case studies will assume familiarity with the features that are used in a repetitive fashion each time a new data set is first analyzed using BugsXLA.

Case Study 3.1 Parallel Groups Clinical Study (Analysis of Covariance)
Chronic obstructive pulmonary disease is a major worldwide cause of morbidity and mortality. It is a severe disease of the lungs, usually associated with smokers. Here, we describe a clinical study investigating the efficacy of a new anti-inflammatory compound in early pharmaceutical research. The study is a simple parallel groups design in which patients were randomized to either a placebo arm or one of four doses of the experimental compound. The outcome of interest was a measure of lung function known as Forced

Expiratory Volume in 1 s, or FEV_1 for short, reported here in liters (L). At the beginning of the study, prior to treatment, a baseline measure of FEV_1 was taken for each patient, since it is known that a large part of the between patients variability in the outcome measure is due to differences in their initial lung function. The traditional response that is modeled is the change from baseline in FEV_1, but we also include the baseline as a covariate in case the correlation between baseline and outcome is less than one (see Chapter 7 of Senn (2007) for a discussion of this matter).

In the absence of informative priors, there is little to be gained from adopting a formal Bayesian analysis of this simple NLM, other than to justify the interpretation usually given to the inferences made, for example, confidence intervals being interpreted as Bayesian credible intervals. This is because the posterior distributions of the parameters will be numerically identical to their equivalent classical confidence distributions (see Box and Tiao, 1973). Provided one is happy to use software designed to produce classical output, it is typically much easier to obtain the results without resorting to WinBUGS or BugsXLA, and simply interpret the output as a Bayesian. It is important that no uniquely frequentist style adjustments have been made, for example, for multiplicity, and any Bayesian interpretation of p-values be carefully done. However, if the plan is to use the NLM as a starting point to explore more sophisticated models, then one might as well adopt the Bayesian computing engine from the outset. We first fit a simple NLM using BugsXLA's default vague priors, comparing the results with those obtained from a classical analysis. We then show how informative priors can be specified via BugsXLA, as well as some other uniquely Bayesian analysis options. Figure 3.1 shows some of the data and completed 'Model Specification' form. This form is obtained via the 'Bayesian Model' icon on the BugsXLA toolbar as discussed in Section 2.2. Treatment is coded as a five level factor, TRT, with the level labeled 'A' being the placebo arm and, for reasons of confidentiality, the four doses of the experimental compound simply being labeled 'B' to 'E'. The model specified is an ANCOVA with one grouping factor and one covariate, and hence can be represented algebraically as before:

$$y_{ij} = \mu + \alpha_i + \beta x_{ij} + e_{ij} \quad i = 1, 2, \ldots, 5 \quad j = 1, 2, \ldots, n_i$$

$$e_{ij} \sim N(0, \sigma^2)$$

where
  y is the response variable FEV1_CFB
  x is the covariate FEV1_BASE (mean centered) with associated regression
    coefficient β
  α_i are the effects of the factor TRT

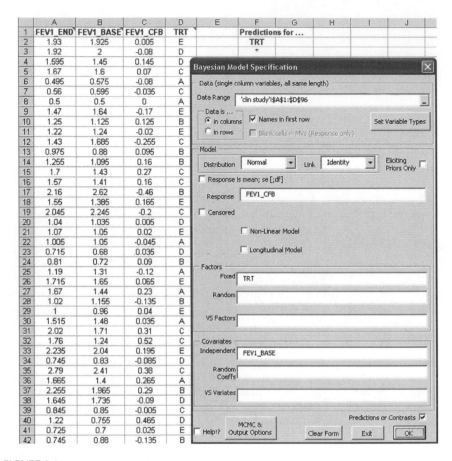

FIGURE 3.1
Data and model specification for clinical study.

Since the effect of a dose level is defined as the mean difference between its response and that of the placebo arm, it is convenient to make level 'A' of TRT the constrained first level so that the free parameters will be the effects of interest. This can be done while defining TRT as a factor via the 'Set Variable Types' button. Figure 3.2 shows the form that appears after clicking on this button.

By default all variables are assumed to be quantitative variates. The variable TRT is defined as a factor by clicking on its name, highlighting it in the list as shown in Figure 3.2, and clicking on the button with the left pointing arrow marks to move it to the list titled 'FACTORS'. The levels for this factor are set by clicking on the button labeled 'Edit Factor Levels', which brings up the form shown in Figure 3.3.

By default all the unique values in the column of data are read into the list shown, but they are in the order they first appear in the data. The buttons labeled 'Move Up' or 'Move Down' could be used to reorder them, but the

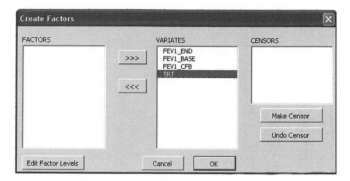

FIGURE 3.2
Creating the TRT factor in the clinical study.

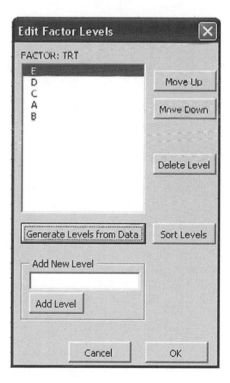

FIGURE 3.3
Editing the levels of the TRT factor in the clinical study.

quickest method is to use the 'Sort Levels' button, which brings factor level 'A' to the top which is what we want. When there are many factors to define, it is possible to highlight more than one variable at a time at each step of this process, making it less tedious, as described in Section 2.4. By making the placebo arm the constrained to zero level of the factor TRT, the model constant parameter, μ, is interpretable as the mean placebo response for subjects with a baseline FEV_1 response equal to the mean observed in this study.

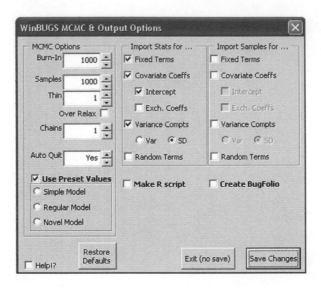

FIGURE 3.4
Setting the MCMC options in the clinical study.

The default MCMC settings (a burn-in of 1000 followed by 1000 samples) are typically not adequate to obtain reliable parameter estimates. They have been chosen to give a "quick and dirty" run of WinBUGS, so that any obvious errors with the data or model specification could be quickly identified and rectified before focusing on the more substantial modeling issues. Better settings are chosen by clicking on the 'MCMC & Output Options' button found at the bottom of the 'Model Specification' form shown in Figure 3.1. This brings up the form shown in Figure 3.4.

By clicking on the 'Use Preset Values' check box, the three preset MCMC options appear as shown in Figure 3.4. Details on the options that can be set via this form can be found in Section 2.5. In this case, we selected the 'Simple Model' setting, which changes the burn-in to 5,000 and the number of samples to 10,000. We discuss some of the other settings on this form later in this case study.

We wanted the estimated predicted means for all levels of TRT. This is possible by selecting the 'Predictions or Contrasts' check box at the bottom of the 'Model Specification' form (see Figure 3.1), which, after clicking on the 'OK' button, brings up the form shown in Figure 3.5. Section 2.6 provides details of how predictions and contrasts are specified. The entries in cells F2:F3 are interpreted by BugsXLA as a request for the predicted means for all levels of the factor TRT. This range is entered into the 'Predictions Range' ref edit box on the 'Predictions & Contrasts' form. The simplest way to do this is, after making sure the ref edit box is active by clicking inside it, to use the mouse to select the range F2:F3 on the worksheet; this should enter the range on the form as shown in Figure 3.5. Since we only require the

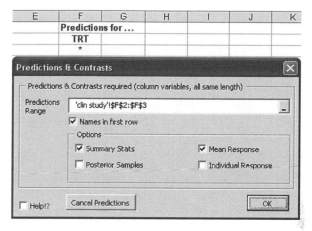

FIGURE 3.5
Specifying the predicted means required in the clinical study

summary statistics for the mean response the default settings on the rest of the form can be left unaltered.

Initially, we ran the analysis using the default priors provided, simply clicking on the 'Run WinBUGS' button at the bottom of the 'Prior Distributions' form. WinBUGS is then started and, in this case, quickly completes the analysis before disappearing and being replaced by the form shown in Figure 3.6. Section 2.10

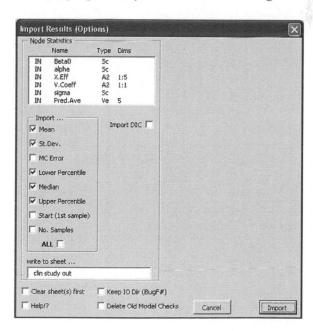

FIGURE 3.6
Importing the results back from WinBUGS in the clinical study.

	A	B	C	D	E	F	G	H	I	J	K
1			Label	Mean	St.Dev.	2.5%	Median	97.5%		WinBUGS Name	
2			CONSTANT	-0.0394	0.0458	-0.1291	-0.0398	0.0520		Beta0	
3			Intercept at 0	0.0448	0.0637	-0.0784	0.0448	0.1693		alpha	
4		TRT	A	0.0000	0.0000					X.Eff[1,1]	
5		TRT	B	0.0922	0.0653	-0.0380	0.0923	0.2207		X.Eff[1,2]	
6		TRT	C	0.1490	0.0586	0.0342	0.1490	0.2638		X.Eff[1,3]	
7		TRT	D	0.0222	0.0645	-0.1060	0.0222	0.1468		X.Eff[1,4]	
8		TRT	E	0.0543	0.0639	-0.0693	0.0543	0.1818		X.Eff[1,5]	
9			FEV1_BASE	-0.0658	0.0418	-0.1470	-0.0661	0.0175		V.Coeff[1,1]	
10			SD(residual)	0.1835	0.0141	0.1587	0.1827	0.2137		sigma	
11	Note: CONSTANT & Factor effects are determined at the mean of the covariate(s).										
12	Interpret these cautiously when Factor x Covariate terms have been fitted.										
13											
14	Model	['clin study'!$A$1:$D$96]									
15	Distribution	Normal									
16	Link	Identity									
17	Response	FEV1_CFB									
18	Fixed	TRT									
19	Covariates	FEV1_BASE									
20											
21	Priors										
22	CONSTANT	N(mu=0.0344, sigma=18.7)									
23	TRT	N(mu=0, sigma=18.7)									
24	FEV1_BASE	N(mu=0, sigma=39.8)									
25	V(residual)	Inv-Gamma(0.001, 0.001)									
26											
27	WinBUGS MCMC Settings										
28	Burn-In: 5000 Samples: 10000 (Thin:1; Chains:1)										
29	Run took 11 seconds										

FIGURE 3.7
Summaries of posterior distributions, model, priors, and MCMC settings for clinical study.

provides details on this form. Here we simply need to click on the 'Import' button, which brings the requested summary statistics calculated in WinBUGS back into Excel.

Figures 3.7 and 3.9 show the results of the analysis as imported back into Excel. Figure 3.7 shows all the parameters in the model and a summary of their marginal posterior distributions. Purely in terms of having a 95% credible interval that does not include zero, only treatment C provides clear evidence of an effect over placebo. The baseline measure of FEV_1 cannot be conclusively regarded as an important covariate, since zero is a credible value. However, there is a reasonably high probability that its regression coefficient has a magnitude of practical importance, and so should not be excluded from the model. When a covariate is included in a model, BugsXLA provides summaries for both the model CONSTANT and the 'Intercept at 0'. These are derived from the same parameter, with the former being the mean response at the constrained levels of the fixed effect terms and the means of the covariates, as discussed previously. The latter term is, as its label suggests, the same but with each covariate set to zero.

Unlike most statistical software packages, BugsXLA explicitly displays the parameters that have been constrained to zero, for example, level A of TRT in Figure 3.7. This is set as the default to avoid any potential mistake in interpretation, particularly for users familiar with software that uses different parameter constraints. For models with lots of fixed effects, and in particular when interactions are specified, this option can lead to a lot of space on

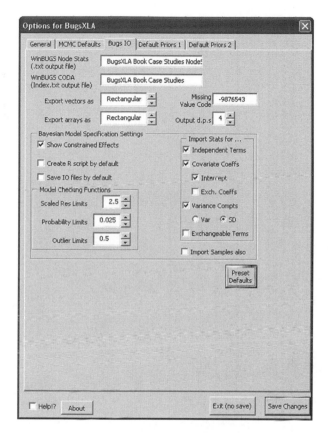

FIGURE 3.8
Form where option to show zero constrained effects can be altered.

the worksheet being taken up displaying these zero constrained parameters. Users can make BugsXLA adopt the more conventional approach of not showing these parameters by altering one of the options found on the 'BugsXLA Options' form that is launched off the main toolbar (see Section 2.12 for more details). Figure 3.8 shows the page on this form where this option can be altered; simply deselect the 'Show Constrained Effects' check box, making sure to click on the 'Save Changes' button afterward. All the remaining case studies in the book have been run with this option deselected.

The Bayesian predicted means for an NLM shown in Figure 3.9 are analogous to the least squares (LS) or fitted means that can be obtained from classical software packages.

It is informative to compare the Bayesian output obtained using BugsXLA's default priors with the output produced by running the equivalent R script, also created by BugsXLA, giving a conventional analysis of these data. This R script is obtained by selecting the 'Make R Script' check box on the 'MCMC & Output Options' form. Figure 3.10 shows this form after making this selection,

N	O	P	Q	R	S	T	U	V
Label	**Mean**	**St.Dev.**	**2.5%**	**Median**	**97.5%**		**WinBUGS Name**	
	Predicted Mean Response							
	Following all at: FEV1_BASE(mean)							
TRT(A)	-0.0397	0.0458	-0.1294	-0.0401	0.0518		Pred.Ave[1]	
TRT(B)	0.0525	0.0469	-0.0393	0.0520	0.1473		Pred.Ave[2]	
TRT(C)	0.1093	0.0347	0.0416	0.1091	0.1772		Pred.Ave[3]	
TRT(D)	-0.0175	0.0447	-0.1047	-0.0177	0.0718		Pred.Ave[4]	
TRT(E)	0.0146	0.0455	-0.0758	0.0148	0.1040		Pred.Ave[5]	

FIGURE 3.9
Predicted means from clinical study.

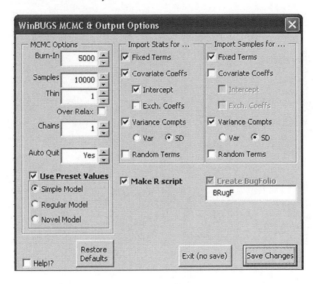

FIGURE 3.10
Setting the output options in the clinical study.

which causes BugsXLA to create a sub-folder with default name 'BRugF' in which the R script will be created. Section 2.5 provides more information regards this feature, and Appendix D describes these R scripts in more detail. The file is called 'BGX R EDA.R' and after reading the data into a data frame named BGX.df, the output obtained from the classical analysis it defines is shown below:

```
> BGX.fm <- lm(FEV1_CFB ~ TRT + FEV1_BASE, data= BGX.df)
> summary(BGX.fm)

Call:
lm(formula = FEV1_CFB ~ TRT + FEV1_BASE, data = BGX.df)

Residuals:
Min            1Q          Median        3Q          Max
-0.45444      -0.08861     -0.01008      0.11124     0.44805
```

```
Coefficients:
                Estimate    Std. Error     t value     Pr(>|t|)
(Intercept)      0.04411      0.06309        0.699       0.4862
TRTB             0.09220      0.06355        1.451       0.1504
TRTC             0.14858      0.05714        2.600       0.0109*
TRTD             0.02209      0.06330        0.349       0.7280
TRTE             0.05446      0.06354        0.857       0.3937
FEV1_BASE       -0.06522      0.04136       -1.577       0.1184

Signif. codes:  0 '***' 0.001 '**' 0.01 '*' 0.05 '.' 0.1 ' ' 1

Residual standard error: 0.1817 on 89 degrees of freedom
Multiple R-squared: 0.1032, Adjusted R-squared: 0.05277
F-statistic: 2.047 on 5 and 89 DF, p-value: 0.07959
```

The estimates of the effects of treatment are very close to the posterior means reported in the Bayesian analysis; the differences are almost solely due to MCMC sampling errors as the priors used would have had little effect. The standard errors are slightly smaller than the equivalent posterior standard deviations. This is to be expected, and is due to the latter representing the dispersion of the posterior distribution, which will be approximately a t-distribution; if the df were much less than 89, then the differences between the classical standard errors and the Bayesian posterior standard deviations would have been greater. The classical point estimate for the residual standard deviation (termed 'residual standard error' in R output) is slightly smaller than the Bayesian posterior mean for this parameter. Again, this is to be expected as the classical point estimate is obtained by maximizing the 'residual likelihood', which is akin to finding the mode of the posterior distribution. If a 95% confidence interval is calculated for this parameter, it will be seen that the limits are very close to the Bayesian credible limits shown. As already mentioned, it is well known that the Bayesian analysis of a NLM with the conventional improper vague priors leads to posterior distributions for each of the parameters that are numerically identical to the classical confidence distributions. This explains why using the very vague default priors provided by BugsXLA gives, within MCMC error, the same credible intervals as their equivalent classically calculated confidence intervals.

This study was actually designed and analyzed using informative priors for some of the parameters. A recent placebo-controlled study in which different treatments were assessed in an identical patient population was used to provide informative priors for both the mean placebo response and the residual standard deviation. The relevant summary statistics from the previous study are

Placebo mean = 0.00, with a standard error of 0.04

Residual variance estimate = 0.026, with 44 df

If we assume a standard vague prior at the start of the previous study, then the posterior distribution for the placebo mean response would be a t-distribution with mean 0, dispersion parameter 0.04^2 and 44 df. It is a sufficiently good approximation to represent this as a Normal distribution with mean 0 and standard deviation 0.04. As previously stated, the posterior for the inverse of the variance parameter would be a scaled chi-square with parameters, df = 44 and $s^2 = 0.026$. In order to justify using these posterior distributions as the priors for the new study, we need to believe that both the true placebo mean change from baseline response and the true residual variance will be the same in this new study as the previous. In this case, due to the extreme similarities between the studies, this belief was considered reasonable. Strictly speaking some account should have been taken for the likely difference that would be seen between the mean baselines for the two studies, but this was assessed has having minimal effect on the placebo mean response.

It is advisable when using informative priors to assess how influential they are on the inferences. This is best done by repeating the analysis with various different priors. A quicker way to obtain some insight into their influence is to produce a plot of the posterior and prior overlaid. BugsXLA will provide such a plot for those parameters it can infer the specified prior. It is first necessary to request that the samples be imported back into Excel. This is done on the 'MCMC & Output Options' form as shown in Figure 3.10. In the frame headed 'Import Samples for …' both the 'Fixed Terms' and the 'Variance Compts' have been checked. This instructs BugsXLA to import the 10,000 samples for each of these terms back into Excel. Figures 3.11 and 3.12 show how the informative priors discussed above are specified using BugsXLA.

On the 'Ind. Factors' page, only the prior for the 'Constant Term' has been altered. In this case, it represents the true placebo mean response at the mean of the baseline covariate. More generally, it is important to check the parameterization of the model being fitted, and be aware of which level of the factors are constrained when interpreting the meaning of the constant term, as well as some of the other parameters. The only parameter on the 'Errors' page when an NLM is being fitted is the residual variance, which is given the informative prior shown in Figure 3.12; note that the Scaled Chi-Square Distribution has been selected before changing the default parameters for the prior. Figure 3.13 shows the output as imported back into Excel.

Comparing this output with that shown in Figure 3.7 when vague priors were used, one can see that the posterior mean for the model constant parameter has been shrunk toward the prior mean of zero, and that its posterior standard deviation has noticeably reduced. The treatment effect estimates have been similarly affected, having a slightly smaller posterior mean due to the affect of the prior for the placebo mean. There is also a clear effect of the informative prior on the posterior of the residual error parameter, with both shrinkage in its posterior mean and a slight reduction in its posterior

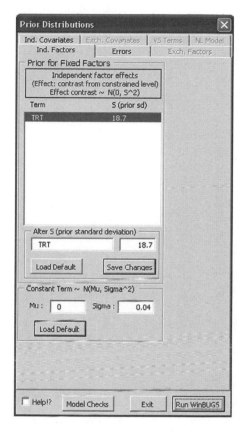

FIGURE 3.11
BugsXLA form for changing priors for fixed effects.

standard deviation, as expected. Figure 3.14 shows how to obtain plots of the posterior and prior distributions for these parameters.

The samples requested for import in Figure 3.10 are saved to a separate sheet in the Excel workbook; by default this sheet has the same name as the sheet containing the data with the suffix 'sams'. As can be seen in Figure 3.14, this sheet contains a column for each of the parameters with imported samples, each column having a header cell plus, in this case, 10,000 rows of values. The header in the first row of each column contains the WinBUGS node name for the parameter, rather than the more informative BugsXLA label (see Appendix C for more details). The easiest way to decode these names is to move the mouse over the cell and read the comment displayed. For example, the comment under cell B1 with value "X.Eff[1,2]" is

```
WinBUGS samples
TRT:B
Normal
0
18.7
```

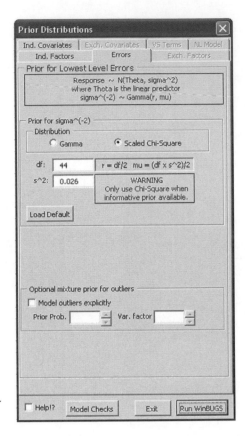

FIGURE 3.12
BugsXLA form for changing priors for residual error terms.

The second line in the comment tells us that this parameter represents the effect of treatment B. The other parts of the comment are used by BugsXLA when plotting these samples, so this comment should not be deleted. An alternative way to decode all the WinBUGS node names is to refer back to the output shown in Figure 3.13 where they are shown in column J in the same row as their BugsXLA labels. To plot one of these columns of data, make any cell in the column active by clicking on a cell, and then from the BugsXLA toolbar click on the 'Post Plots' icon. This icon is the one highlighted in Figure 3.14, and has a bell-shaped histogram with a curve overlaid as its picture. The form shown on the right of Figure 3.14 will appear. In this example, the column headed 'sigma' was selected for plotting, and so the Posterior Plotting Tool form shows the information retrieved for this parameter: the BugsXLA label, "SD(residual)," and the prior, "ScChiSq(df=44, s^2=0.026) for Inv-V." Clicking on the button labeled 'Update Graph' produces a plot similar to that shown in Figure 3.15.

This plot was made slightly different to the default via the 'User Specified' options on the Posterior Plotting Tool form. It can be seen that although the prior is clearly informative, since it is not close to being flat across the

	A	B	C	D	E	F	G	H	I	J	K
1			Label	Mean	St.Dev.	2.5%	Median	97.5%		WinBUGS Name	
2			CONSTANT	-0.0179	0.0297	-0.0768	-0.0179	0.0403		Beta0	
3			Intercept at 0	0.0615	0.0560	-0.0501	0.0624	0.1700		alpha	
4		TRT	A	0.0000	0.0000					X.Eff[1,1]	
5		TRT	B	0.0709	0.0532	-0.0333	0.0710	0.1758		X.Eff[1,2]	
6		TRT	C	0.1267	0.0453	0.0371	0.1269	0.2159		X.Eff[1,3]	
7		TRT	D	3.111E-4	0.0526	-0.1034	-1.291E-4	0.1026		X.Eff[1,4]	
8		TRT	E	0.0327	0.0523	-0.0686	0.0322	0.1358		X.Eff[1,5]	
9			FEV1_BASE	-0.0620	0.0399	-0.1408	-0.0621	0.0168		V.Coeff[1,1]	
10			SD(residual)	0.1762	0.0110	0.1564	0.1755	0.1992		sigma	
11	Note: CONSTANT & Factor effects are determined at the mean of the covariate(s).										
12		Interpret these cautiously when Factor x Covariate terms have been fitted.									
13											
14	Model	['clin study'!$A$1:$D$96]									
15	Distribution	Normal									
16	Link	Identity									
17	Response	FEV1_CFB									
18	Fixed	TRT									
19	Covariates	FEV1_BASE									
20											
21	Priors										
22	CONSTANT	N(mu=0, sigma=0.04)									
23	TRT	N(mu=0, sigma=18.7)									
24	FEV1_BASE	N(mu=0, sigma=39.8)									
25	V(residual)	Inv-ChiSqr(44) x (44 x 0.026)									
26											
27	WinBUGS MCMC Settings										
28	Burn-In: 5000 Samples: 10000 (Thin:1; Chains:1)										
29	Run took 13 seconds										

FIGURE 3.13
Summaries of posterior distributions, model, priors, and MCMC settings for clinical study using informative priors.

range of credible values, the information in the new study has been influential in determining the posterior. The equivalent plot for the model constant parameter is shown in Figure 3.16. The posterior for the model constant has a mean fairly close to the prior, but has a smaller variance due to the additional information in the new study. Figure 3.17 shows this posterior and prior plot for the effect of treatment C.

The prior is indistinguishable from the horizontal axis due to its being so dispersed relative to the posterior. It is important not to conclude from this plot that the posterior for this parameter is uninfluenced by any of the priors specified. In fact, for this simple model, it can be shown analytically that the informative prior for the model constant has a direct impact on the posteriors for all of the treatment effects; this is intuitively reasonable since a treatment's effect is defined as the mean difference from the placebo response. Figure 3.17 also shows how to obtain more refined percentiles of the posterior distribution than can be routinely imported back from WinBUGS; these are obtained simply by selecting the 'Percentiles' check box before clicking on the 'Display' button. It is also possible to specify up to three "tail area" probability calculations. In this case the posterior probability that the effect of treatment C is greater than zero, that is, superior to placebo, is 99.8%. An effect size of 0.075 L was considered clinically meaningful for this population, and the posterior probability that the effect of treatment C is greater than or equal to this amount is 87%.

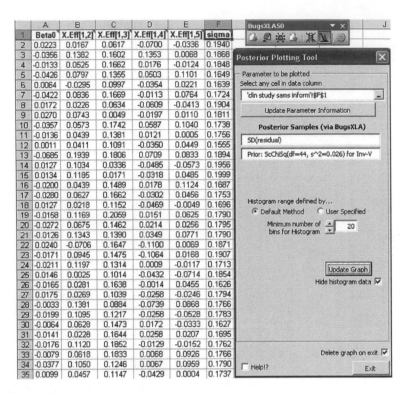

FIGURE 3.14
BugsXLA's posterior plotting tool.

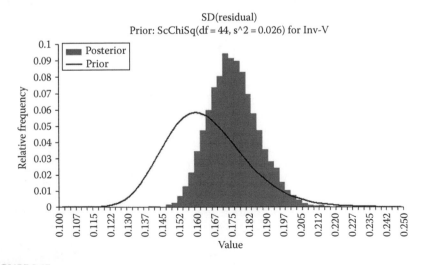

FIGURE 3.15
BugsXLA histogram of posterior samples with informative prior overlaid for the residual standard deviation parameter in the clinical study.

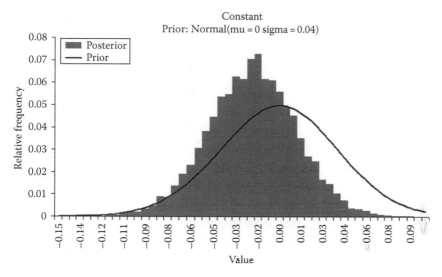

FIGURE 3.16

BugsXLA histogram of posterior samples with informative prior overlaid for the model constant parameter in the clinical study.

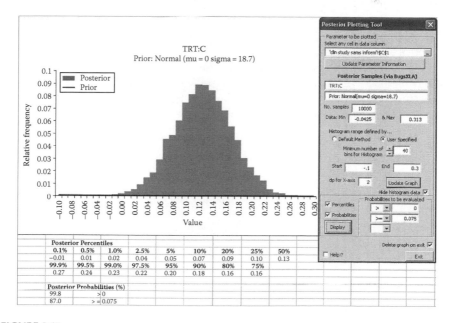

FIGURE 3.17

BugsXLA histogram of posterior samples with prior overlaid for the effect of treatment C in the clinical study. Posterior percentiles and probabilities of effect being greater than specified values also shown.

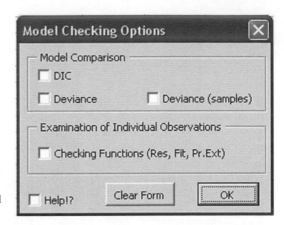

FIGURE 3.18
BugsXLA form for specifying model checking options.

BugsXLA has not been designed to provide a comprehensive range of model checking features, and the user is strongly encouraged to use other software to do this. As described in Section 2.5, BugsXLA can create an R script that can be used as a starting point for such an exploratory analysis. However, it is possible to obtain additional output from WinBUGS to aid in assessing the model fit. Figure 3.12 shows the form where the prior distributions can be altered. At the bottom of this form is a button labeled 'Model Checks', which when clicked brings up the form shown in Figure 3.18.

The model comparison options, DIC and deviance, will be discussed later in the book. Here we explain what is produced when the checking functions to help assess the influence of individual observations are requested. On importing the results back into Excel, an additional worksheet is produced with the name 'Mdl Chks(1)', or next smallest integer if this sheet already exists. On this sheet, for each observation in the data set, the posterior mean for each of the following functions is shown:

Fit: the fitted value or predicted mean response

RawRes: the raw residual, that is, difference between observed and fitted value

ScRes: the scaled residual, that is, the raw residual divided by the residual standard deviation

These are all Bayesian versions of standard classical model checking statistics, which can be interpreted in the conventional manner. BugsXLA also computes what has been referred to as "the chance of getting a more extreme observation: $\min(p(Y_i < y_i), p(Y_i \geq y_i))$," where the y_i are the observations and $p(Y_i)$ is their predictive distribution (see Chapter 9 of Spiegelhalter et al. (1996) or Gelfand et al. (1992)). This probability, labeled **PrExt** in the output, is similar in nature to the classical p-value, and so small values can be interpreted as evidence that the observation is in some way discordant with

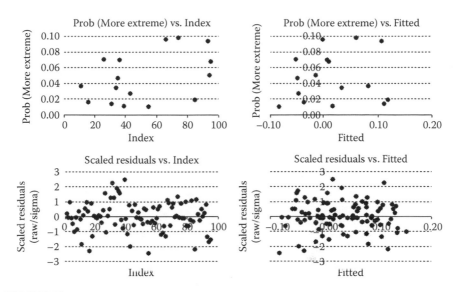

FIGURE 3.19

BugsXLA's model checking plots of the Bayesian "prob. extreme values" and the scaled residuals for the clinical study.

the model. It is important to be aware that no cross-validation approach is used in computing this value, and so it is conservative in its ability to detect discordant observations. The observed values, the raw residuals, the scaled residuals and the "Prob[More Extreme]," PrExt, are each plotted both against the observation's index and the fitted values. Figure 3.19 shows the plots for the latter two model checking functions for the clinical study.

By default, only those observations with a probability of a more extreme result less than 0.1 are plotted in the graphs on the top row in Figure 3.19. Inspection of the plots shows a good correlation between low probabilities in the top graph and relatively large-scaled residuals in the bottom graph, for example, compare probabilities of less than 0.02 with absolute scaled residuals greater than 2. For models with Normal errors, it is unlikely that these probabilities of being extreme will pick up anything that the scaled residuals will miss. These additional Bayesian diagnostics may prove more valuable in more complex models where scaled residuals are difficult to define. Although these plots do not raise any serious concerns with the adequacy of the assumption of Normal errors, we will come back to this example in Chapter 10 when illustrating how to model heavy-tailed error distributions.

4

Generalized Linear Models

Generalized linear models (GLMs) are a natural extension of the NLM that permit a wide range of non-Normal responses to be modeled as flexibly as in the Normal case. The most common examples are Binomial and Poisson data, with these often being modeled using logistic and log-linear regression, respectively. As discussed in Chapter 3 for the NLM, all the location parameters in a GLM are also fixed constants. GLMs were introduced by Nelder and Wedderburn (1972), with McCullagh and Nelder (1989) being the standard reference book on the subject. However, more recent books have been written that are much more accessible to the applied statistician, for example, Myers et al. (2002), Dobson (2002), or Faraway (2006). Chapter 16 of Gelman et al. (2004) covers GLMs from a Bayesian perspective. There are three components to a GLM:

1. The linear predictor, $\eta = X \beta$
2. The link function, $g(\mu) = \eta$
3. The error distribution, specifying the distribution of the response, y, given the explanatory variables, X, with mean $E(y \mid X) = \mu$. This error distribution can also depend upon a further dispersion parameter, φ.

Here, X is the design matrix of explanatory variables, and β the vector of parameters defining their effects on the response. Note that this general notation still permits the fitting of categorical factors, as well as continuous variates, via the specification of dummy variables (see, e.g., Dobson (2002)). It should be clear that when $g(.)$ is the identity link function, that is, $\mu = \eta$, and the error distribution is Normal, then we are back to the NLM, with the dispersion parameter φ being σ. More generally, a non-linear link function is required in order to allow a relatively simple error distribution to be a credible component of the model, as well as to avoid difficulties in estimating the parameters when the mean, μ, is constrained to a subset of the whole real line, for example, [0, 1] for the Binomial proportion or [0, ∞) for the Poisson count response. Note that when a non-linear link function is used, GLMs are not "linear models" in the usual statistical sense, since the mean is no longer a weighted linear combination of the parameters. They are linear in the generalized sense that the mean is related to such a linear combination via the link function. This non-linearity has implications for the interpretation of the explanatory variables' effects, as well as other inferences of interest, particularly when additional random effects are included in the model as discussed in Chapter 6.

4.1 Binomial Data

To help illustrate how a GLM is formulated, we show how the equivalent ANCOVA model to that discussed in Chapter 3 could be fitted to a Binomial response. We again consider an experiment in which three treatments are being compared, with some continuous baseline measurement considered a potential explanatory variable. The response after treatment is now a Binomial proportion of some event that is of interest. This model can be represented algebraically:

$$\eta_{ij} = \mu + \alpha_i + \beta x_{ij} \quad i = 1, 2, 3 \quad j = 1, 2, \dots, n_i$$

$$g(\pi_{ij}) = \eta_{ij}$$

$$r_{ij} \sim \text{Bin}(\pi_{ij}, m_{ij})$$

where μ, α_i, β, x_{ij}, and n_i are as defined before, but noting that all parameters are operating on the link scale. The pair (r_{ij}, m_{ij}) is the response, that is, the number of events and tests, respectively, on the jth experimental unit receiving the ith treatment, and π_{ij} is the true event rate given the treatment applied and value of x_{ij}. A common choice of link function for a Binomial response is the logit, $g(\pi) = \log(\pi/(1 - \pi))$, which transforms the linear predictor from the whole real line to [0, 1], and leads to the logistic regression model. Note that the Binomial Distribution does not require a dispersion parameter, that is, $\varphi = 1$, although over-dispersion (as explained in, e.g., Brown and Prescott (2006)) is very common in practice. Modeling over-dispersed data is one of the reasons for including random effects as discussed in Chapter 6. It should be apparent how any "fixed effects" linear model, however complex, that attempts to asses the effects of explanatory variables on a Normal response could be modified to model a Binomial response. Note that due to the link function, the parameters no longer represent the effects of the variables as linear changes in the true mean response. When the logit link is used, the model assumes that these represent linear changes in the log odds, where odds are the ratio of the true event rate to its complement, $\pi/(1 - \pi)$. The implications for interpretation will be discussed in the example that follows shortly.

As well as the logit, BugsXLA allows the user to specify either of the other two common link functions for Binomial data: the probit and complementary log-log (cloglog). The less commonly used generalized logit link, introduced by Prentice (1976), is also available. BugsXLA determines default priors for the parameters in the same way as described in Chapter 3 for the NLM (refer to Section 2.7 for details of these default priors).

Case Study 4.1 Respiratory Tract Infections

Smith et al. (1995) discuss the meta-analysis of 22 randomized trials to investigate the clinical benefits of selective decontamination of the digestive tract on respiratory infections acquired in intensive care units. The authors compared Bayesian approaches to various alternatives, and in particular the method adopted in the original analysis by a Collaborative Group (1993) of digestive tract trialists. In each trial, intensive care patients were randomized, in a blinded fashion, to either a treatment or control group. Here we show how a fixed effects model can be fitted using BugsXLA. We come back to this example in Chapter 6, where a more appropriate random-effects model is fitted.

Figure 4.1 shows the data and completed BugsXLA model specification form. The variables study and trt are both defined as factors, with the latter

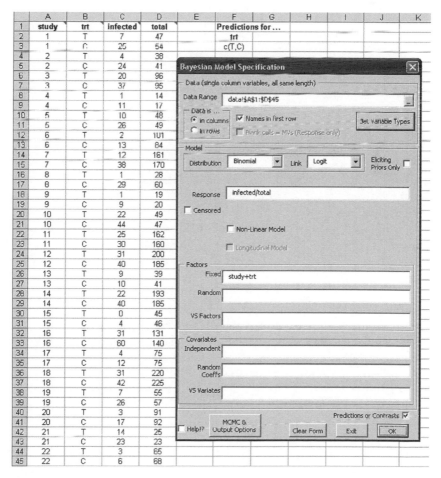

FIGURE 4.1
Data and model specification for respiratory tract infections meta-analysis.

having levels 'T' and 'C' to represent the treated and control groups, respectively. The variables infected and total are variates that contain the Binomial response; the number of patients with respiratory tract infections and total number in the trial for each study-by-treatment group combination. Note how the Binomial response is defined as the proportion, 'infected/total', on the BugsXLA model specification form. If the user wished to apply a link function other than the logit, then this can be selected via the drop-down menu. Both of the factors are included in the fixed effects part of the model specification by including the '+' operator between them. The model specified is an analysis of variance with two grouping factors, and hence can be represented algebraically:

$$\eta_{ij} = \mu + \alpha_i + \gamma_j \quad i = 1, 2, \ldots, 22 \quad j = 1, 2$$

$$\text{logit}(\pi_{ij}) = \eta_{ij}$$

$$r_{ij} \sim \text{Bin}(\pi_{ij}, m_{ij})$$

where
   r and m are the response variables infected and total, respectively
   α_i are the effects of the factor study
   γ_j are the effects of the factor trt

The simple model defaults for the MCMC settings were chosen (a burn-in of 5,000 followed by 10,000 samples), the contrast between the treated and control groups' means was requested (see cells F2:F3 in Figure 4.1), and the default priors were unchanged (see cells A33:B35 in Figure 4.2). The model checking functions and Deviance Information Criterion (DIC) statistic were also requested. Refer back to Case Study 3.1 for details of how to set all these options. Figures 4.2 through 4.4 show the results of the analysis as imported back into Excel. Figure 4.2 shows all the parameters in the model and a summary of their marginal posterior distributions.

Inspection of the 95% credible interval for the effect of treatment clearly shows that selective decontamination of the digestive tract reduces the incidence of respiratory tract infections. However, due to the treatment parameter measuring the effect on the logit scale, it is difficult to interpret the values shown in Figure 4.2 directly. Figure 4.3 shows the output produced to summarize the treatment contrast requested. When a contrast is specified for a Binomial response, the contrast can only be estimated on the odds scale and not the probability scale; this is because there is no one-to-one mapping of a difference on the logit scale to anything meaningful on the probability scale. The contrast has been back-transformed from the logit scale, so that it is now the ratio of the odds of infection for the treated group to the odds of infection in the control group that is being estimated, that is,

	A	B	C	D	E	F	G	H	I	J	K
1			Label	Mean	St.Dev.	2.5%	Median	97.5%		WinBUGS Name	
2			CONSTANT	-0.3314	0.2160	-0.7449	-0.3353	0.1033		Beta0	
3		study	2	0.1962	0.3236	-0.4537	0.1999	0.8303		X.Eff[1,2]	
4		study	3	-0.0491	0.2667	-0.5805	-0.0472	0.4639		X.Eff[1,3]	
5		study	4	0.3063	0.4421	-0.5796	0.3059	1.1660		X.Eff[1,4]	
6		study	5	0.2917	0.3077	-0.3268	0.2940	0.8844		X.Eff[1,5]	
7		study	6	-1.6620	0.3519	-2.3710	-1.6570	-0.9863		X.Eff[1,6]	
8		study	7	-0.9825	0.2650	-1.5210	-0.9793	-0.4667		X.Eff[1,7]	
9		study	8	-0.0382	0.3166	-0.6713	-0.0332	0.5686		X.Eff[1,8]	
10		study	9	-0.3185	0.4410	-1.1960	-0.3123	0.5319		X.Eff[1,9]	
11		study	10	1.7310	0.3140	1.1250	1.7320	2.3430		X.Eff[1,10]	
12		study	11	-0.8108	0.2604	-1.3280	-0.8078	-0.3170		X.Eff[1,11]	
13		study	12	-0.6940	0.2517	-1.1870	-0.6922	-0.2066		X.Eff[1,12]	
14		study	13	-0.4105	0.3476	-1.1090	-0.4069	0.2614		X.Eff[1,13]	
15		study	14	-0.8536	0.2586	-1.3660	-0.8522	-0.3517		X.Eff[1,14]	
16		study	15	-2.4650	0.5780	-3.7070	-2.4300	-1.4260		X.Eff[1,15]	
17		study	16	0.1157	0.2514	-0.3885	0.1199	0.6008		X.Eff[1,16]	
18		study	17	-1.3940	0.3469	-2.0960	-1.3850	-0.7326		X.Eff[1,17]	
19		study	18	-0.8692	0.2506	-1.3680	-0.8656	-0.3862		X.Eff[1,18]	
20		study	19	-0.0859	0.3003	-0.6878	-0.0803	0.4884		X.Eff[1,19]	
21		study	20	-1.3650	0.3193	-2.0050	-1.3590	-0.7474		X.Eff[1,20]	
22		study	21	2.2120	0.4161	1.4230	2.2040	3.0500		X.Eff[1,21]	
23		study	22	-1.9470	0.4141	-2.8050	-1.9350	-1.1760		X.Eff[1,22]	
24		trt	T	-1.0700	0.0888	-1.2420	-1.0710	-0.8962		X.Eff[2,2]	
25											
26	Model	[data!$A$1:$D$45]									
27	Distribution	Binomial									
28	Link	Logit									
29	Response	infected/total									
30	Fixed	study+trt									
31											
32	Priors										
33	CONSTANT	N(mu=0, sigma=100)									
34	study	N(mu=0, sigma=100)									
35	trt	N(mu=0, sigma=100)									
36											
37	WinBUGS MCMC Settings										
38	Burn-In: 5000 Samples: 10000 (Thin:1; Chains:1)										
39	Run took 20 seconds										

FIGURE 4.2
Summaries of posterior distributions, model, priors, and MCMC settings for respiratory tract infections meta-analysis.

N	O	P	Q	R	S	T	U	V
Label	Mean	St.Dev.	2.5%	Median	97.5%		WinBUGS Name	
Predicted Probability								
Predicted Odds								
study(1) contr[trt T / C]								
	0.3442	0.0306	0.2887	0.3427	0.4081		Pred.Odds[1]	

FIGURE 4.3
Summary of treatment contrast for the respiratory tract infections meta-analysis.

the odds-ratio $(\pi_{i2}/(1-\pi_{i2}))/(\pi_{i1}/(1-\pi_{i1}))$. BugsXLA indicates this by the label, "contr[trt T/C]." Although, for a given treatment group, the probability, or odds, of infection differs between studies in this model, it should be clear that the difference between the log-odds of infection for the two treatment groups does not; the effect of study is additive on the logit scale. Hence, the odds-ratio, being the back-transformed difference between log-odds, is also the same for all studies. In Figure 4.3 it can be seen that BugsXLA states

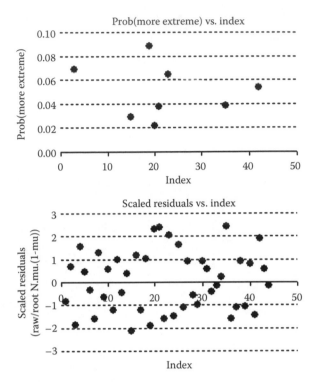

FIGURE 4.4

Bayesian "prob. extreme values" and the scaled residuals for the respiratory tract infections meta-analysis.

that the contrast is for "study(1)," the constrained level of the study factor (see Section 2.12). Although, as just discussed, the odds ratio is the same for all studies in this example, and hence the specific study can be ignored, this would not have been the case if a study-by-treatment interaction had been included in the model. The posterior mean (and 95% credible interval) for the odds ratio is 0.34 (0.29, 0.41). This estimate is very similar to the 95% confidence interval obtained via the Mantel–Haenszel–Peto method, given in Smith et al. (1995) as 0.36 (0.31, 0.43).

Figure 4.4 shows two of the model checking plots produced by BugsXLA for this analysis. Refer to Section 2.9 and the discussion in Case Study 3.1 for more on these plots. Although these plots do not indicate any extreme outliers—the largest scaled residual is 2.38 (with 'more extreme' probability of 0.04), and the smallest 'more extreme' probability is 0.02 (with scaled residual 2.32)—due to the relatively large number of moderately extreme observations, these plots do suggest that there is more variation in these data than the model adequately explains. Some caution is required when interpreting these plots, since the model induces dependencies between these observation level values, for example, the pairs of values in each study will be very highly correlated, negatively in the case of the scaled residuals. We will come back to this example in Chapter 6 when random effects are discussed, and use another "goodness of fit" diagnostic, the Bayesian DIC, to help identify the most appropriate model for these data. The next case study explains how the DIC is obtained via BugsXLA, and how it can be interpreted.

Case Study 4.2 Binary Dose–Response Data

We use data from a small dose–response study to illustrate how the different link functions that BugsXLA offers can be compared. We show how the DIC can be used to help compare between models, with a cautionary example of the dangers of using it blindly; refer to Spiegelhalter et al. (2002) for details of the DIC and discussion of its validity as a model comparison statistic (see also comments in Section 2.9). The data were first published by Bliss (1935) and show the number of beetles killed after 5 h exposure to carbon disulfide at eight different concentrations. BugsXLA offers four link functions for binomial data: logit, probit, cloglog, and generalized logit (see Section 2.3.2 for definitions). These link functions essentially represent the assumption that we are modeling a latent variable with a logistic, Normal, extreme value or generalized logistic error distribution, respectively. Figure 4.5 shows the data and model specification when the generalized logit link is used.

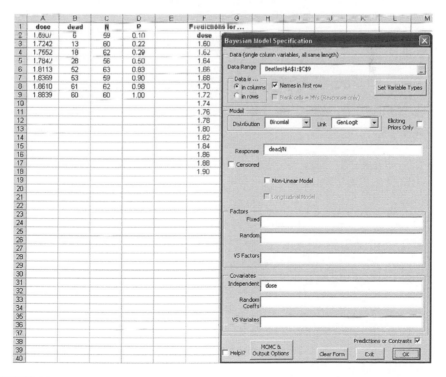

FIGURE 4.5
Data and model specification for binary dose response with generalized logit link.

All the variables are continuous variates so there is no need to utilize the 'Set Variable Types' facility. The values in column D, labeled P, are the crude estimates of the proportion killed obtained from the ratio dead/N. These values will be used later when plotting the fitted models. The predicted means defined in column F will also be used later to create the plot. The model specified is a generalized logistic regression, and can be represented algebraically:

$$\eta_i = \mu + \beta x_i \quad i = 1, 2, \ldots, 8$$

$$\text{logit}(\pi_i^*) = \eta_i$$

$$\pi_i = (\pi_i^*)^m$$

$$r_i \sim \text{Bin}(\pi_i, n_i)$$

where
  r and n are the response variables dead and N, respectively
  x is the covariate dose with regression coefficient β
  m is the power parameter that generalizes the logit link; when m equals one this becomes the standard logit link

The regular model defaults for the MCMC settings were chosen (a burn-in of 10,000 followed by 10,000 samples after a one-fifth thinning), predicted means were requested as discussed above, and the default priors were unchanged (see cells A21:B23 in Figure 4.6). Finally, before clicking on the 'Run WinBUGS' button, the DIC statistics were requested from the 'Model Checking Options' form obtained via the 'Model Checks' button, as explained in Case Study 3.1 and Section 2.9. Figure 4.6 shows all the parameters in the model and a summary of their marginal posterior distributions.

Inspection of the credible interval for the parameter 'GenLogit m' shows that there is strong evidence that the standard logit link is inadequate since one is not a credible value. As well as showing all the parameters in the model and a summary of their marginal posterior distributions, Figure 4.6 also shows the Bayesian DIC statistics:

Dbar: the posterior mean of the deviance
Dhat: a point estimate of deviance utilizing the parameters' posterior means
pD: the effective number of parameters; pD = Dbar − Dhat
DIC: the Deviance Information Criterion; DIC = Dbar + pD

Refer to Spiegelhalter et al. (2002) for the detailed definition of these terms. The DIC will be used to help compare the different link functions used to

	A	B	C	D	E	F	G	H	I	J	K
1			Label	Mean	St.Dev.	2.5%	Median	97.5%		WinBUGS Name	
2			GenLogit m	0.2607	0.1384	0.0817	0.2342	0.6201		m.gnlgt	
3			CONSTANT	-2.7130	1.9980	-7.9740	-2.3030	-0.0204		Beta0	
4			Intercept at 0	-143.5000	55.9600	-291.2000	-131.5000	-73.1200		alpha	
5			dose	78.6500	30.1800	40.7500	72.2000	158.5000		V.Coeff[1,1]	
6	Note: CONSTANT & Factor effects are determined at the mean of the covariate(s).										
7	Interpret these cautiously when Factor x Covariate terms have been fitted.										
8											
9				Y							
10			Dbar	32.8910							
11			Dhat	38.2940							
12			pD	-5.4030							
13			DIC	27.4870							
14	Model	[Beetles!$A$1:$C$9]									
15	Distribution	Binomial									
16	Link	GenLogit									
17	Response	dead/N									
18	Covariates	dose									
19											
20	Priors										
21	GenLogit m	Gamma(0.25, 0.25)									
22	CONSTANT	N(mu=0, sigma=100)									
23	dose	N(mu=0, sigma=1480)									
24											
25	WinBUGS MCMC Settings										
26	Burn-In: 10000 Samples: 10000 (Thin:5; Chains:1)										
27	Run took 14 seconds										

FIGURE 4.6

Summaries of posterior distributions, model, priors, and MCMC settings for binary dose response with generalized logit link.

model these data, with smaller values of DIC indicating a better model in terms of its ability to make short-term predictions. The WinBUGS website provides the following guidance when comparing the DIC values obtained from different models fitted to the same data: a difference of greater than 10 between DIC values is clear evidence that one model is better than another, while differences between 5 and 10 can be considered substantial.

Inspection of the DIC output in Figure 4.6 reveals a rather strange phenomenon, a negative pD value. Since this quantity essentially represents the effective number of parameters in the model, it should not be negative, and in this case should be somewhere between two and three. The Bugs website (http://www.mrc-bsu.cam.ac.uk/bugs/) has a page dedicated to the DIC, which includes answers to frequently asked questions such as "Can pD be negative?". There you will find the technical details of how this can happen, but essentially it is due to the point estimates of one or more of the parameters not providing a good estimate of the deviance when used in calculating Dhat. This occurs in some cases when the posterior distributions are very non-Normal. Since pD equals Dbar – Dhat, this "error" is transmitted to this value also. When using BugsXLA there is nothing one can do to fix this problem, other than consider alternative prior distributions that are both credible and result in a pD value that makes sense. Of the four statistics related to the deviance shown in Figure 4.6, only the posterior mean of the deviance, Dbar, can be considered reliable. Note that this problem with the DIC statistics does not affect the reliability of the parameter estimates themselves.

TABLE 4.1

DIC Statistics Used to Compare Link Functions

Link	GenLogit	GenLogit	Logit	cLogLog	Probit
Prior SDs	Default	30	Either	Either	30
Dbar	32.9	32.5	39.4	31.6	38.3
Dhat	38.3	31.3	37.4	29.6	36.3
pD	−5.4	1.2	2.0	2.0	2.0
DIC	27.5	33.8	41.4	33.6	40.3

The same model was then rerun using each of the other three link functions in turn. When using the probit link with the default prior distributions, WinBUGS crashed. Noting that the default prior distributions are incredibly dispersed on these link-scales, it is reasonable to assume that the problem is numerical. By reducing the prior standard deviation to 30 for both the model constant and the regression coefficient, while still representing very dispersed distributions, was enough to allow WinBUGS to complete the analysis for all the link functions. Note that neither the logit nor the cloglog links are sensitive to the change from the default to these prior distributions, and the probit is insensitive to increases in the standard deviation up to the point that WinBUGS fails to complete the analysis. However, the parameter estimates of the generalized logit link are sensitive to the prior distribution chosen, and this is discussed in more detail later. Table 4.1 shows the DIC statistics for the various analyses undertaken.

The pD values obtained when using any of the other three link functions are two, which is the number of parameters in the model as expected. Although the pD for the generalized logit link with a less dispersed prior is positive, given that it is still not between two and three, and being aware of the problem observed with the default priors, we should be very cautious about the reliability of Dhat, pD, and DIC in this case. This illustrates one of the worrying issues with the DIC. It is only because this is a relatively simple model in which we have a good idea of the value pD should take, that we can be alerted to a potential issue with these statistics. More generally, and in particular when fitting complex random effects models as discussed in Chapters 5 and 6, it will not be easy to spot that the value of pD is misleading. It is recommended that, until the practical performance of the DIC has been reported in many more published applications with much stronger advice on when it can be relied upon, the DIC should not be used as the sole evidence for selecting a particular model. However, experience to date suggests that the DIC statistics are useful indicators of the relative goodness of fit for most of the common models. In this particular case, it is probably reasonable to replace the DIC for the generalized logit link by the value Dbar plus three; the usual formula is Dbar + pD, and we know that there are three parameters in the generalized logit model (pD might still

O	P	Q	R	S	T	U	V
Mean	St.Dev.	2.5%	Median	97.5%		WinBUGS Name	
Predicted Probability							
dose(1.6)							
0.0254	0.0132	0.0055	0.0234	0.0566		Pred.Ave[1]	
0.0349	0.0162	0.0093	0.0328	0.0722		Pred.Ave[2]	
0.0481	0.0195	0.0158	0.0459	0.0919		Pred.Ave[3]	
0.0666	0.0230	0.0269	0.0644	0.1171		Pred.Ave[4]	
0.0925	0.0264	0.0460	0.0905	0.1493		Pred.Ave[5]	
0.1290	0.0292	0.0767	0.1271	0.1907		Pred.Ave[6]	
0.1805	0.0305	0.1244	0.1789	0.2439		Pred.Ave[7]	
0.2531	0.0303	0.1955	0.2521	0.3141		Pred.Ave[8]	
0.3544	0.0312	0.2945	0.3534	0.4174		Pred.Ave[9]	
0.4906	0.0377	0.4224	0.4886	0.5718		Pred.Ave[10]	
0.6582	0.0442	0.5737	0.6584	0.7430		Pred.Ave[11]	
0.8266	0.0366	0.7468	0.8296	0.8893		Pred.Ave[12]	
0.9400	0.0199	0.8968	0.9422	0.9729		Pred.Ave[13]	
0.9830	0.0107	0.9563	0.9851	0.9976		Pred.Ave[14]	
0.9949	0.0051	0.9809	0.9964	0.9999		Pred.Ave[15]	
0.9983	0.0023	0.9915	0.9992	1.0000		Pred.Ave[16]	

FIGURE 4.7
Predicted means imported into Excel for binary dose response with generalized logit link, edited in preparation for plotting.

be less than three, but it should be quite close). Using these values in the comparison of the models shows that the logit and probit models are clearly inferior, being more than five different in their DIC values from the rest. The cloglog has the smallest DIC, although not much smaller than the generalized logit. Unless other evidence was available to differentiate between these two models, the cloglog should probably be preferred due to it having one fewer parameter, greater precedence, and being more robust to the prior distributions specified.

We show how the predicted means requested can be used to produce a plot that helps to understand the model fits better. Figure 4.7 shows the predicted means for the model shown in Figure 4.6; the output has been edited in preparation for plotting by removing all the rows in between the values. A plot of the fitted model is created by first selecting from the Excel charting options the XY (scatter) plot with points connected by smoothed lines without markers. Into this blank plot, one can add a series, specifying the cells F3:F18 in Figure 4.5 as the X values, and cells O4:O19 in Figure 4.7 as the Y values, naming this series 'GenLogit(V)' to denote the very vague prior used here. The plot should then be formatted to your preferred style. An additional series can be added for each of the other models, editing the line colors and styles to make identification easier. The data can be included on this plot by adding another series, naming it 'data', with X and Y values in cells A2:A9 and D2:D9 of Figure 4.5, respectively. This data series should be formatted as points without lines. Figure 4.8 shows the final plot.

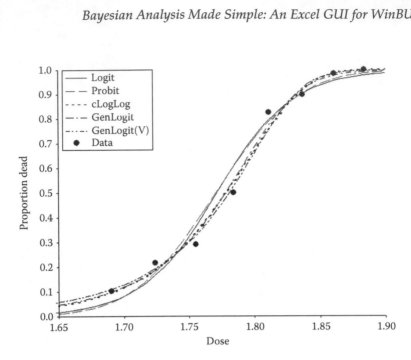

FIGURE 4.8
Fitted models and crude estimates (data) for binary dose–response data.

The models fall roughly into two groups, with the logit and probit forming one group, and the cloglog and both generalized logits forming the other. The main differences occur at the lower half of the dose response, where the logit and probit models appear to be a noticeably poorer fit to the data. If we zoom in on this part of the dose response, the differences are more apparent, as shown in Figure 4.9; this is done simply by altering the scale for both axes on the Excel graph.

It was stated earlier that the parameter estimates of the generalized logit model were sensitive to the priors chosen. Table 4.2 shows the two sets of parameter estimates obtained. Although the different prior distributions have led to quite markedly different marginal posterior distributions, the two fitted models shown in Figures 4.8 and 4.9 appear very similar, as do their Dbar values in Table 4.1. This suggests that the parameters in the model could be highly correlated, so that a change in one can be compensated by a change in the other. One way to assess this is to rerun the analysis and request that the posterior samples be returned for all the parameters. This is done by selecting the 'Fixed Terms' and 'Covariate Coeffs' check boxes under 'Import Samples for ...' on the 'MCMC & Output Options' form as explained in Section 2.5 and illustrated in Case Study 3.1. It is then a simple matter to produce an XY (scatter) plot of the two columns of imported samples for parameters m (WinBUGS node name m.gnlgt) and dose (WinBUGS node name V.Coeff[1,1]). After changing the default format of the plot produced, the graph shown in Figure 4.10 can be produced. This clearly shows a very strong correlation between these two parameters,

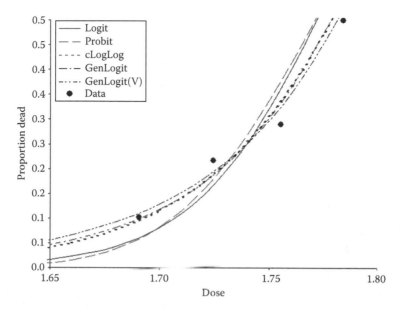

FIGURE 4.9

Close up of lower part of fitted models and crude estimates (data) for binary dose–response data.

TABLE 4.2

Posterior Means (Standard Deviations) for Parameters of Generalized Logit Model Using Different Prior Distributions

Prior SDs	Default	30
m	0.26 (0.14)	0.38 (0.14)
Intercept	−144 (56)	−102 (21)
Dose	79 (30)	56 (12)

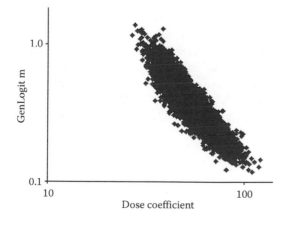

FIGURE 4.10

Scatter plot of samples from posterior distributions of generalized logit m and dose coefficient parameters. Note that both axes are on the log-scale.

which explains why the marginal posterior distributions can change without greatly impacting the model fit.

Case Study 4.3　Binary Classification of Sexual Orientation Using Hormones
Logistic regression can be used for probabilistic binary classification. Details on various ways to tackle classification and discrimination problems can be found in Hand (1981). We explain the general principles behind using logistic regression via a specific case study that also illustrates a difficulty that sometimes arises, and discuss a uniquely Bayesian solution. Margolese (1970) describe a study in which 26 healthy males, 11 heterosexual and 15 homosexual, had 24 h urine samples taken and analyzed for concentrations of the hormones androsterone (androgen) and etiocholanolone (estrogen). The objective of the study was to determine whether the two hormone variables could be used to predict sexual orientation. Figure 4.11 shows the data and model specification form.

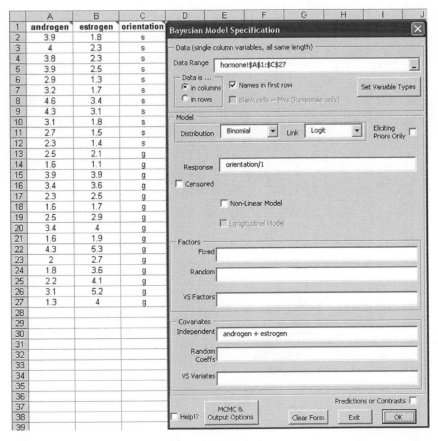

FIGURE 4.11
Data and model specification for binary classification problem.

The binary response variable, orientation, is defined as a factor with levels s (heterosexual) and g (homosexual). Note the notation, 'orientation/1', used when the response is binary. The first level, s, is taken to be the event that is being modeled. The model specified is a logistic regression, and can be represented algebraically:

$$\eta_i = \mu + \beta x_i + \gamma z_i \quad i = 1, 2, \ldots, 26$$

$$\text{logit}(\pi_i) = \eta_i$$

$$r_i \sim \text{Bern}(\pi_i)$$

where
  r is the coded binary response, s = 1, g = 0
  x and z are the two hormone covariates with regression coefficients β and γ, respectively

The simple model defaults for the MCMC settings were chosen (a burn-in of 5,000 followed by 10,000 samples) and the default priors were unchanged. On running WinBUGS, the program failed, displaying the dreaded trap error window. Since this is a very simple data set and model, a failure like this should be a surprise and indicate something peculiar might exist in these data. A reasonable next step is to tighten the very highly dispersed default vague priors, without representing unreasonably precise prior beliefs. We will come back to this matter, but first we fit an equivalent classical model as part of a more exploratory analysis of the data. BugsXLA can create the code necessary for the R package to run this analysis, which is obtained by requesting R scripts as explained in Section 2.5.4 and illustrated in Case Study 3.1. After rerunning the analysis, and exiting WinBUGS after it crashes, the file 'BGX R EDA.R' can be found in the specified sub-folder. This script file can be run from within the R package. After reading the data into a data frame named BGX.df, the output obtained from the classical analysis it specifies is shown below (note that the R script is modeling the event 'orientation = g' due to it sorting the levels alphabetically):

```
> orientation.bern <- (BGX.df$orientation ==
            levels(BGX.df$orientation)[1])
> BGX.fm <- glm(orientation.bern ~ androgen + estrogen,
            family =binomial(link ='logit'), data = BGX.df)
Warning messages:
1: glm.fit: algorithm did not converge
2: glm.fit: fitted probabilities numerically 0 or 1 occurred

> summary(BGX.fm)
```

```
Call:
glm(formula = orientation.bern ~ androgen + estrogen,
    family = binomial(link = "logit"), data = BGX.df)

Deviance Residuals:
Min          1Q           Median      3Q          Max
-3.380e-05   -2.107e-08   2.107e-08   2.107e-08   2.759e-05

Coefficients:
             Estimate    Std. Error   z value   Pr(>|z|)
(Intercept)    84.49    136095.02     0.001     1
androgen     -100.91     92755.62    -0.001     1
estrogen       90.22     75910.98     0.001     1

(Dispersion parameter for binomial family taken to be 1)

Null deviance: 3.5426e+01 on 25 degrees of freedom
Residual deviance: 2.3229e-09 on 23 degrees of freedom
AIC: 6

Number of Fisher Scoring iterations: 25
```

It is clear that the classical likelihood based fitting algorithm also has problems with this model and data set. R provides two warning messages indicating that there were problems with convergence. The reported standard errors for the parameters are very high, which, despite the incredibly small residual deviance, lead to the superficial conclusion that none of the predictors are statistically significant. A simple plot of the data reveals the problem. Figure 4.12 was obtained using the R command:

```
plot(estrogen ~ androgen, data = BGX.df, pch =

    as.character(orientation))
```

It is clear from Figure 4.12 that the two groups are linearly separable so that a perfect fit is possible, that is, it is possible to draw a straight line on the graph that will completely separate the two groups. This is a known problem for methods based on the likelihood alone, and so alternative classical methods have been proposed to deal with this issue (see Section 2.8 of Faraway (2006) for more information). Now that we know that the problem is caused by the likelihood not being able to rule out infinitely large values for the regression coefficients, it makes sense to assess whether prior distributions that down weight incredibly large values would help. Ideally subject matter experts should be involved in helping to elicit reasonable prior beliefs; O'Hagan et al. (2006) discuss scientifically rigorous ways of doing this. Since this is not possible here, we take a more pragmatic approach and through trial and error identify the most dispersed prior distribution that BugsXLA

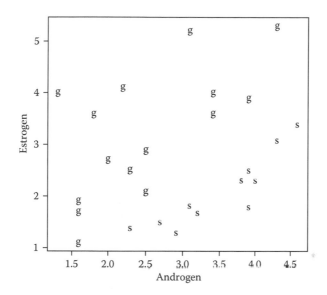

FIGURE 4.12
Scatter plot of estrogen vs. androgen identified by sexual orientation for binary classification problem.

offers that does not cause WinBUGS to crash. Given this completely post-hoc approach, it is essential that the sensitivity of any inferences drawn from the analysis be thoroughly assessed. Figure 4.13 shows the priors chosen (cells A16:B18), as well as the MCMC settings, the model and a summary of the parameters' marginal posterior distributions.

	A	B	C	D	E	F	G	H	I	J	K
1			Label	Mean	St.Dev.	2.5%	Median	97.5%		WinBUGS Name	
2			CONSTANT	-3.1470	1.7270	-7.0700	-2.9120	-0.3793		Beta0	
3			Intercept at 0	-6.4550	3.9550	-15.0200	-6.0800	0.3485		alpha	
4			androgen	7.2210	2.3080	3.3520	7.0060	12.2600		V.Coeff[1,1]	
5			estrogen	-6.4670	2.1550	-11.2000	-6.2680	-2.8970		V.Coeff[2,1]	
6	Note: CONSTANT & Factor effects are determined at the mean of the covariate(s).										
7	Interpret these cautiously when Factor x Covariate terms have been fitted.										
8											
9	Model	[hormone|$A$1:$C$27]									
10	Distribution	Bernoulli									
11	Link	Logit									
12	Response	orientation/1 : Pr[orientation = s]									
13	Covariates	androgen + estrogen									
14											
15	Priors										
16	CONSTANT	N(mu=0, sigma=100)									
17	androgen	N(mu=0, sigma=5)									
18	estrogen	N(mu=0, sigma=5)									
19											
20	WinBUGS MCMC Settings										
21	Burn-In: 5000 Samples: 10000 (Thin:1; Chains:1)										
22	Run took 11 seconds										

FIGURE 4.13
Summaries of posterior distributions, model, priors, and MCMC settings for binary classification problem.

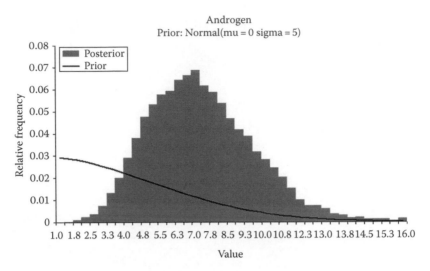

FIGURE 4.14
Posterior and prior overlaid for the androgen parameter in the binary classification problem.

The magnitude of the regression coefficients' posterior means are large both in absolute and relative terms (compared to their posterior standard deviations). When prior distributions are chosen this way, it is advisable to view the posterior and prior overlaid using BugsXLA's posterior plotting tool as explained in Section 2.11 and illustrated in Case Study 3.1. Figure 4.14 shows the plot for the androgen parameter. Given that the androgen values have a range of approximately three, a prior that is consistent with regression coefficients as large as 10 on the logit scale does not seem unduly informative. In fact, it is quite likely that these tightened priors are still too vague, leading to posterior credible intervals that include values that may be too extreme to be truly credible.

4.2 Poisson Data

The simple ANCOVA model for Poisson data can be represented algebraically:

$$\eta_{ij} = \mu + \alpha_i + \beta x_{ij} \quad i = 1, 2, 3 \quad j = 1, 2, \dots, n_i$$

$$\log(\lambda_{ij}) = \eta_{ij}$$

$$y_{ij} \sim \text{Pois}(\lambda_{ij})$$

where μ, α_i, β, x_{ij}, n_i, and y_{ij} are as defined in Chapter 3, but noting that all parameters are operating on the log scale, this being the only link function that BugsXLA offers for Poisson data. The response, y_{ij}, is typically an observed count with a true mean rate λ_{ij}. BugsXLA determines default priors using the same general approach as for other GLMs.

Case Study 4.4 Control of Cockchafer Larvae

Bartlett (1936) discusses the analysis of an experiment designed to assess the efficacy of five treatments for the control of cockchafer larvae, a pest that causes extensive and lethal damage to the roots of young trees. Eight "blocks" of land, labeled A–H, each had the five treatments randomly assigned to plots in a randomized block design. Larvae were divided into two groups by age, labeled a or b, in order to assess any differential effects of treatment by age. The response was the count of larvae still living after an unspecified time following treatment. Unfortunately, due to time constraints, only in blocks A and B were the counts over the whole plots; in the remaining blocks sample quadrant counts over a quarter of the plot were completed. Figure 4.15 shows some of the data and completed BugsXLA model specification form.

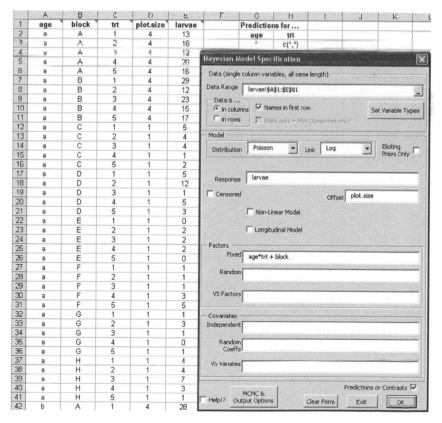

FIGURE 4.15
Data and model specification for cockchafer larvae experiment.

The variables age, block, and trt are all defined as factors. The variable larvae is a variate that contains the Poisson count response. An additional variate, plot.size, is defined and included as an offset in the model, as explained in Brown and Prescott (2006), for example, to account for the differences in the counts due purely to the area over which they were made. The interaction between age and trt, as well as their main effects, are specified by using the '*' operator; 'age*trt' is a convenient shorthand for 'age + trt + age:trt,' where the ':' operator defines an interaction term. The model specified is a slightly more complicated analysis of variance than in previous examples, and can be represented algebraically:

$$\eta_{ijk} = \mu + \alpha_i + \gamma_j + (\alpha\gamma)_{ij} + \omega_k + \log(\text{plot.size}) \quad i = 1, 2 \quad j = 1, 2, \ldots, 5 \quad k = 1, 2, \ldots, 8$$

$$\log(\lambda_{ijk}) = \eta_{ijk}$$

$$y_{ijk} \sim \text{Pois}(\lambda_{ijk})$$

where
  y is the response variable larvae
  α_i are the effects of the factor age
  γ_j are the effects of the factor trt
  $(\alpha\gamma)_{ij}$ are the effects of the interaction between age and trt
  ω_k are the effects of the factor block

The simple model defaults for the MCMC settings were chosen (a burn-in of 5,000 followed by 10,000 samples), all pair wise contrasts between the treatments for both age groups were requested (see cells G2:H3 in Figure 4.15), and the default priors were unchanged (see cells A28:B32 in Figure 4.16). Figures 4.16 and 4.17 show the results of the analysis as imported back into Excel. Figure 4.16 shows all the parameters in the model and a summary of their marginal posterior distributions.

Inspection of the 95% credible intervals for the main and interaction effects of treatment shows that they all contain zero, and hence, a naïve interpretation of these results would be that the experiment has failed to show that any of the treatments clearly affect larvae survival. However, just as when interpreting the equivalent effect estimates from a non-Bayesian analysis, it is not appropriate to make such a broad inference in this way. The treatment factor consists of more than two levels, so there are numerous ways to define a set of contrasts that account for the variation between their mean responses. It cannot be concluded that because all the contrasts in any particular chosen set are relatively close to zero, this implies that all possible contrasts of interest will be similarly close to zero. A classical analysis of variance often utilizes a test based on "additional sums of squares"

	A	B	C	D	E	F	G	H	I	J	K
1			Label	Mean	St.Dev.	2.5%	Median	97.5%		WinBUGS Name	
2			CONSTANT	1.2770	0.1443	0.9837	1.2810	1.5560		Beta0	
3		age	b	0.9155	0.1542	0.6227	0.9124	1.2190		X.Eff[1,2]	
4		trt	2	-0.0726	0.1895	-0.4425	-0.0753	0.3053		X.Eff[2,2]	
5		trt	3	-0.1047	0.1913	-0.4760	-0.1070	0.2734		X.Eff[2,3]	
6		trt	4	-0.1625	0.1973	-0.5427	-0.1643	0.2299		X.Eff[2,4]	
7		trt	5	-0.2495	0.2014	-0.6597	-0.2448	0.1409		X.Eff[2,5]	
8		block	B	0.4878	0.0869	0.3195	0.4862	0.6566		X.Eff[3,2]	
9		block	C	-0.3912	0.1806	-0.7604	-0.3882	-0.0447		X.Eff[3,3]	
10		block	D	0.3620	0.1327	0.0943	0.3631	0.6212		X.Eff[3,4]	
11		block	E	-0.9438	0.2317	-1.4170	-0.9377	-0.5140		X.Eff[3,5]	
12		block	F	-0.2384	0.1696	-0.5772	-0.2343	0.0838		X.Eff[3,6]	
13		block	G	-0.3402	0.1775	-0.6983	-0.3379	-2.358E-4		X.Eff[3,7]	
14		block	H	0.3212	0.1352	0.0563	0.3221	0.5829		X.Eff[3,8]	
15		age x trt	b, 2	-0.3619	0.2308	-0.8229	-0.3609	0.0929		X.Eff[4,4]	
16		age x trt	b, 3	0.0283	0.2248	0.4057	0.0274	0.4677		X.Eff[4,6]	
17		age x trt	b, 4	-0.0353	0.2322	-0.4912	-0.0351	0.4176		X.Eff[4,8]	
18		age x trt	b, 5	-0.1654	0.2419	-0.6298	-0.1693	0.3213		X.Eff[4,10]	
19											
20	Model	[larvae!$A$1:$E$81]									
21	Distribution	Poisson									
22	Link	Log									
23	Response	larvae									
24	Offset	plot.size									
25	Fixed	age*trt + block									
26											
27	Priors										
28	CONSTANT	N(mu=1.82, sigma=113)									
29	age	N(mu=0, sigma=113)									
30	trt	N(mu=0, sigma=113)									
31	block	N(mu=0, sigma=113)									
32	age x trt	N(mu=0, sigma=113)									
33											
34	WinBUGS MCMC Settings										
35	Burn-In: 5000 Samples: 10000 (Thin:1; Chains:1)										
36	Run took 36 seconds										

FIGURE 4.16
Summaries of posterior distributions, model, priors, and MCMC settings for cockchafer larvae experiment.

or an equivalent likelihood ratio test (as explained in, e.g., Brown and Prescott (2006)) to assess the "statistical significance" of the terms in the model. These tests bring their own issues: those designed to assess the "overall significance" of a factor (e.g., F-tests) can fail to detect one good treatment if tested alongside many ineffective ones, and so called "multiple-comparisons" suffer from the problem of multiplicity. Although there are Bayesian methods that could be used to mimic the classical significance tests (e.g., Bayes factors as discussed in Gelman et al. (2004)), we take the view that in most cases, and certainly in designed experiments such as this one, the model, which includes the prior distribution, should be specified to represent all reasonable beliefs about the effects being estimated. For example, Chapter 8 explains how mixture priors can be specified so that a point mass of prior probability is assigned to an effect of size zero, effectively providing a hybrid analysis between effect estimation and testing. Here we take a much simpler approach and now inspect all the pair wise contrasts, shown in Figure 4.17.

The contrasts have been back-transformed from the log scale, so that they are now ratios of mean counts that are being estimated; BugsXLA indicates

Label	Mean	St.Dev.	2.5%	Median	97.5%		WinBUGS Name
Predicted Mean Response							
Following all at: block(A)							
age(a) contr[trt 1 / 2]	1.0950	0.2094	0.7372	1.0780	1.5570		Pred.Ave[1]
age(a) contr[trt 1 / 3]	1.1310	0.2175	0.7610	1.1130	1.6100		Pred.Ave[2]
age(a) contr[trt 1 / 4]	1.1990	0.2382	0.7949	1.1790	1.7210		Pred.Ave[3]
age(a) contr[trt 1 / 5]	1.3100	0.2702	0.8693	1.2770	1.9360		Pred.Ave[4]
age(a) contr[trt 2 / 3]	1.0530	0.2088	0.7003	1.0330	1.5190		Pred.Ave[5]
age(a) contr[trt 2 / 4]	1.1160	0.2253	0.7458	1.0940	1.6240		Pred.Ave[6]
age(a) contr[trt 2 / 5]	1.2190	0.2556	0.8036	1.1870	1.8020		Pred.Ave[7]
age(a) contr[trt 3 / 4]	1.0810	0.2207	0.7161	1.0590	1.5710		Pred.Ave[8]
age(a) contr[trt 3 / 5]	1.1810	0.2477	0.7713	1.1560	1.7440		Pred.Ave[9]
age(a) contr[trt 4 / 5]	1.1150	0.2356	0.7246	1.0920	1.6410		Pred.Ave[10]
age(b) contr[trt 1 / 2]	1.5580	0.2077	1.1960	1.5420	2.0180		Pred.Ave[11]
age(b) contr[trt 1 / 3]	1.0870	0.1308	0.8536	1.0780	1.3690		Pred.Ave[12]
age(b) contr[trt 1 / 4]	1.2280	0.1518	0.9586	1.2180	1.5540		Pred.Ave[13]
age(b) contr[trt 1 / 5]	1.5270	0.2032	1.1740	1.5110	1.9670		Pred.Ave[14]
age(b) contr[trt 2 / 3]	0.7053	0.0952	0.5369	0.6992	0.9030		Pred.Ave[15]
age(b) contr[trt 2 / 4]	0.7967	0.1101	0.6000	0.7906	1.0320		Pred.Ave[16]
age(b) contr[trt 2 / 5]	0.9909	0.1453	0.7394	0.9805	1.3100		Pred.Ave[17]
age(b) contr[trt 3 / 4]	1.1380	0.1439	0.8831	1.1290	1.4460		Pred.Ave[18]
age(b) contr[trt 3 / 5]	1.4150	0.1909	1.0890	1.4000	1.8300		Pred.Ave[19]
age(b) contr[trt 4 / 5]	1.2540	0.1749	0.9520	1.2420	1.6340		Pred.Ave[20]

FIGURE 4.17
Summary of treatment contrasts for the cockchafer larvae experiment.

this by the label, "contr[trt 1/2]" etc. For reasons previously discussed (see Case Study 4.1), the fact that these are all calculated for block(A) is irrelevant. Inspection of the credible intervals suggest that some of the treatments were effective at reducing the number of larvae in age group b, but there is no evidence of any treatment effects in the other age group. In a classical analysis of such data some adjustment for multiplicity would need to be made due to the numerous significance tests being made. In the Bayesian approach to inference, if the prior and model fairly represent our beliefs then no such adjustments are necessary as the posterior distribution will contain all relevant information. However, if the posterior distributions will be used to make decisions, then the multiplicity issue is not so easily sidestepped. Assuming one is unable to undertake a fully Bayesian utilitarian approach, as discussed in Lindley (1985) or Raiffa (1968), then the False Discovery Rate (see Benjamini and Hochberg, 1995) is a method that is closest in spirit to the Bayesian approach. BugsXLA does not offer any features to undertake such analyses, and so the user would need to take the output provided and apply these methods using other software. Some have argued (see, e.g., Gelman (2005)) that the fixed effects assumption is rarely appropriate, and that adopting a random effects model will ensure appropriate adjustments are made to the effect estimates so that the multiplicity issue is addressed. This approach is discussed further in Case Study 5.1.

4.3 Survival or Reliability Data

Survival and reliability data, although typically continuous in nature, are usually poorly modeled by a Normal error distribution. It is common to use right-skewed distributions such as the Log-Normal, Weibull, or Gamma. It is also common for such data to be censored, for example, instead of observing the actual response, we only observe that the response must be greater than some value. Such censoring often occurs in time to event studies when the event, for example, death, has not occurred in the individual during the course of the study. Refer to Collett (2003) or Wolstenholme (1999) for good introductions to the modeling of these types of data in the medical and engineering fields, respectively.

Case Study 4.5 Mice Survival Times

The WinBUGS examples include a data set, originally published by Grieve (1987), from a photocarcinogenecity study in mice. In this study, 80 mice were randomly assigned to one of four treatments: three being a type of control (irradiated, vehicle, or positive), and the other being the test compound. The response was the survival time in weeks, with some values being censored due to some mice still being alive at the end of the study. We use this example to show how BugsXLA can be used to fit error distributions more appropriate to survival or reliability data, as well as how it easily handles censored data. We provide another example of the use of the DIC, once again emphasizing some caution in its use. Initially, we repeat the analysis given in the WinBUGS examples and fit a Weibull error distribution; Figure 4.18 shows some of the data and the model specification form.

BugsXLA requires censored data to be specified in essentially the same way as WinBUGS. The response variable, Time, contains all the uncensored data, with blanks or 'NA' in the cells where the data are censored. A second variable, Tcen, referred to as a 'censor' by BugsXLA, contains the censoring values with blanks or 'NA' where the data are uncensored. BugsXLA can handle three types of censoring:

Right censored (>= value)   denoted by 'GE', for example, GE 20
Left censored (<= value)   denoted by 'LE', for example, LE 20
Interval censored (>= min, <= max) denoted by 'IN', for example, IN 15 25

In this example, all the censored values are right censored, and so the values in Tcen need the GE prefix. The cells could be manually edited, or this prefix can be automatically added at the same time as the variable Tcen is defined as type censor. Censors are defined via the 'Set Variable Types' button on the model specification form, and so are set at the same time as any factors, such

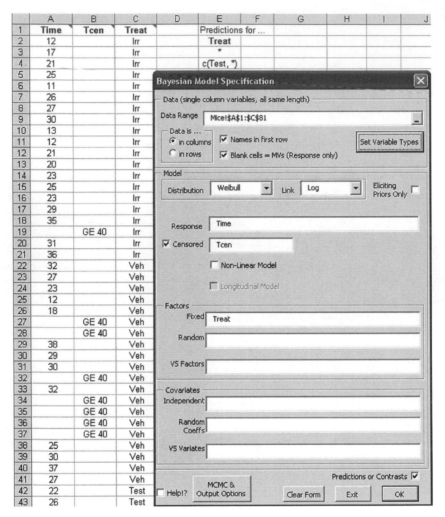

The table on the left of the figure contains the following data:

	A	B	C
1	Time	Tcen	Treat
2	12		Irr
3	17		Irr
4	21		Irr
5	25		Irr
6	11		Irr
7	26		Irr
8	27		Irr
9	30		Irr
10	13		Irr
11	12		Irr
12	21		Irr
13	20		Irr
14	23		Irr
15	25		Irr
16	23		Irr
17	29		Irr
18	35		Irr
19		GE 40	Irr
20	31		Irr
21	36		Irr
22	32		Veh
23	27		Veh
24	23		Veh
25	12		Veh
26	18		Veh
27		GE 40	Veh
28		GE 40	Veh
29	38		Veh
30	29		Veh
31	30		Veh
32		GE 40	Veh
33	32		Veh
34		GE 40	Veh
35		GE 40	Veh
36		GE 40	Veh
37		GE 40	Veh
38	25		Veh
39	30		Veh
40	37		Veh
41	27		Veh
42	22		Test
43	26		Test

FIGURE 4.18
Data and model specification for mice survival study.

as Treat in this case. The variate to be converted to type censor must be high-lighted on the next form that appears, and the 'Make Censor' button clicked. Figure 4.19 shows the forms that will then be visible. Since all the values in Tcen are right censored the first option is selected and the 'OK' button completes the task. You will notice that the values on the spreadsheet now have the GE prefix. Note that even if the cell entries are manually edited, it is necessary to define the variable as type censor via this process. In which case, one should select the 'Mixed censoring (manual edit)' option on the 'Create Censor' form shown in Figure 4.19.

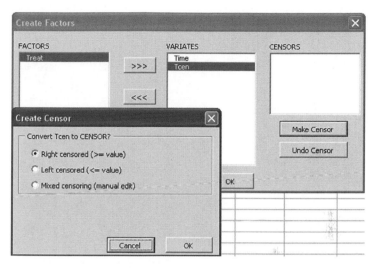

FIGURE 4.19
Forms showing how Tcen variable is defined as type censor.

The model specified is a one-way analysis of variance, and can be represented algebraically:

$$\eta_i = \mu + \alpha_i \quad i = 1, 2, 3, 4$$

$$\log(\lambda_i) = \eta_i$$

$$\theta_i = \log(2)\, \lambda_i^{-r}$$

$$y_i \sim \text{Weib}(r,\ \theta_i)\, I[a_i, b_i]$$

where
  y is the response variable Time
  α_i are the effects of the factor Treat
  λ is the median of the Weibull Distribution
  I[a, b] is an indicator function denoting that the response is in the interval
    [a, b]

For uncensored data the interval is unconstrained, that is, $[0, \infty]$. For the censored data in this example, the a_i are the censoring values in Tcen, and the b_i are all ∞. Infinity is approximated by a very large value in the code generated by BugsXLA.
    The simple model defaults for the MCMC settings were chosen (a burn-in of 5,000 followed by 10,000 samples), predicted means for all levels of the

	A	B	C	D	E	F	G	H	I	J	K
1			Label	Mean	St.Dev.	2.5%	Median	97.5%		WinBUGS Name	
2			Weibull r	3.1400	0.3304	2.5130	3.1320	3.8080		r.weib	
3			CONSTANT	3.1800	0.0827	3.0100	3.1820	3.3380		Beta0	
4		Treat	Veh	0.3763	0.1257	0.1356	0.3725	0.6378		X.Eff[1,2]	
5		Treat	Test	0.1162	0.1101	-0.0983	0.1186	0.3365		X.Eff[1,3]	
6		Treat	Pos	-0.1123	0.1123	-0.3355	-0.1146	0.1095		X.Eff[1,4]	
7											
8				Y							
9			Dbar	494.8240							
10			Dhat	489.7390							
11			pD	5.0850							
12			DIC	499.9090							
13	Model	[Micel$A$1:$C$81]									
14	Distribution	Weibull									
15	Link	Log									
16	Response	Time									
17	Censor	Tcen									
18	Fixed	Treat									
19											
20	Priors										
21	CONSTANT	N(mu=3.13, sigma=34.2)									
22	Treat	N(mu=0, sigma=34.2)									
23	Weibull r	Gamma(1, 0.001)									
24											
25	WinBUGS MCMC Settings										
26	Burn-In: 5000 Samples: 10000 (Thin:1; Chains:1)										
27	Run took 18 seconds										

FIGURE 4.20
Summaries of posterior distributions, DIC, model, priors, and MCMC settings for the mice survival study.

factor Treat as well as all pair wise contrasts involving the test compound were requested (see cells E2:E4 in Figure 4.18), and the default priors were unchanged (see cells A21:B23 in Figure 4.20). The default prior for the Weibull shape parameter, r, is a highly dispersed Exponential Distribution, defined as a Gamma Distribution to provide more flexibility for the case when a more informative prior for this parameter is appropriate. The model DIC statistics were also requested. Figures 4.20 and 4.21 show the results of the analysis as imported back into Excel.

As well as showing all the parameters in the model and a summary of their marginal posterior distributions, Figure 4.20 also shows the Bayesian

N		O	P	Q	R	S	T	U	V
Label		Mean	St.Dev.	2.5%	Median	97.5%		WinBUGS Name	
		Predicted Median (not mean)							
Treat(Irr)		24.1300	2.0010	20.2800	24.1000	28.1500		Pred.Ave[1]	
Treat(Veh)		35.1900	3.2610	29.6100	34.9100	42.3400		Pred.Ave[2]	
Treat(Test)		27.1100	2.2430	22.9000	27.0100	31.8200		Pred.Ave[3]	
Treat(Pos)		21.5700	1.7620	18.3200	21.5100	25.4000		Pred.Ave[4]	
contr[Treat Test / Irr]		1.1300	0.1250	0.9064	1.1260	1.4000		Pred.Ave[5]	
contr[Treat Test / Veh]		0.7768	0.0953	0.6001	0.7721	0.9777		Pred.Ave[6]	
contr[Treat Test / Pos]		1.2650	0.1424	0.9981	1.2570	1.5630		Pred.Ave[7]	

FIGURE 4.21
Summary of predicted means and treatment contrasts for the mice survival study.

DIC statistics. Note that the pD value of approximately five equals the number of parameters in the model as one would expect; the parameters are the Weibull shape, the model constant, plus the three parameters that define the effects of the four level Treat factor. Figure 4.21 shows the predicted means and contrasts requested.

Note that the contrasts are estimated ratios due to the log link. It is clear that the test compound has a negative effect on the survival time relative to the vehicle, indicating that the compound is carcinogenetic: test/vehicle ratio of 0.78 (sd = 0.10). However, there is reasonably good evidence that the compound has a smaller negative effect on survival time than the positive control: test/positive ratio of 1.27 (sd = 0.14).

We now compare these results with those obtained from fitting three other error distributions: Normal, Log-Normal, and Gamma. The only difference to how these models are specified using BugsXLA is to change the distribution via the drop down menu on the model specification form. However, it is important to recognize that although the default priors are chosen to be vague in each case, one should be aware of the possibility that inferences may be more sensitive to priors depending upon the error distribution. It is also important to be aware of how BugsXLA parameterizes the linear predictor in each case:

Normal: identity link for mean
Log-Normal: identity link for mean of log-response; effectively modeling
               the median
Weibull: log link for median
Gamma: log link for mean

It follows that the effects are additive only for the Normal model, and multiplicative for the others. The scale parameter for the Log-Normal model is given the same default Inverse-Gamma(0.001, 0.001) prior distribution as is the case for the Normal model, although this parameter is the variance of the log-response in the Log-Normal case. The prior for the Gamma shape parameter is itself given a Gamma(0.1, 0.1) by default. Refer to Appendix A for details of the parameterizations that BugsXLA uses for each of these distributions. Table 4.3 shows some output after fitting each of these models that will be used to compare them.

A simple comparison of the DIC would suggest that the Gamma model provides a much poorer fit to these data than any of the others. Also, following the guidance that differences between 5 and 10 can be considered substantial, the Normal should be ruled out, supporting the usual assumption that Normal errors are inadequate to model survival data. There is little difference between the Weibull and the Log-Normal DIC values, and so unless there were other reasons to differentiate between these models, it would be wise to assess the sensitivity of the inferences to the assumed error distribution.

Inspection of the pD values suggests there may be a problem with the DIC calculations for the Gamma model. Although different error distributions

TABLE 4.3

Comparison of Models for the Mice Survival Study

			Predictions		Contrasts	
	DIC	pD	Vehicle	Test	Test/Vehicle	Test/Pos
Weibull	500	5.1	35.2 (3.3)	27.1 (2.2)	0.78 (0.10)	1.27 (0.14)
Log-Normal	501	5.0	34.0 (3.4)	24.4 (2.3)	0.73 (0.10)	1.13 (0.15)
Gamma	515	9.7	36.2 (3.7)	26.7 (2.6)	0.74 (0.10)	1.18 (0.16)
Normal	506	4.9	33.7 (2.1)	26.0 (2.1)	NA	NA

Note that the Log-Normal Distribution is fitted to the untransformed response, not the Normal to the logged response. The DIC values are only directly comparable if the response is identical in all models fitted. Refer to p. 281 of Ntzoufras (2009) for details of how the deviance would need to be adjusted if the response variable was transformed.

are fitted, each model has the same number of parameters, and so the pD should be approximately five in every case. It can be seen that this is true for all bar the Gamma model where the pD is over nine. To investigate this issue further, it is necessary to take a closer look at the Gamma model, and in particular the samples generated by WinBUGS for the deviance node. By turning off the 'Auto Quit' from WinBUGS option on the 'MCMC & Output Options' form (see Section 2.5), it is possible to use the features of WinBUGS to assess convergence (see Section 1.2 for details). Although it appears that the MCMC chain has converged satisfactorily, the density plot for the deviance node is clearly bimodal. A plot of the posterior density of the deviance can also be obtained via BugsXLA by selecting the 'Deviance (samples)' check box on the 'Model Checking Options' form; this instructs BugsXLA to import back into Excel the samples for the deviance node. It is then possible to use the 'Post Plots' utility, as explained in Case Study 3.1, to plot the posterior distribution of the deviance. Figure 4.22 shows the plot that BugsXLA produces; note that no prior distribution is overlaid since the prior inferred for the deviance from those specified for the model's parameters has not been determined.

Since the DIC assumes that the posterior mean of the deviance is a good estimate, the bimodal distribution seen in Figure 4.22 invalidates its use as a good measure of model fit. It is not understood why this bimodality occurs for the Gamma model in this case. Further analysis with the censored data excluded did not show the same anomaly with the Gamma model. Also, further investigation using different parameterizations of the Gamma model, not available via BugsXLA, did reduce the size of the second mode, although the deviance remained bimodal. These additional analyses suggest that when censored data exists the performance of the MCMC algorithm in WinBUGS for the Gamma model is quite sensitive to the way it is parameterized.* This example also suggests that it is important when using the DIC not only to

---

* This problem may be peculiar to WinBUGS, as I have been informed that the Gamma model as parameterized here works fine when using the JAGS software (see Plummer 2010).

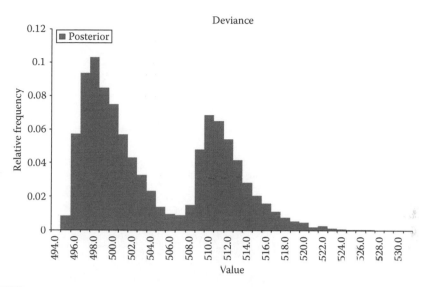

FIGURE 4.22

Posterior distribution of the deviance node, Gamma model, for the mice survival study.

assess convergence of the MCMC chain, but also to check the shape of the posterior distribution of the deviance. As was stated in Case Study 4.2, the DIC should not be used as the sole evidence for selecting a particular model. As always, the best advice is to assess the sensitivity of the main inferences of interest to as wide a range of plausible assumptions as time permits. In this particular case, inspection of the contrasts shown in Table 4.3, indicates that the three usual candidates for survival data, Weibull, Log-Normal, and Gamma, give similar results.

4.4 Multivariate Categorical Data

When the response of interest is categorical, that is, it can take one of a discrete number of possible categories, the multivariate extension of the Binomial model can be used. If the individual responses are being modeled, then BugsXLA refers to this as a 'Categorical' distribution, as opposed to when tabulated count data are being modeled, in which case it is referred to as a 'Multinomial' distribution. This is analogous to the distinction between Bernoulli and Binomial models for response data that can only take one of two categories. In some cases, the categories have a natural ordering such that the response provides a semiquantitative measure of some quantity of interest. BugsXLA refers to these as 'Ordered Cat' or 'Ordinal' depending upon whether individual or tabulated responses are being modeled. Refer to Agresti (2007) for a good introduction to the modeling of both unordered and ordered categorical data.

Case Study 4.6 Cold Feet Incidence in Hypertension Trial
Brown and Prescott (2006) discuss a hypertension trial in which one of the outcome measures was the incidence of cold feet in the patients. This adverse event was recorded on an ordered categorical scale of 1–5 with the following meaning:

1. None
2. Occasionally
3. On most days
4. Most of the time
5. All of the time

This was a randomized double blind trial of three treatments for hypertension in which a new drug Carvedilol, labeled A here, was compared with two standard drugs, Nifedipine and Atenolol, labeled B and C, respectively, here. Patients were randomly assigned to treatment and as well as providing a pre-treatment baseline measure of cold feet, gave a post-treatment measure on each of four visits to their medical centre; visits are labeled 3–6 representing the actual week post treatment. As is common in such studies, some patients failed to attend at every visit, and others dropped out completely before the end of the study. Here, we show how to apply a Bayesian fixed effects ordinal logistics regression under the proportional odds assumption (see, e.g., Agresti (2007)). Figure 4.23 shows some of the data and the model specification form.

All the variables, including the response, CF_end, and baseline, CF_pre, are defined as factors with their levels sorted into ascending order. When fitting an ordered categorical model using BugsXLA, the response must be a factor with the levels either in ascending or descending order of magnitude of the response. Since BugsXLA parameterizes the proportional odds model such that the probability of being in a higher category is logit-linked to the linear predictor, a positive effect of a covariate implies that it increases the magnitude of the response when the response's levels are in ascending order, and decreases the magnitude when they are in descending order. BugsXLA explicitly shows the order when reporting the results (see cell B19 in Figure 4.24), and it is advisable to check this before interpreting the output. The model specified can be represented algebraically:

$$\eta_{cijk} = \mu_c + \alpha_i + \gamma_j + \omega_k \quad c = 1, 2, 3, 4 \quad i = 1, 2, 3, 4 \quad j = 1, 2, 3 \quad k = 1, 2, \ldots, 5$$

$$\text{logit}(\pi_{cijk}^*) = \eta_{cijk}$$

$$\pi_{cijk}^* = \pi_{(c+1)ijk} + \pi_{(c+2)ijk} \cdots + \pi_{(5)ijk}$$

$$y_{ijk} \sim \text{Cat}(\pi_{(1:5)ijk})$$

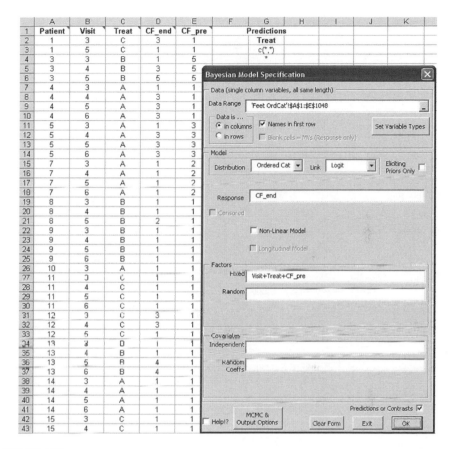

	A	B	C	D	E	F	G	H	I	J	K
1	Patient	Visit	Treat	CF_end	CF_pre		Predictions				
2	1	3	C	3	1		Treat				
3	1	5	C	1	1		c(",")				
4	3	3	B	1	5		*				
5	3	4	B	3	5						
6	3	5	B	5	5						
7	4	3	A	1	1						
8	4	4	A	3	1						
9	4	5	A	3	1						
10	4	6	A	3	1						
11	5	3	A	1	3						
12	5	4	A	3	3						
13	5	5	A	3	3						
14	5	6	A	3	3						
15	7	3	A	1	2						
16	7	4	A	1	2						
17	7	5	A	1	2						
18	7	6	A	1	2						
19	8	3	B	1	1						
20	8	4	B	1	1						
21	8	5	B	2	1						
22	9	3	B	1	1						
23	9	4	B	1	1						
24	9	5	B	1	1						
25	9	6	B	1	1						
26	10	3	A	1	1						
27	11	3	C	1	1						
28	11	4	C	1	1						
29	11	5	C	1	1						
30	11	6	C	1	1						
31	12	3	C	3	1						
32	12	4	C	3	1						
33	12	5	C	1	1						
34	13	3	D	1	1						
35	13	4	B	1	1						
36	13	5	B	4	1						
37	13	6	B	4	1						
38	14	3	A	1	1						
39	14	4	A	1	1						
40	14	5	A	1	1						
41	14	6	A	1	1						
42	15	3	C	1	1						
43	15	4	C	1	1						

Dialog box (Bayesian Model Specification):

- Data (single column variables, all same length)
- Data Range: 'Feet OrdCat'!$A$1:$E$1048
- Data is ... ● in columns ☑ Names in first row   ○ in rows ☐ Blank cells = MVs (Response only)   [Set Variable Types]
- Model
  - Distribution: Ordered Cat ▼   Link: Logit ▼   Eliciting Priors Only ☐
  - Response: CF_end
  - ☐ Censored
  - ☐ Non-Linear Model
  - ☐ Longitudinal Model
- Factors
  - Fixed: Visit+Treat+CF_pre
  - Random:
- Covariates
  - Independent:
  - Random Coeffs:
- Predictions or Contrasts ☑
- ☐ Help!? | MCMC & Output Options | Clear Form | Exit | OK

FIGURE 4.23
Data and model specification for cold feet analysis.

where
    y is the response factor CF_end
    μ_c are the "cut-points" that split the latent variable, implied by the logistic model, into the five ordered categories. The cut-points are constrained:
    $\mu_1 < \mu_2 < \mu_3 < \mu_4$
    α_i are the effects of the factor Visit
    γ_j are the effects of the factor Treat
    ω_k are the effects of the factor CF_pre
    π^*_{cijk} is the probability of the response being in a category higher than c
    $\pi_{(c)ijk}$ is the probability of the response being in category c
    $Cat(\pi_{(1:5)ijk})$ is a categorical distribution with probabilities given by the five values $\pi_{(1:5)ijk}$

The simple model defaults for the MCMC settings were chosen (a burn-in of 5,000 followed by 10,000 samples), all pair wise contrasts (odds ratios)

	A	B	C	D	E	F	G	H	I	J	K
1			Label	Mean	St.Dev.	2.5%	Median	97.5%		WinBUGS Name	
2		Cut point	1	2.3560	0.2304	1.8960	2.3540	2.7920		CF.a[1]	
3		Cut point	2	2.9770	0.2417	2.4910	2.9740	3.4360		CF.a[2]	
4		Cut point	3	4.1530	0.2766	3.6080	4.1490	4.6910		CF.a[3]	
5		Cut point	4	5.0290	0.3158	4.4050	5.0260	5.6440		CF.a[4]	
6		Visit	4	0.0109	0.2589	-0.5007	0.0106	0.5114		X.Eff[1,2]	
7		Visit	5	0.4273	0.2462	-0.0537	0.4273	0.9133		X.Eff[1,3]	
8		Visit	6	0.2269	0.2579	-0.2803	0.2276	0.7363		X.Eff[1,4]	
9		Treat	B	0.7409	0.2101	0.3330	0.7405	1.1600		X.Eff[2,2]	
10		Treat	C	-0.5427	0.2564	-1.0510	-0.5432	-0.0405		X.Eff[2,3]	
11		CF_pre	2	0.9765	0.3746	0.2271	0.9834	1.7000		X.Eff[3,2]	
12		CF_pre	3	2.9780	0.2638	2.4680	2.9790	3.4930		X.Eff[3,3]	
13		CF_pre	4	1.5320	0.7031	0.1043	1.5500	2.8510		X.Eff[3,4]	
14		CF_pre	5	5.5040	0.4834	4.5900	5.4900	6.4930		X.Eff[3,5]	
15											
16	Model	['Feet OrdCat'!$A$1:$E$1048]									
17	Distribution	Ordered Cat									
18	Link	Logit									
19	Response	CF_end (1,2,3,4,5)									
20	Fixed	Visit+Treat+CF_pre									
21											
22	Priors										
23	Cat. Consts	N(mu=0, sigma=100)									
24	Visit	N(mu=0, sigma=100)									
25	Treat	N(mu=0, sigma=100)									
26	CF_pre	N(mu=0, sigma=100)									
27											
28	WinBUGS MCMC Settings										
29	Burn-In: 5000 Samples: 10000 (Thin:1; Chains:1)										
30	Run took 32.4 minutes										

FIGURE 4.24
Summaries of posterior distributions, model, priors, and MCMC settings for the cold feet analysis.

as well as predicted means (probabilities) for all levels of the factor Treat were requested (see cells G2:G4 in Figure 4.23), and the default priors were unchanged (see cells A23:B26 in Figure 4.24). Figures 4.24 and 4.25 show the results of the analysis as imported back into Excel.

As always when a link other than the identity is used, it is difficult to interpret the parameter estimates directly. However, inspection of the treatment effect estimates does indicate that there are real differences in the incidence of cold feet in the different treatment groups. Note that WinBUGS is significantly slower in analyzing this problem than the previous examples; the time taken for the analysis is shown in row 30 in Figure 4.24. This is due to the relative complexity of the ordered categorical model and the fact that over 1000 observations are included in this data set. Figure 4.25 shows the predicted probabilities and odds ratios requested.

Inspection of the odds ratios shows that all three of the 95% credible intervals exclude one, providing strong evidence that the treatments are each having a different effect on the incidence of cold feet. Since a higher category in the response implies more frequent incidence of the adverse event, the treatments can be ordered C, A, and B in terms of being better as measured by this outcome. If we simply take the predicted probabilities shown for category 1 (no cold feet during the treatment period), then these are 95% (C), 91% (A), and 83% (B). Note that these predictions are for visit 3 and a baseline

N		O	P	Q	R	S	T	U	V
Label		**Mean**	**St.Dev.**	**2.5%**	**Median**	**97.5%**		**WinBUGS Name**	
		Predicted Probability							
		Following all at: Visit(3) CF_pre(1)							
Treat(A) :1		0.9116	0.0187	0.8694	0.9133	0.9422		Pred.Ave[4,1]	
Treat(A) :2		0.0387	0.0087	0.0246	0.0377	0.0583		Pred.Ave[4,2]	
Treat(A) :3		0.0336	0.0080	0.0206	0.0329	0.0517		Pred.Ave[4,3]	
Treat(A) :4		0.0092	0.0028	0.0048	0.0089	0.0158		Pred.Ave[4,4]	
Treat(A) :5		0.0068	0.0022	0.0035	0.0065	0.0121		Pred.Ave[4,5]	
Treat(B) :1		0.8320	0.0294	0.7700	0.8336	0.8842		Pred.Ave[5,1]	
Treat(B) :2		0.0697	0.0132	0.0464	0.0688	0.0977		Pred.Ave[5,2]	
Treat(B) :3		0.0654	0.0134	0.0422	0.0646	0.0945		Pred.Ave[5,3]	
Treat(B) :4		0.0188	0.0052	0.0103	0.0181	0.0307		Pred.Ave[5,4]	
Treat(B) :5		0.0141	0.0041	0.0075	0.0136	0.0235		Pred.Ave[5,5]	
Treat(C) :1		0.9462	0.0134	0.9167	0.9475	0.9684		Pred.Ave[6,1]	
Treat(C) :2		0.0240	0.0063	0.0138	0.0234	0.0379		Pred.Ave[6,2]	
Treat(C) :3		0.0203	0.0055	0.0114	0.0197	0.0324		Pred.Ave[6,3]	
Treat(C) :4		0.0055	0.0018	0.0027	0.0052	0.0097		Pred.Ave[6,4]	
Treat(C) :5		0.0040	0.0014	0.0019	0.0038	0.0073		Pred.Ave[6,5]	
		Predicted Odds							
		Following all at: Visit(3) CF_pre(1)							
contr[Treat A / B] :1		0.4873	0.1035	0.3136	0.4769	0.7172		Pred.Odds[1]	
contr[Treat A / C] :1		1.7780	0.4649	1.0420	1.7220	2.8600		Pred.Odds[2]	
contr[Treat B / C] :1		3.7140	0.9058	2.2920	3.5940	5.7940		Pred.Odds[3]	

FIGURE 4.25
Summary of predicted means and treatment contrasts for the cold feet analysis.

cold feet value of 1; other values for these factors could have been set when specifying the predictions in Figure 4.23. In this fixed effects analysis, we have ignored the possibility of any additional variation between subjects that would lead to correlation between responses obtained from the same subject at different visits, this being an example of how over-dispersion could occur. In Chapter 6, we come back to this example and show how this can be modeled using BugsXLA.

Sometimes it is possible to tabulate the ordered categorical responses so that instead of analyzing the individual response for each patient at each visit, we analyze the counts. In the example discussed above, this is not possible since the baseline response is included in the analysis. In general, whenever we have individual observation level covariates then tabulation is not possible. However, in order to illustrate how BugsXLA analyses such data, we reanalyze this example excluding the baseline covariate. Figure 4.26 shows the tabulated data and model specification form.

When analyzing tabulated data BugsXLA refers to the distribution as 'Ordinal' as opposed to 'Ordered Cat'. The response is now multivariate, and so the five variables that contain the counts must be entered into the 'Response' field separated by commas; these variables are defined as variates. The order in which the response variates are listed is important since, like the factor levels in the ordered categorical response, they must be in

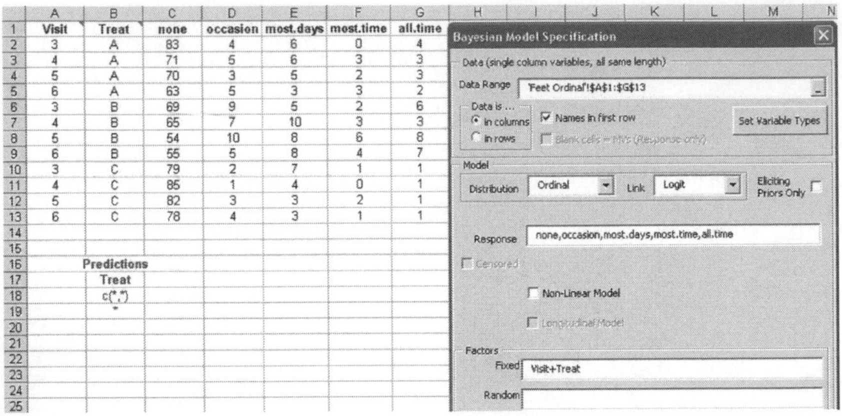

	A	B	C	D	E	F	G	H	I	J	K	L	M	N
1	Visit	Treat	none	occasion	most.days	most.time	all.time							
2	3	A	83	4	6	0	4							
3	4	A	71	5	6	3	3							
4	5	A	70	3	5	2	3							
5	6	A	63	5	3	3	2							
6	3	B	69	9	5	2	6							
7	4	B	65	7	10	3	3							
8	5	B	54	10	8	6	8							
9	6	B	55	5	8	4	7							
10	3	C	79	2	7	1	1							
11	4	C	85	1	4	0	1							
12	5	C	82	3	3	2	1							
13	6	C	78	4	3	1	1							
14														
15														
16		Predictions												
17		Treat												
18		c("","")												
19														
20														
21														
22														
23														
24														
25														

Bayesian Model Specification

Data (single column variables, all same length)

Data Range: `Feet Ordinal'!$A$1:$G$13`

Data is ... • in columns ☑ Names in first row ○ in rows ☐ Blank cells = MVs (Response only)   Set Variable Types

Model

Distribution: Ordinal   Link: Logit   Eliciting Priors Only ☐

Response: `none,occasion,most.days,most.time,all.time`

☐ Censored

☐ Non-Linear Model

☐ Longitudinal Model

Factors

Fixed: `Visit+Treat`

Random:

FIGURE 4.26
Tabulated data and model specification for cold feet analysis.

either ascending or descending order of magnitude. The model specified is only slightly different to when individual observations are analyzed and can be represented algebraically:

$$\eta_{cij} = \mu_c + \alpha_i + \gamma_j \quad c = 1, 2, 3, 4 \quad i = 1, 2, 3, 4 \quad j = 1, 2, 3$$

$$\mathrm{logit}(\pi^*_{cij}) = \eta_{cij}$$

$$\pi^*_{cij} = \pi_{(c+1)ij} + \pi_{(c+2)ij} \dots + \pi_{(5)ij}$$

$$y_{(1:5)ij} \sim \mathrm{Multi}(\pi_{(1:5)ij}, n_{ij})$$

$$n_{ij} = y_{(1)ij} + y_{(2)ij} \dots + y_{(5)ij}$$

where
  $y_{(1:5)}$ is the multivariate response of counts in each of the five categories
  μ_c, α_i, γ_j, π^*_{cij} and $\pi_{(c)ij}$ being the same as before
  $\mathrm{Multi}(\pi_{(1:5)})$ is a multinomial distribution with probabilities given by the five values $\pi_{(1:5)}$
  n_{ij} is the number of observations in the ith visit by patients receiving the jth treatment

The output from this analysis is not shown, but even though the baseline covariate has been excluded, the results are fairly similar. The biggest difference is the speed with which WinBUGS can analyze data that has been tabulated in this way. The MCMC run was completed more than 80 times faster than when the individual responses were analyzed.

Case Study 4.7 Alligator Feeding Habits

Agresti (2007) analyses data collected on the feeding habits of alligators in four lakes of Florida, United States. The response was multinomial, being one of the five categories: fish, invertebrate, reptile, bird, or other. Here we reanalyze these data with the aim being to assess whether there are any differences in the feeding habits assignable to the lake or size of the alligator. Each alligator was recorded as being either greater than or less than 2.3 m long. Ideally, the actual length of the alligator should be used as a continuous covariate, since we inevitably lose information by dichotomizing a continuous measurement. However, in this case, since BugsXLA v5.0 does not include the functionality to fit continuous covariates in multinomial models, this variable would need to be converted into a factor anyway. Unless there was a very good reason for choosing the split point, the sensitivity of the results to changing it should be assessed. Also, depending upon how much variation exists between the lengths of the alligators, a more refined analysis might be possible by splitting this variable into more than two categories. Since no structure is imposed on this factor's effects, unlike when a continuous covariate is modeled, the relationship between the variable and the response is allowed to change dramatically between levels. This feature of the model may not matter when there are sufficient data at each level to estimate the effects precisely, but more generally one needs to be wary of making many relatively small categories. Figure 4.27 shows the data and the completed model specification form.

The variables Size and Lake are both defined as factors. Note that in this study, it was possible to group the data so that counts could be formed. BugsXLA refers to this as multinomial data. As when fitting an ordinal model (see discussion at end of Case Study 4.6) the response consists of the five variables containing the count data. This is entered in the 'Response'

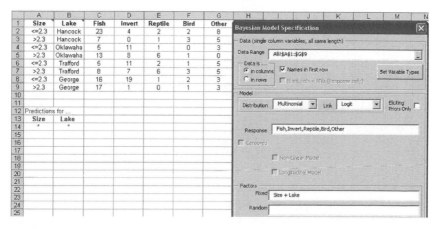

FIGURE 4.27
Data and model specification for alligator feeding habits.

field as a comma separated list as shown in Figure 4.27. If the response was in a single column, defined as a factor with a level for each type of food, with each row representing a separate observation and the explanatory factors similarly expanded, then BugsXLA could undertake the same analysis using the 'Categorical' error distribution. However, it is strongly recommended that such data be converted into counts and analyzed as discussed here, since WinBUGS is approximately 30 times slower in running the categorical model.

BugsXLA uses the multinomial-Poisson transformation when modeling multinomial data in order to obtain faster MCMC sampling. The idea is to pretend that the counts in the categories are actually independent Poisson variables, with additional parameters added to account for the finite sample sizes; see Baker (1995) for details. The model specified here can be represented algebraically:

$$\eta_{cij} = \nu_{ij} + \mu_c + \alpha_{ic} + \gamma_{jc} \quad c = 1, 2, \dots, 5 \quad i = 1, 2 \quad j = 1, 2, 3, 4$$

$$\log(\lambda_{cij}) = \eta_{cij}$$

$$y_{cij} \sim \text{Pois}(\lambda_{cij})$$

where
  y_{cij} is the cth food category count for the ith alligator size in the jth lake
  ν_{ij} are the additional parameters needed to make the multinomial-Poisson transformation work; they result in $E(\lambda_{\cdot ij}) = y_{\cdot ij}$ where the '·' notation means sum over that level
  μ_c are the effects that account for inherent differences in the amount of each food type eaten by alligators in Florida
  α_{ic} are the effects associated with the factor Size
  γ_{jc} are the effects associated with the factor Lake

Note that in the multinomial model the effects of each factor are actually interactions with an implicit factor that defines the food categories. The objective is to determine how each of the explanatory factors alters the relative number of alligators preferring each of the food types. The regular model defaults for the MCMC settings were chosen (a burn-in of 10,000 followed by 10,000 samples after a one-fifth thinning), predicted means (probabilities) for all combinations of the factors Size and Lake were requested (see cells A13:B14 in Figure 4.27), and the default priors were unchanged (see cells A30:B32 in Figure 4.28). Figures 4.28 and 4.29 show the results of the analysis as imported back into Excel.

Effectively it is the log-ratio of the probabilities of preferring each food type relative to fish that is modeled. Fish is taken as the reference since

	A	B	C	D	E	F	G	H	I	J	K
1			Label	Mean	St.Dev.	2.5%	Median	97.5%		WinBUGS Name	
2		Category	Invert/Fish	-1.8510	0.5606	-3.0620	-1.8110	-0.8597		CF.Eff[2]	
3		Category	Reptile/Fish	-2.6390	0.7059	-4.1550	-2.5830	-1.4060		CF.Eff[3]	
4		Category	Bird/Fish	-2.1800	0.5953	-3.4640	-2.1530	-1.1140		CF.Eff[4]	
5		Category	Other/Fish	-0.7707	0.3557	-1.4840	-0.7659	-0.0824		CF.Eff[5]	
6		Size	>2.3:Invert/Fish	-1.5190	0.4063	-2.3390	-1.5070	-0.7285		X.Eff[1,7]	
7		Size	>2.3:Reptile/Fish	0.3757	0.6073	-0.7768	0.3570	1.6130		X.Eff[1,8]	
8		Size	>2.3:Bird/Fish	0.6844	0.6715	-0.6009	0.6806	2.0270		X.Eff[1,9]	
9		Size	>2.3:Other/Fish	-0.3522	0.4554	-1.2510	-0.3480	0.5295		X.Eff[1,10]	
10		Lake	Oklawaha:Invert/Fish	2.7350	0.6839	1.4520	2.7120	4.1480		X.Eff[2,7]	
11		Lake	Oklawaha:Reptile/Fish	1.3520	0.8491	-0.1971	1.3210	3.1630		X.Eff[2,8]	
12		Lake	Oklawaha:Bird/Fish	-1.8060	1.4310	-5.0620	-1.6410	0.5369		X.Eff[2,9]	
13		Lake	Oklawaha:Other/Fish	-0.9386	0.7760	-2.5700	-0.9084	0.4675		X.Eff[2,10]	
14		Lake	Trafford:Invert/Fish	2.9150	0.6938	1.6240	2.8840	4.3650		X.Eff[2,12]	
15		Lake	Trafford:Reptile/Fish	1.8490	0.8268	0.3235	1.8040	3.5770		X.Eff[2,13]	
16		Lake	Trafford:Bird/Fish	0.3746	0.8237	-1.2580	0.3807	1.9560		X.Eff[2,14]	
17		Lake	Trafford:Other/Fish	0.7018	0.5735	-0.4369	0.7051	1.8290		X.Eff[2,15]	
18		Lake	George:Invert/Fish	1.7570	0.6302	0.5910	1.7340	3.0930		X.Eff[2,17]	
19		Lake	George:Reptile/Fish	-1.6370	1.4450	-4.9400	-1.5110	0.7851		X.Eff[2,18]	
20		Lake	George:Bird/Fish	-0.7850	0.8384	-2.4970	-0.7646	0.8209		X.Eff[2,19]	
21		Lake	George:Other/Fish	-0.8780	0.5780	-2.0640	-0.8587	0.2139		X.Eff[2,20]	
22											
23	Model	[Alli!$A$1:$G$9]									
24	Distribution	Multinomial									
25	Link	Logit									
26	Response	Fish,Invert,Reptile,Bird,Other									
27	Fixed	Size + Lake									
28											
29	Priors										
30	Cat. Consts	N(mu=1.34, sigma=104)									
31	Size	N(mu=0, sigma=104)									
32	Lake	N(mu=0, sigma=104)									
33											
34	WinBUGS MCMC Settings										
35	Burn-In: 10000 Samples: 10000 (Thin:5; Chains:1)										
36	Run took 42 seconds										

FIGURE 4.28
Summaries of posterior distributions, model, priors, and MCMC settings for alligator feeding habits.

it is the first variable in the list entered as the response. This can also be thought of as the log(odds) in favor of a particular food type given the choice is restricted to this food or fish. This can be seen as a generalized version of the logit link used for Binomial data, and is why the link is referred to as 'logit' rather than 'log'. BugsXLA reflects this interpretation of each effect by adding a suffix of the form 'Invert/Fish' to each parameter's label. The parameters defining the effects of the factors Size and Lake are contrasts on the logit-scale and hence represent log(odds-ratio)s relative to the 'zero constrained' level of the factor. As always when a link other than the identity is used, it is difficult to interpret these parameter estimates directly. However, a cursory inspection of the credible intervals indicate that there are some real differences in the feeding habits of alligators of different sizes and in different lakes. Figure 4.29 shows some of the predicted probabilities imported back into Excel. The output has been edited in preparation for plotting using Excel's charting features. The labels have been simplified and gaps added to break up the line plots into appropriate groups. Note also that two additional columns of data have been added, W and X, which are the differences between columns O and Q (mean and 2.5%ile) and columns S and O (97.5%ile and mean), respectively.

N	O	P	Q	R	S	T	U V	W	X
Label	Mean	St.Dev.	2.5%	Median	97.5%		WinBUGS Name		
	Predicted Probability								
SIZE <=2.3									
LAKE HANCOCK									
Fish	0.5369	0.0710	0.3961	0.5370	0.6733		Pred.Ave[1,1]	0.1408	0.1364
Invert	0.0937	0.0432	0.0273	0.0877	0.1941		Pred.Ave[1,2]	0.0664	0.1004
Reptile	0.0463	0.0284	0.0086	0.0404	0.1166		Pred.Ave[1,3]	0.0377	0.0703
Bird	0.0690	0.0356	0.0180	0.0627	0.1545		Pred.Ave[1,4]	0.0510	0.0855
Other	0.2541	0.0628	0.1428	0.2501	0.3903		Pred.Ave[1,5]	0.1113	0.1362
LAKE OKLAWAHA									
Fish	0.2558	0.0722	0.1310	0.2491	0.4138		Pred.Ave[2,1]	0.1248	0.1580
Invert	0.6011	0.0887	0.4208	0.6043	0.7657		Pred.Ave[2,2]	0.1803	0.1646
Reptile	0.0791	0.0435	0.0191	0.0708	0.1857		Pred.Ave[2,3]	0.0600	0.1066
Bird	0.0094	0.0112	0.0002	0.0054	0.0400		Pred.Ave[2,4]	0.0092	0.0307
Other	0.0546	0.0346	0.0104	0.0471	0.1399		Pred.Ave[2,5]	0.0442	0.0853
LAKE TRAFFORD									
Fish	0.1838	0.0569	0.0887	0.1785	0.3081		Pred.Ave[3,1]	0.0951	0.1243
Invert	0.5147	0.0864	0.3462	0.5155	0.6796		Pred.Ave[3,2]	0.1685	0.1649
Reptile	0.0904	0.0454	0.0249	0.0825	0.1994		Pred.Ave[3,3]	0.0655	0.1090
Bird	0.0369	0.0264	0.0059	0.0301	0.1053		Pred.Ave[3,4]	0.0309	0.0685
Other	0.1742	0.0612	0.0745	0.1685	0.3107		Pred.Ave[3,5]	0.0997	0.1365
LAKE GEORGE									
Fish	0.4523	0.0689	0.3177	0.4517	0.5893		Pred.Ave[4,1]	0.1346	0.1370
Invert	0.4130	0.0706	0.2774	0.4119	0.5548		Pred.Ave[4,2]	0.1356	0.1418
Reptile	0.0115	0.0123	0.0003	0.0075	0.0439		Pred.Ave[4,3]	0.0112	0.0324
Bird	0.0294	0.0204	0.0046	0.0244	0.0817		Pred.Ave[4,4]	0.0248	0.0524
Other	0.0939	0.0387	0.0331	0.0889	0.1835		Pred.Ave[4,5]	0.0608	0.0896
SIZE >2.3									
LAKE HANCOCK									
Fish	0.5670	0.0913	0.3794	0.5703	0.7376		Pred.Ave[5,1]	0.1876	0.1706
Invert	0.0232	0.0144	0.0050	0.0198	0.0606		Pred.Ave[5,2]	0.0182	0.0374
Reptile	0.0723	0.0472	0.0124	0.0618	0.1897		Pred.Ave[5,3]	0.0599	0.1174
Bird	0.1428	0.0711	0.0389	0.1309	0.3105		Pred.Ave[5,4]	0.1039	0.1677
Other	0.1947	0.0703	0.0813	0.1865	0.3543		Pred.Ave[5,5]	0.1134	0.1596

FIGURE 4.29
Predicted probabilities that an alligator prefers each food type. Output has been edited in preparation for plotting.

The plot shown in Figure 4.30 can be produced by first selecting the range of cells N3:O53, which includes the labels and the posterior means. Note that Figure 4.29 does not show all the data being plotted. From the Excel charting options select the line chart with markers, removing the default legend and formatting the plot to your preferred style. The lines joining points in the same lake-size combination are included purely to aid the eye, and could be removed if these offended the purist. Add custom 'Y Error Bars' specifying the range X3:X53 for the positive errors and W3:W53 for the negative errors. Note that it is important that the ranges used for the error bars are the same length and aligned with the original range of cells plotted; Excel will not complain if these are not aligned, but the error bars plotted will be incorrect. The advantage of the plot produced is that it makes interpretation of the output much easier. For example, it appears that the larger alligators in Lake Trafford were virtually indifferent to the types of food they ate. More generally, it appears that the larger alligators had a strong preference for fish, while the smaller alligators ate both fish and invertebrates in relatively large numbers.

BugsXLA v5.0 does not allow contrasts to be specified for multinomial models, so if one wanted to make formal inferences on other quantities of interest, the advanced techniques for users familiar with WinBUGS discussed in Chapter 11 could be used. However, sometimes it is possible to make these additional inferences by manipulating the output provided by BugsXLA.

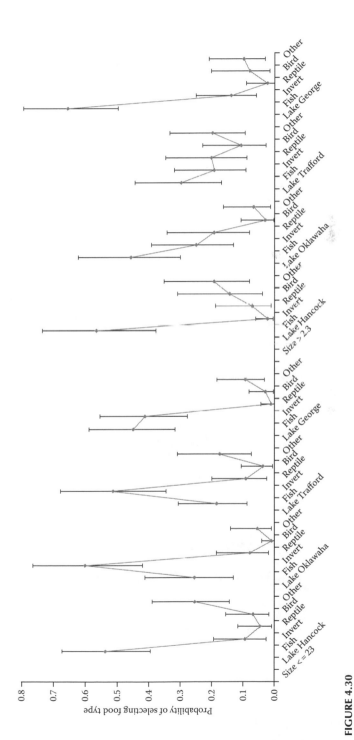

FIGURE 4.30

Plot showing predicted probabilities that an alligator prefers each food type. Error bars represent 95% credible interval estimates.

For example, using such an approach we can obtain estimates comparable with the parameterization adopted by Agresti (2007). He constrained the parameters for each food choice to sum to zero, for example, his parameters defining the effects of Lake related to those defined using BugsXLA via the formula:

$$\gamma_{jc}^* = \gamma_{jc} - \frac{\gamma_{.c}}{4}$$

The posterior distribution for these constrained parameters can be obtained by first requesting that the samples be imported for 'Fixed Terms' on the 'MCMC & Output Options' form as explained in Section 2.5 and illustrated in Case Study 3.1. After importing these samples into Excel, there will be 16 columns on the sheet labeled 'Alli sams', one for each of the 'X.Eff' fixed effects terms shown in Figure 4.28. Comparing the labels in columns B:C with the WinBUGS names in column J, it can be seen that parameters defining the effects of Lake have WinBUGS names beginning 'X.Eff[2,'. It can also be seen that of these terms, those relating to the relative number of invertebrates eaten have an index value for the second dimension of 7, 12, and 17 for Lakes Oklawaha, Trafford, and George, respectively. Note that the effects

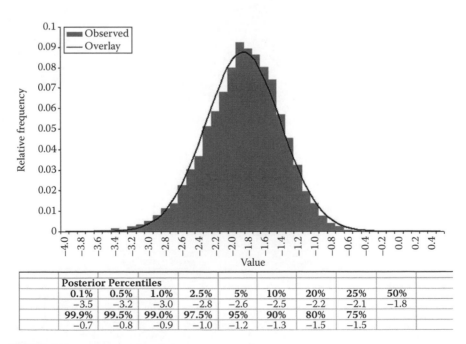

Hancock: Invert/Fish
Data fitted normal(mu = −1.86 sigma = 0.455)

Posterior Percentiles									
0.1%	0.5%	1.0%	2.5%	5%	10%	20%	25%	50%	
−3.5	−3.2	−3.0	−2.8	−2.6	−2.5	−2.2	−2.1	−1.8	
99.9%	99.5%	99.0%	97.5%	95%	90%	80%	75%		
−0.7	−0.8	−0.9	−1.0	−1.2	−1.3	−1.5	−1.5		

FIGURE 4.31
Plot of posterior distribution for one of the sum-to-zero constrained effects of Lake on alligator feeding habits.

associated with Lake Hancock, the first level of Lake, have been constrained to zero. Posterior samples for the constrained parameters can be obtained by entering the right hand side of the equation above as an Excel formula in an empty cell on sheet 'Alli sams', and copying this formula down the column utilizing the fact that Excel uses relative referencing by default. For example, if the 10,000 samples for each of X.Eff[2,7], X.Eff[2,12] and X.Eff[2,17] were in columns E, I, and M, respectively, then in the second row of an empty column enter the formula:

$$= 0 - \frac{(E2 + I2 + M2)}{4}$$

This will calculate the first sample from the posterior distribution for the sum-to-zero constrained effect of Lake Hancock:Invert/Fish. By copying the formula to all the cells in the column down to row 10,001 generates 10,000 samples from this posterior distribution. These values can be summarized using Excel's built in functions for calculating the mean, standard deviation and quantiles. Alternatively, a name for this column of data can be entered in the first row and BugsXLA's posterior plotting tool used to produce a histogram and percentiles as explained in Section 2.11 and illustrated in Case Study 6.3. Figure 4.31 shows such a plot, with the best fitting Normal Distribution overlaid, and percentiles underneath. This whole process can be repeated for each of the parameters in turn. Clearly this approach can be generalized to obtain the posterior distribution for any function of the parameters that can be expressed as a formula in Excel.

5

Normal Linear Mixed Models

Mixed models are so called because they contain both fixed and random effects. In classical statistics, a fixed effect is an unknown constant that is estimated from the data, while a random effect is not estimated directly, but instead the parameters of its underlying distribution are estimated. It is only fairly recently that accessible texts on mixed models have been available, with the classical approach being well covered by books such as Brown and Prescott (2006) or McCulloch and Searle (2001). Texts aimed at the applied data analyst explaining the Bayesian approach are typically harder to find, although books such as Gelman et al. (2004) and Ntzoufras (2009), which cover the normal linear mixed model (NLMM) under the more general heading of 'hierarchical models', are clearly aimed at applied statisticians. In the Bayesian approach to statistical inference, the terms "fixed" and "random" are strictly speaking inappropriate, since all parameters in the model are thought of as random variables. It is convenient to continue using these terms since there are Bayesian equivalents, with effects being defined as fixed or random depending upon the nature of their prior distribution. As stated previously, a Bayesian considers a factor as having fixed effects if the parameters defining them are completely independent, implying that knowledge of the true value of any of them tells us nothing about the true values of any of the others. The parameters defining the effects of a random factor are not considered independent, implying that we learn about all the parameters even if we only obtain information on some of them directly. Unlike in the classical approach to statistical inference, there is no need to imagine a population from which the levels of the factor were randomly chosen in order to justify specifying it as random. It is sufficient that we believe that the effects of each level should be "similar," as implied by the random effects prior that we now come to discuss.

Chapter 3 explains how independent priors for fixed effects are formally specified in BugsXLA. If a factor is defined as random in BugsXLA, then the effects, α_i, associated with the levels of this factor are given the prior:

$$\alpha_i \sim N(0, \tau^2)$$

where τ^2 is the variance component associated with this term, which has itself a prior distribution. It is this hierarchy in the parameters that is referred to when these models are termed Hierarchical Linear Models. There is still some debate about the most appropriate default vague prior to use for τ^2 (see Chapter 5 of Spiegelhalter et al. (2004) for a thorough discussion). BugsXLA offers the following options:

$$\tau \sim \text{Half-N(K)}$$

The Half-Normal prior on the standard deviation of the effects is the default, as this has been recommended by both Spiegelhalter et al. (2004) and Gelman (2006).

$$\tau \sim \text{Half-C(K)}$$

The Half-Cauchy prior on the standard deviation has been recommended by Gelman (2006) for cases when the factor only has a few levels, for example, less than six. Note that this is a very heavy tailed distribution, equivalent to a Half-t distribution with one degree of freedom.

$$\tau \sim \text{U}(0,K)$$

The Uniform prior on the standard deviation has been recommended by Gelman (2006) for cases when the factor has many levels. This distribution has the slight disadvantage of having to define an upper limit, beyond which the value of τ is forced to have zero posterior probability.

In all of the above prior distributions, K is a fixed value with a default chosen by BugsXLA to represent little information relative to the variability in the response. In all cases, this value can be either altered from the default directly or inferred via an elicitation of beliefs regarding the factor's effects; this is discussed in more detail later.

$$\tau^{-2} \sim \text{Ga}(r,\mu) \quad \text{or} \quad \text{Inverse-Gamma for } \tau^2$$

The Inverse-Gamma prior is included as this is the standard distribution to use for the residual variance parameter, and was used in the past for these hierarchical variance parameters. Using the value 0.001 for both r and μ is now discredited as a default uninformative prior. The Inverse-Gamma is only recommended when its parameters can be chosen to represent an informative prior.

$$\tau^{-2} \sim \text{ScChiSq}(df,s^2)$$

See the discussion in Chapter 3 on when this distribution might be an appropriate representation of an informative prior.

Random effects can also be specified for the regression coefficients associated with continuous covariates; typically this occurs when a covariate-by-factor interaction is fitted, with the belief that the coefficients should be similar for each level of the factor. Refer to Case Study 5.4

and Section 2.7 for an explanation of how BugsXLA allows a prior to be specified in these cases.

Case Study 5.1 Latin Square Industrial Experiment

Davies (1954) describes an experiment in which four materials, labeled A, B, C, and D, were fed into a wear-testing machine. The loss of weight in 0.1 mm units was measured. The machine contained four testing positions, enabling four samples to be tested at each run. Past experience suggested that systematic differences might exist both between the wear seen at each of the positions and also between different runs of the machine. A Latin square design was used to test each material four times each, giving a total of 16 measurements from four runs of the machine. It might be reasonable to fit an NLMM with the effect of materials regarded as fixed, and the effects of position and run regarded as random. Figure 5.1 shows the data and BugsXLA model specification form.

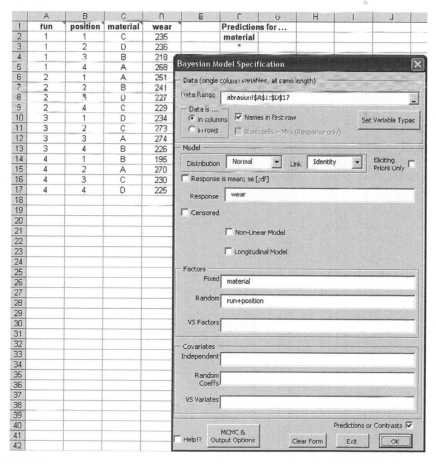

FIGURE 5.1
Data and model specification for abrasion study.

The variables run, position, and material are all defined as factors, with their levels sorted into ascending order. Since for a Latin square design the blocking factors are crossed, the random effects part of the model is specified simply by entering the names of the factors with the '+' operator between them. This model can be represented algebraically:

$$y_{ijk} = \mu + \alpha_i + r_j + p_k + e_{ijk} \quad i = 1,2,3,4 \quad j = 1,2,3,4 \quad k = 1,2,3,4$$

$$r_j \sim N\left(0, \sigma_3^2\right)$$

$$p_k \sim N\left(0, \sigma_2^2\right)$$

$$e_{ijk} \sim N\left(0, \sigma_1^2\right)$$

where
   y is the response variable wear
   α_i are the effects of the factor material
   r_j are the effects of the factor run
   p_k are the effects of the factor position
   $\sigma_1^2, \sigma_2^2, \sigma_3^2$ are the variance components

The simple model defaults for the MCMC settings were chosen (a burn-in of 5,000 followed by 10,000 samples), the predicted means for all levels of material were requested (see cells F2:F3 in Figure 5.1), and the default priors were unchanged (see cells A18:B22 in Figure 5.3). Figure 5.2 shows the page labeled 'Exch. Factors' on the 'Prior Distributions' form where the priors for the random effects can be altered.

The default value for the Half-Normal parameter defining the prior for the run term is 111, which is very vague relative to the variation in the data. The form provides a crude elicitation process by which the prior can be inferred, either by specifying the maximum credible value for the parameter τ directly or via the maximum credible difference between the effects of two randomly chosen levels of the factor. A more sophisticated elicitation tool can be utilized by clicking on the button labeled GFI (Graphical Feedback Interface). Refer to Section 2.8 for details on how to use the GFI and how BugsXLA infers a prior distribution from the elicitation process provided here.

Figures 5.3 and 5.4 show the results of the analysis as imported back into Excel. The summaries of the parameters' posterior distributions show that there are clear differences between the wear rates of the four materials, since all of the contrasts have 95% credible intervals that exclude zero. It would require subject matter expertise to assess whether these differences

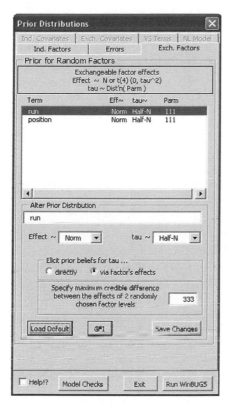

FIGURE 5.2
Prior for random effects in abrasion study.

	A	B	C	D	E	F	G	H	I	J	K
1			Label	Mean	St.Dev.	2.5%	Median	97.5%		WinBUGS Name	
2			CONSTANT	266.5000	14.8700	237.4000	266.1000	298.5000		Beta0	
3		material	B	-45.6000	7.0980	-59.8100	-45.6000	-30.7500		X.Eff[1,2]	
4		material	C	-23.9900	7.2850	-39.0400	-24.0400	-9.4790		X.Eff[1,3]	
5		material	D	-35.1800	7.1650	-49.6600	-35.1800	-20.8900		X.Eff[1,4]	
6			SD(run)	15.5400	14.8800	1.6620	11.4800	54.8900		sigma.Z[1]	
7			SD(position)	18.2100	14.3800	2.9700	14.3600	55.6200		sigma.Z[2]	
8			SD(residual)	9.4400	3.4090	5.0840	8.6460	18.2500		sigma	
9											
10	**Model**	[abrasion!$A$1:$D$17]									
11	Distribution	Normal									
12	Link	Identity									
13	Response	wear									
14	Fixed	material									
15	Random	run+position									
16											
17	**Priors**										
18	CONSTANT	N(mu=240, sigma=2000)									
19	material	N(mu=0, sigma=2000)									
20	run	N(0,tau^2); tau ~ Half-N(sigma=111)									
21	position	N(0,tau^2); tau ~ Half-N(sigma=111)									
22	V(residual)	Inv-Gamma(0.001, 0.001)									
23											
24	**WinBUGS MCMC Settings**										
25	Burn-In: 5000 Samples: 10000 (Thin:1; Chains:1)										
26	Run took 12 seconds										

FIGURE 5.3
Summaries of posterior distributions, model, priors, and MCMC settings for abrasion study.

N	O	P	Q	R	S	T	U	V
Label	Mean	St.Dev.	2.5%	Median	97.5%		WinBUGS Name	
	Predicted Mean Response							
	Following all at: run(dist. mean) position(dist. mean)							
material(A)	266.5000	14.8700	237.4000	266.1000	298.5000		Pred.Ave[1]	
material(B)	220.9000	14.7600	192.2000	220.4000	252.2000		Pred.Ave[2]	
material(C)	242.5000	14.7700	213.6000	242.1000	274.0000		Pred.Ave[3]	
material(D)	231.3000	14.8200	202.0000	230.9000	262.5000		Pred.Ave[4]	

FIGURE 5.4
Predicted means from abrasion study.

were of practical importance. Inspection of the summarized posteriors for the three variance components suggests that both the run and position might be important sources of variation relative to the residual. However, as should be expected from such a small experiment, the hierarchical standard deviation parameters, SD(run) and SD(position), are estimated with relatively low precision. A more formal assessment is not warranted here as the experiment was designed with a strong prior belief that these terms were important, and so the method of analysis should not be altered post-hoc without very good reasons. Figure 5.4 show the predicted means for the four treatments.

As one should expect, the differences between the posterior means of the predicted mean response for each type of material are consistent with the effects of material shown in Figure 5.3. Note that the uncertainty in these predicted means is much greater than that in the effect estimates. This is to be expected, as these include the uncertainty in the mean response averaged over all possible runs and positions of the machine, not just those runs and positions included in the study. It is important to note that BugsXLA assumes that the predicted mean of interest is centered at the means of the underlying distributions for the random effects, and not just the means of the specific levels of these factors included in the model. Note also the two or more trailing zeroes in each of the imported values. This is due to WinBUGS only reporting its output to four significant figures, and so this needs to be considered when interpreting these summaries. This issue does not arise when working with imported samples as WinBUGS provides these in double precision.

It is again informative to compare the results of this Bayesian analysis with that obtained from an equivalent likelihood analysis computed by R. Using the R code created by BugsXLA (refer to Case Study 3.1 for details of how to obtain), the summary of this likelihood analysis is shown below:

```
> library(lme4)
> BGX.fm<-lmer(wear~material+(1|run)+(1|position),data=BGX.df)
> summary(BGX.fm)

Linear mixed model fit by REML
Formula: wear ~ material+(1|run)+(1|position)
```

```
    Data: BGX.df
AIC    BIC    logLik  deviance  REMLdev
114.3  119.7  -50.13  120.4      100.3
Random effects:
Groups      Name           Variance    Std. Dev.
run         (Intercept)     66.896       8.1790
position    (Intercept)    107.062      10.3471
Residual                    61.250       7.8262
Number of obs: 16, groups: run, 4; position, 4
Fixed effects:
                Estimate        Std. Error    t value
(Intercept)      265.750          7.668        34.66
materialB        -45.750          5.534        -8.27
materialC        -24.000          5.534        -4.34
materialD        -35.250          5.534        -6.37
Correlation of Fixed Effects:
                (Intr)          matrlB          matrlC
materialB       -0.361
materialC       -0.361          0.500
materialD       -0.361          0.500           0.500
```

The estimates of the effects of material are very close to the posterior means reported in the Bayesian analysis, the differences are mainly due to MCMC sampling errors as the priors used would have had little effect. The standard errors are smaller than the equivalent posterior standard deviations. This is at least partly due to the latter representing the dispersion of the posterior distribution, which will be approximately a t-distribution, as discussed in Chapter 3. Unlike in the fixed effects analysis, the lme4 package in R does not provide degrees of freedom for these effect estimates, so it is not straightforward to compute interval estimates from the summary provided. These are absent due to controversy on how to calculate values that provide the best approximation in the general case. If this same analysis is run in SAS Proc Mixed, it states that there are 6 degrees of freedom associated with these estimates, the same as would exist for a completely fixed effects analysis. The 95% credible intervals reported in the Bayesian analysis are slightly wider than the equivalent SAS confidence intervals. Refer to Box and Tiao (1973) for a discussion of how the Bayesian approach recovers information from different strata in an NLMM.

Again as discussed in Chapter 3, the likelihood estimates of the variance parameters are likely to be smaller than the Bayesian posterior means due to the skewed nature of the posterior distribution, and the maximum likelihood being more analogous to posterior modes. Typically, the posterior median is closer to the classical point estimate for variance components. It is not easy to obtain interval estimates for these parameters using R, although a SAS Proc Mixed analysis can provide interval estimates for the variance components, which can be square-rooted to give the same for the standard deviations:

SAS 95% confidence limits for standard deviations of random effects

run	4.2–53
position	5.5–54
residual	5.0–17

Comparison with the Bayesian credible intervals shows that the limits are very similar for the residual, and also very close at the upper end for the two random factors. The lower Bayesian credible limits for run and position are somewhat smaller than the equivalent values given by SAS. The close match for the residual interval estimates is due to the likelihood dominating the very weak prior used, even for a study as small as this one. There is much less information in the likelihood for the hierarchical variance parameters, and hence the Half-Normal prior, with more weight closer to zero, is influential in shrinking the credible values toward zero. In general, the posterior distributions for hierarchical variance parameters are likely to be sensitive to the choice of prior in cases where there are only a few levels of the factor being studied, as in this case. Often this is not an issue, since these variance parameters are usually not of interest in themselves, and a well-designed study should make the inferences of interest insensitive to variation in their true values.

In the frequentist approach to statistical inference, when more than one treatment comparison is being made, strictly speaking, some formal adjustment for multiplicity is necessary. This issue was briefly discussed at the end of Case Study 4.4, where it was mentioned that in the Bayesian approach to inference no adjustment is necessary providing we believe that the model appropriately reflects the nature of the variation and correlation in the data. Some have argued (see, e.g., Gelman (2005)) that this implies that the fixed effects assumption is rarely appropriate and that adopting a hierarchical random effects model, with informative priors as necessary, should ensure adjustments are made to the effect estimates that address the multiplicity issue. This method for dealing with multiplicity is peculiar to the Bayesian approach, and like all modeling assumptions, is not something to be adopted blindly. We illustrate the general principle using this case study, with the caveat that the necessary subject matter expertise is not available to comment on the implied prior beliefs regarding the similarity between the materials' effects. Due to the small size of the experiment, there is insufficient information in the data alone to illustrate this approach when factors run and position are included in the model, that is, unless a strong prior belief of similarity between the materials' effects is included, the effect estimates hardly change. For the purposes of this illustration the model is simplified to a one-way analysis of variance, that is, the factors run and position are not fitted. We suppose that prior to the experiment being run, the wear rate of the four materials were considered to be "similar" and exchangeable as discussed in Section 1.1. It is not a simple task to explain this concept to a layman, but this is necessary in order to obtain agreement that the way this belief is to be modeled is reasonable. In the simplest formulation, one's beliefs concerning the materials' effects should be reasonably represented by them behaving as if being drawn from a Normal Distribution with a standard deviation that determines the degree of similarity. The data will be used to estimate this standard deviation parameter, but any additional prior information can be incorporated as usual in the Bayesian paradigm. To fit this model using BugsXLA, all that needs to be changed on the 'Model Specification' form is to remove material as a fixed effect and add it as a random effect. Table 5.1 shows all pair wise contrasts of the effects of material, facilitating a comparison between the standard fixed effects and the random effects analysis (the simple model MCMC settings with default priors were used).

Note that the estimated contrasts have all shrunk toward zero, this being a direct consequence of the prior assumption that the materials have effects with "similar" magnitudes. The shrinkage is more noticeable for the larger effects, but it does not change their order. Like in the classical approaches to dealing with the multiplicity issue, the Bayesian estimates, due to the shrinkage that has occurred, are all "less statistically significant" than when the standard fixed effects analysis is undertaken. Unlike in most classical approaches, it is straightforward to obtain "multiplicity adjusted" Bayesian interval estimates as they are inferred directly from the model in the usual way. In cases where there are many treatments, with prior knowledge of how they might be formed into "similar subgroups," a more complex hierarchical model with random effects to model both the between and within subgroup similarities could be specified. By allowing more flexibility in

TABLE 5.1

Pair Wise Contrasts (Posterior Means and Standard Deviations) for Abrasion Study, Comparing the Differences When the Effects of Material Are Modeled as Fixed or Random in a One-Way ANOVA Model (Run and Position Factors Removed from Original Model)

	Effects of Material Modeled as	
	Fixed	Random
Materials (A – B)	46 (12)	39 (13)
Materials (A – C)	24 (12)	21 (12)
Materials (A – D)	35 (12)	30 (12)
Materials (B – C)	−22 (12)	−19 (11)
Materials (B – D)	−10 (12)	−9 (11)
Materials (C – D)	−11 (12)	−10 (11)

how one's beliefs can be incorporated into the formal model, the Bayesian approach to dealing with multiplicity is much more satisfying than the largely ad-hoc classical methods. However, to coin a phrase, "with power comes responsibility," and so, unless the additional structure representing the prior beliefs are well founded, the sensitivity to these assumptions should be assessed.

Case Study 5.2 Variance Components Estimation
When developing a new product it is important to run experiments to determine the major sources of variation in the product's critical to quality characteristics. Here we describe one such experiment that was run during the development of a needle-free device for the delivery of DNA vaccine drugs. The device delivers gold micro-particles coated with DNA-based vaccine into the immune network of the outer epidermal layer of the skin. Each device contains enough drug product to enable approximately 8000 shots to be delivered over its lifetime. The purpose of this experiment was to assess the variation that occurred in numerous measurable quality characteristics over the lifetime of the product, as well as the variation due to other features of the sampling and measurement process. In this experiment, a device was repeatedly fired into a series of gels designed to mimic the outer layers of the skin. At each of five different time points throughout the lifetime of the device, a set of six gels were separately shot into. From each gel three slides were made, and the image from these slides digitized. Two different analysts viewed each of the slides and recorded their measurements in a completely blinded study. Here we show how the measurement of the image mean observed depth, in micrometers, could be analyzed by Bayesian methods using BugsXLA.

Figure 5.5 shows some of the data and BugsXLA model specification form. The factor ShotSet refers to the set of shots taken at each time point, with levels labeled with the letters C through to G. The factors GelNum, SlideNum, and Analyst should be self explanatory, and depth is the response variable. The model fitted is partially nested with factor SlideNum nested within GelNum, which is nested within the ShotSet factor. Analyst is included as a completely crossed factor, with an Analyst-by-ShotSet interaction term also fitted. It was not believed appropriate to consider the levels of the ShotSet factor as exchangeable in the sense implied by the "random effects" assumption. This is because it was believed likely that the performance of the device over its lifetime would be smoothly changing, and unlikely that it would change in a random manner between sampled time points with no correlation between them. Although some type of longitudinal model would be needed to properly model the time course, in the absence of a strong rationale for a particular model, it was believed adequate to adopt the "fixed effects" assumption for the purpose of this experiment. Since the main effect of ShotSet is included as a fixed effect, the ShotSet-by-GelNum interaction term included as random effects, using the ':' operator, models

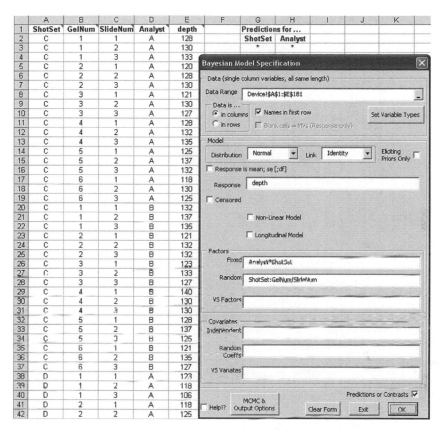

FIGURE 5.5
Data and model specification for device study.

the variation between gels within each time point. The '/' operator is used to include SlideNum as nested within this interaction term, that is, nested within GelNum that is nested within ShotSet. This model can be represented algebraically:

$$y_{ijkl} = \mu + \alpha_i + \gamma_j + (\alpha\gamma)_{ij} + (\gamma g)_{jk} + (\gamma g s)_{jkl} + e_{ijkl} \quad i=1,2 \quad j=1,2,\ldots,5$$

$$k=1,2,\ldots,6 \quad l=1,2,3$$

$$(\gamma g)_{jk} \sim N(0, \sigma_3^2)$$

$$(\gamma g s)_{jkl} \sim N(0, \sigma_2^2)$$

$$e_{ijkl} \sim N(0, \sigma_1^2)$$

where

    y is the response variable depth

    α_i are the effects of the factor analyst

    γ_j are the effects of the factor ShotSet

    $(\alpha\gamma)_{ij}$ are the interaction effects between analyst and factor ShotSet

    $(\gamma g)_{jk}$ are the nested effects of GelNum within ShotSet

    $(\gamma gs)_{jkl}$ are the nested effects of SlideNum within GelNum

    $\sigma_1^2, \sigma_2^2, \sigma_3^2$ are the variance components

The simple model defaults for the MCMC settings were chosen (a burn-in of 5,000 followed by 10,000 samples), the predicted means for all combinations of the levels of ShotSet and Analyst were requested (see cells G2:H3 in Figure 5.5), and the default priors were unchanged (see cells A24:B30 in Figure 5.6).

    Inspection of the credible intervals for the fixed effects terms shows that there are some real differences between the mean depths, both over the lifetime of the device and between the analysts. The nature of these differences is best seen by plotting the predicted means, as shown in Figure 5.7. The default output from BugsXLA has been slightly reformatted to aid interpretation and facilitate the production of the Excel plot shown in Figure 5.7; the number of decimal places has been reduced, and a blank line has been inserted so that a break occurs between analysts in the plot. The plot is not a feature of BugsXLA but simply an Excel line plot of cells N4:O14, with error bars added using the posterior standard deviation values in cells P4:P14. It is clear from this plot that although both analysts have observed essentially the same time course pattern, there is a shift of approximately

	A	B	C	D	E	F	G	H	I	J	K
1			Label	Mean	St.Dev.	2.5%	Median	97.5%		WinBUGS Name	
2			CONSTANT	128.6000	2.9360	123.1000	128.5000	134.6000		Beta0	
3		Analyst	B	1.9970	1.0560	-0.0331	2.0030	4.0660		X.Eff[1,2]	
4		ShotSet	D	-8.9510	4.3310	-18.0800	-8.7810	-0.9597		X.Eff[2,2]	
5		ShotSet	E	-10.4600	4.1340	-18.6900	-10.4600	-2.4690		X.Eff[2,3]	
6		ShotSet	F	-7.4470	4.2280	-15.7200	-7.4740	1.0800		X.Eff[2,4]	
7		ShotSet	G	-11.4500	3.9650	-19.0700	-11.3900	-3.5760		X.Eff[2,5]	
8		Analyst x ShotSet	B, D	1.9730	1.4810	-0.9010	1.9650	4.8790		X.Eff[3,4]	
9		Analyst x ShotSet	B, E	3.7780	1.4930	0.8014	3.7820	6.6860		X.Eff[3,6]	
10		Analyst x ShotSet	B, F	4.8900	1.4950	1.9680	4.8900	7.8150		X.Eff[3,8]	
11		Analyst x ShotSet	B, G	1.5450	1.4970	-1.3730	1.5390	4.4700		X.Eff[3,10]	
12			V(ShotSet x GelNum)	45.7400	17.1600	22.0000	42.6000	87.9800		sigma2.Z[1]	
13			V(ShotSet x GelNum x SlideNum)	17.3200	4.2900	10.2900	16.8000	27.1000		sigma2.Z[2]	
14			V(residual)	9.9350	1.5710	7.3270	9.7850	13.4100		sigma2	
15											
16	Model	[Device!$A$1:$E$181]									
17	Distribution	Normal									
18	Link	Identity									
19	Response	depth									
20	Fixed	Analyst*ShotSet									
21	Random	ShotSet:GelNum/SlideNum									
22											
23	Priors										
24	CONSTANT	N(mu=123, sigma=874)									
25	Analyst	N(mu=0, sigma=874)									
26	ShotSet	N(mu=0, sigma=874)									
27	Analyst x ShotSet	N(mu=0, sigma=874)									
28	ShotSet x GelNum	Norm(0,tau^2); tau ~ Half-N(sigma=43.7)									
29	ShotSet x GelNum x SlideNum	Norm(0,tau^2); tau ~ Half-N(sigma=43.7)									
30	V(residual)	Inv-Gamma(0.001, 0.001)									
31											
32	WinBUGS MCMC Settings										
33	Burn-In: 5000 Samples: 10000 (Thin:1; Chains:1)										
34	Run took 72 seconds										

FIGURE 5.6

Summaries of posterior distributions, model, priors, and MCMC settings for device study.

N	O	P	Q	R	S	T	U	V
Label	Mean	St.Dev.	2.5%	Median	97.5%		WinBUGS Name	
	Predicted Mean Response							
	Following all at: GelNum(dist. mean) SlideNum(dist. mean)							
Analyst(A) ShotSet(C)	128.6	2.9	123.1	128.5	134.6		Pred.Ave[1]	
Analyst(A) ShotSet(D)	119.6	3.2	113.1	119.6	126.0		Pred.Ave[2]	
Analyst(A) ShotSet(E)	118.1	3.0	112.0	118.2	124.0		Pred.Ave[3]	
Analyst(A) ShotSet(F)	121.1	3.1	115.1	121.1	127.5		Pred.Ave[4]	
Analyst(A) ShotSet(G)	117.1	2.9	111.6	117.1	122.7		Pred.Ave[5]	
Analyst(B) ShotSet(C)	130.6	2.9	125.0	130.5	136.5		Pred.Ave[6]	
Analyst(B) ShotSet(D)	123.6	3.3	117.0	123.6	130.0		Pred.Ave[7]	
Analyst(B) ShotSet(E)	123.9	3.0	117.8	124.0	129.7		Pred.Ave[8]	
Analyst(B) ShotSet(F)	128.0	3.1	121.9	128.0	134.4		Pred.Ave[9]	
Analyst(B) ShotSet(G)	120.7	2.9	115.2	120.7	126.3		Pred.Ave[10]	

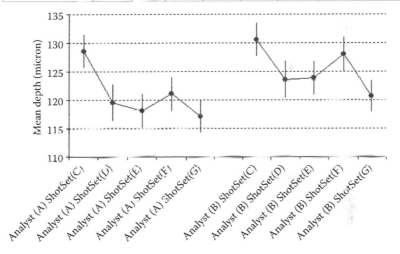

FIGURE 5.7
Predicted means with error bars (posterior standard deviation) from device study.

5 microns between their readings, with the possibility that this shift only occurred after the first set of shots. It also appears that the first set of shots (labeled 'C') is markedly different from the remaining shots, being a little over 5 microns higher. If it was important to report these differences more precisely, then the specific contrasts of interest should be specified and estimated directly; in this case this was not required. It is important to note that the posterior standard deviations reported and plotted in Figure 5.7 should not be used when assessing differences between the levels of the factors. This is because the model implies that the observations have a fairly complicated correlation structure that needs to be accounted for in such comparisons. Any contrasts of interest should be explicitly specified when determining the analysis to be run and then these correlations will be appropriately accounted for in the estimates provided by WinBUGS. Note also that this issue is not peculiar to the Bayesian approach, but holds for standard errors produced for predicted or fitted means by conventional software packages.

Referring back to Figure 5.6 it can be seen that the largest source of random variation is that due to the differences between gels, that is, the nested GelNum factor labeled 'V(ShotSet × GelNum)' in the output. Hence, the most impactful way to improve the precision of the measurement process would be to identify and eliminate the major causes of these differences. If such improvements were not possible, then the sampling scheme should focus on optimizing the number of gels analyzed in order to maximize the precision. When fitting relatively complex hierarchical models such as this one, it is particularly important to assess the sensitivity of the inferences to the priors, even when default "vague" priors are adopted. Ultimately, the only way to do this thoroughly is to rerun the analysis under different prior assumptions. However, BugsXLA offers a useful plotting tool where by the prior and posterior for each parameter can be overlaid (refer to Section 2.11 for details and Case Study 3.1 for an example). Figure 5.8 shows this plot for the nested GelNum variance component.

It can be seen that over the posterior credible range of values, the specified Half-Normal prior distribution is relatively flat, indicating that the information in the likelihood is likely to be dominant in forming the posterior. In general, it is possible that the prior distributions for other parameters are strongly influencing the posterior for the parameter being assessed, and so this "one parameter at a time" analysis cannot be taken as sufficient to assess sensitivity to the priors. However, these plots are likely to identify influential prior distributions in the majority of the less complex models that are regularly used to tackle real problems. When informative priors are used it is not necessarily a "bad thing" if these are influential on the posterior distribution; the purpose of using an informative prior is sometimes to incorporate

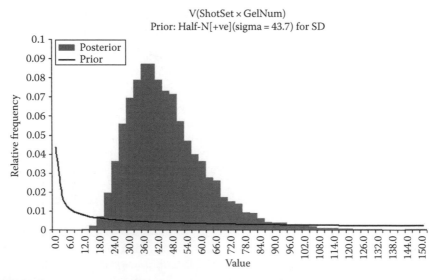

FIGURE 5.8
Posterior and prior distributions for nested GelNum variance component from device study.

appropriate knowledge about those parameters on which it is expected that the current study will not provide much additional information. Examples include information available on historical controls, variance components not of direct interest or non-linear model parameters that can be constrained via a prior without unduly influencing the inferences of interest. But it is important to be aware if one's prior is influential, and justify the validity of the conclusions in the light of this fact.

Variance component estimation is an area that can easily lead to difficulties in the non-Bayesian approaches. Some methods can lead to negative estimates of variance, while obtaining interval estimates can be problematic, particularly when relying on asymptotic results in cases where the effective sample size is small. Lawson (2008) discusses these issues in the context of the gauge repeatability and reproducibility study, and demonstrates how BugsXLA can be used to analyze such a study. The Bayesian approach to variance component estimation is not free of controversy, and the main issue lies in the choice of a default prior distribution for the hierarchical variance components, for example, σ_2^2 and σ_3^2 in Case Study 5.2. Further reference to this controversy, as well as the list of default priors that BugsXLA offers, is provided at the beginning of this chapter.

Case Study 5.3 Meta-Analysis Using Study Level Summaries

Fluticasone Propionate (FP) is an inhaled corticosteroid (ICS) widely used for the treatment of asthma. Any new treatment aimed at the same population for which FP is routinely prescribed, will very likely need either to be non-inferior when administered alone or additive in its effect when co-administered with FP or another equivalent ICS. In order to assimilate the available knowledge on the performance of FP in relevant clinical trials, a non-linear mixed effects meta-analysis of published studies, modeling the mean and dispersion jointly, was undertaken. A reduced data set, excluding all bar the placebo and 200 mg FP daily dose, is reanalyzed here using a simpler NLMM, which still shows some of the major findings uncovered in the original analysis. The response of interest was a measure of lung function known as Forced Expiratory Volume in one second, or FEV_1 for short, reported here in liters (L). As is typically the case, it was not possible to obtain the patient level data for these studies, and so the analysis was based on the study level summaries defined below:

- **CFB.mn:** Mean change from baseline FEV_1 for each treatment arm in the study
- **CFB.se:** Standard error of each mean change from baseline FEV_1
- **CFB.df:** Degrees of freedom associated with each standard error
- **FPmg:** A two-level factor denoting the treatment associated with each arm; either 200 mg daily dose of FP or placebo.
- **STUDY:** A 27-level factor allowing between study differences to be assessed.

- **YR2K**: A two-level factor indicating whether the study was published since the year 2000. There was a belief that the introduction of centralized spirometry at around this time could have had some effect.

- **FORM**: A two-level factor indicating which formulation was used; either a metered dose inhaler (MDI) or dry powder inhaler (DPI).

- **ICS.USE**: A three-level factor identifying whether the patients had previously been using any ICS treatments; some studies only recruited ICS users (ALL), some only recruited ICS naïve patients (NONE), and some had no restrictions on this matter (MIX).

- **BASE.PC**: Mean percent predicted FEV_1 for subjects in the study taken prior to being randomized to treatment; a measure of the asthma severity of the patients included in the study.

Figure 5.9 shows the data and BugsXLA model specification form. Note that the check box labeled 'Response is mean; se [;df]' has been selected. This informs BugsXLA that the response will consist of two or three variates each separated by a semi-colon. The first two variates will contain the means and their standard errors in that order. If a third variate is specified then this must contain the degrees of freedom associated with each standard error. If a third variate is not specified then the analysis is undertaken assuming that the standard errors are known without error. In this particular example, the degrees of freedom are in each case so large that there is little to be gained by including them in the analysis, but they are specified here simply for illustration. Since we are interested in assessing how the effect of FP might be different depending upon the drug formulation, the type of patients included in the study, or the date of the study, interactions between the treatment factor and these other factors have been specified in the fixed effects part of the model. Note how brackets have been used to simplify the specification of this part of the model. Similarly, we are also interested in whether the effect of FP differs depending upon the severity of the asthma patients in the study. To assess this, the BASE.PC covariate is included with an interaction term in the model. Finally, we not only model the between study effect on the mean response, but also fit a study-by-treatment interaction to allow for differences in the true effect of FP between studies. Both of the terms included to account for differences between studies are treated as random effects. This model can be represented algebraically:

$$\eta_{ijklm} = \mu + \alpha_i + \gamma_j + \delta_k + \zeta_l + (\alpha\gamma)_{ij} + (\alpha\delta)_{ik} + (\alpha\zeta)_{il} + s_m + (\alpha s)_{im} + \beta x_{ijklm} + (\alpha\beta)_i x_{ijklm}$$

$$i = 1,2 \quad j = 1,2,3 \quad k = 1,2 \quad l = 1,2 \quad m = 1,2, \ldots, 27$$

$$y_{ijklm} \sim N(\eta_{ijklm}, \omega_{ijklm}^2)$$

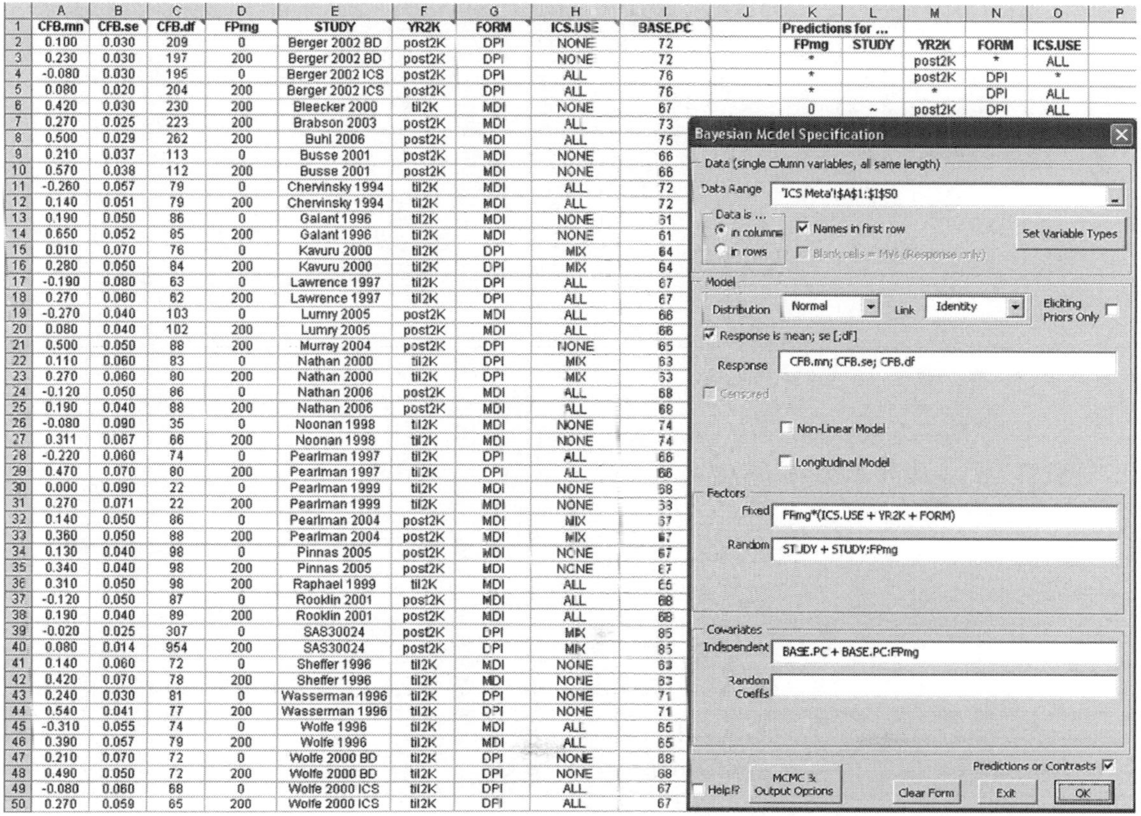

	A CFB.mm	B CFB.se	C CFB.df	D FPmg	E STUDY	F YR2K	G FORM	H ICS.USE	I BASE.PC
2	0.100	0.030	209	0	Berger 2002 BD	post2K	DPI	NONE	72
3	0.230	0.030	197	200	Berger 2002 BD	post2K	DPI	NONE	72
4	-0.080	0.030	195	0	Berger 2002 ICS	post2K	DPI	ALL	76
5	0.080	0.020	204	200	Berger 2002 ICS	post2K	DPI	ALL	76
6	0.420	0.030	230	200	Bleecker 2000	tii2K	MDI	NONE	67
7	0.270	0.025	223	200	Brabson 2003	post2K	MDI	ALL	73
8	0.500	0.029	262	200	Buhl 2006	post2K	MDI	ALL	75
9	0.210	0.037	113	0	Busse 2001	post2K	MDI	NONE	66
10	0.570	0.038	112	200	Busse 2001	post2K	MDI	NONE	66
11	-0.260	0.057	79	0	Chervinsky 1994	tii2K	MDI	ALL	72
12	0.140	0.051	79	200	Chervinsky 1994	tii2K	MDI	ALL	31
13	0.190	0.050	86	0	Galant 1996	tii2K	MDI	NONE	61
14	0.650	0.052	85	200	Galant 1996	tii2K	MDI	NONE	64
15	0.010	0.070	76	0	Kavuru 2000	DPI	MIX	MIX	64
16	0.280	0.050	84	200	Kavuru 2000	tii2K	MDI	MIX	67
17	-0.190	0.080	63	0	Lawrence 1997	tii2K	DPI	ALL	67
18	0.270	0.080	62	200	Lawrence 1997	tii2K	DPI	ALL	66
19	-0.270	0.040	103	0	Lumry 2005	post2K	MDI	ALL	66
20	0.080	0.040	102	200	Lumry 2005	post2K	MDI	ALL	65
21	0.500	0.050	88	0	Murray 2004	post2K	DPI	NONE	63
22	0.110	0.060	83	0	Nathan 2000	tii2K	DPI	MIX	53
23	0.270	0.060	80	200	Nathan 2000	tii2K	MDI	MIX	68
24	-0.120	0.050	86	0	Nathan 2006	tii2K	MDI	ALL	68
25	0.190	0.040	88	200	Nathan 2006	tii2K	MDI	ALL	74
26	-0.080	0.090	35	0	Noonan 1998	tii2K	MDI	NONE	74
27	0.311	0.067	66	200	Noonan 1998	tii2K	MDI	NONE	66
28	-0.220	0.070	74	0	Pearlman 1997	tii2K	DPI	ALL	66
29	0.470	0.070	80	200	Pearlman 1997	tii2K	MDI	ALL	58
30	0.000	0.090	22	0	Pearlman 1999	post2K	MDI	NONE	37
31	0.270	0.071	22	200	Pearlman 1999	post2K	MDI	MIX	67
32	0.140	0.050	66	0	Pearlman 2004	post2K	MDI	NONE	67
33	0.360	0.050	88	200	Pearlman 2004	post2K	MDI	NONE	65
34	0.130	0.040	98	0	Pinnas 2005	post2K	MDI	ALL	68
35	0.340	0.040	98	200	Pinnas 2005	tii2K	MDI	ALL	68
36	0.310	0.050	98	200	Raphael 1999	tii2K	MDI	ALL	85
37	-0.120	0.050	87	0	Rooklin 2001	post2K	MDI	MIX	63
38	0.190	0.040	89	200	Rooklin 2001	post2K	DPI	NONE	53
39	-0.020	0.025	307	0	SAS30024	post2K	MDI	NONE	71
40	0.080	0.014	954	200	SAS30024	tii2K	DPI	ALL	65
41	0.140	0.060	72	0	Sheffer 1996	tii2K	MDI	NONE	65
42	0.420	0.070	78	200	Sheffer 1996	tii2K	MDI	NONE	71
43	0.240	0.030	81	0	Wasserman 1996	tii2K	DPI	NONE	65
44	0.540	0.041	77	200	Wasserman 1996	tii2K	DPI	ALL	65
45	-0.310	0.055	74	0	Wolfe 1996	tii2K	MDI	ALL	68
46	0.390	0.057	79	200	Wolfe 1996	tii2K	MDI	NONE	68
47	0.210	0.070	72	0	Wolfe 2000 BD	tii2K	DPI	ALL	67
48	0.490	0.050	72	200	Wolfe 2000 BD	tii2K	MDI	ALL	67
49	-0.080	0.060	68	0	Wolfe 2000 ICS	tii2K	DPI	ALL	67
50	0.270	0.059	65	200	Wolfe 2000 ICS	tii2K	DPI	ALL	67

Bayesian Model Specification

Data (single column variables, all same length)

Data Range: `ICS Meta!$A$1:$I$50`

Data is ... (•) in columns ( ) in rows   [X] Names in first row   [ ] Blank cells = NAs (Response only)

Set Variable Types

Model

Distribution: Normal   Link: Identity

[X] Response is mean; se [.df]

Response: CFB.mm; CFB.se; CFB.df

[ ] Censored   [ ] Non-Linear Model   [ ] Longitudinal Model

Factors

Fixed: FPmg*(ICS.USE + YR2K + FORM)

Random: STUDY + STUDY:FPmg

Covariates

Independent: BASE.PC + BASE.PC:FPmg

[ ] Random Coeffs   [ ] Eliciting Priors Only

Predictions for ... FPmg   STUDY   YR2K   FORM   ICS.USE

HelpR   MCMC & Output Options   Clear Form   Exit   Predictions or Contrasts [X]   OK

FIGURE 5.9
Data and model specification for FP meta-analysis.

$$se^2_{ijklm} \sim Ga\left(0.5\ df_{ijklm}, 0.5\ df_{ijklm}/\omega^2_{ijklm}\right)$$

$$s_m \sim N(0,\ \sigma^2_3)$$

$$(\alpha s)_{im} \sim N(0,\ \sigma^2_2)$$

where

y is the response variable CFB.mn

se is the estimated standard error CFB.se

df is the degrees of freedom CFB.df

x is the covariate BASE.PC with associated regression coefficients β and $(\alpha\beta)_i$, allowing different coefficients for the two levels of the factor FPmg

α_i are the effects of the factor FPmg

γ_j are the effects of the factor ICS.USE

δ_k are the effects of the factor YR2K

ζ_l are the effects of the factor FORM

s_m are the effects of the factor STUDY

$(\alpha\gamma)_{ij}$, $(\alpha\delta)_{ik}$, $(\alpha\zeta)_{il}$ and $(\alpha s)_{im}$ are the interaction effects between FPmg and the other factors

σ^2_2, σ^2_3 are the variance components

The simple model defaults for the MCMC settings were chosen (a burn-in of 5,000 followed by 10,000 samples). In order to help interpret the interaction effects, predicted means were requested for all combinations of the two-factor interactions involving treatment (see cells K2:O5 in Figure 5.9). The final row of cells K2:O6 defining the predictions required, specifies the predicted placebo mean response from a future study in which the DPI formulation is used in an ICS experienced patient population, assuming the mean percent predicted FEV_1 for this study will be the same as the mean for the studies included in this meta-analysis. The '~' symbol, used here in the STUDY column, tells BugsXLA that the prediction is for a level of the factor not studied in the current data set. This can only be used for random effects, and is typically used as described above to specify that we wish to make predictions for a future realization of a level of this factor. In this particular case, the resultant predictive distribution obtained is the appropriate prior distribution for the placebo mean response in a future study considered "exchangeable" with those included in the meta-analysis. This would be useful, for example, if we planned to study a different treatment for asthma in a placebo controlled study, enabling us to justify randomizing fewer subjects to the placebo arm due to the knowledge quantified in the prior derived here.

Since this is a more complicated model than the previous NLMM case studies, with an "unbalanced data set," the regular model defaults for the MCMC

	A	B	C	D	E	F	G	H	I	J	K
1			Label	Mean	St.Dev.	2.5%	Median	97.5%		WinBUGS Name	
2			CONSTANT	0.0559	0.0584	-0.0617	0.0560	0.1700		Beta0	
3			Intercept at 0	0.5482	0.3853	-0.2043	0.5438	1.3230		alpha	
4		FPmg	200	0.3855	0.0572	0.2651	0.3873	0.4935		X.Eff[1,2]	
5		ICS.USE	MIX	-0.1101	0.0790	-0.2664	-0.1104	0.0459		X.Eff[2,2]	
6		ICS.USE	ALL	-0.2982	0.0611	-0.4173	-0.2988	-0.1759		X.Eff[2,3]	
7		YR2K	post2K	0.1121	0.0615	-9.592E-3	0.1117	0.2325		X.Eff[3,2]	
8		FORM	DPI	0.0870	0.0614	-0.0332	0.0869	0.2068		X.Eff[4,2]	
9		FPmg x ICS.USE	200, MIX	-0.0419	0.0722	-0.1876	-0.0410	0.0997		X.Eff[5,4]	
10		FPmg x ICS.USE	200, ALL	0.1389	0.0580	0.0360	0.1349	0.2647		X.Eff[5,6]	
11		FPmg x YR2K	200, post2K	-0.1404	0.0585	-0.2534	-0.1423	-0.0182		X.Eff[6,4]	
12		FPmg x FORM	200, DPI	-0.0981	0.0585	-0.2103	-0.0992	0.0226		X.Eff[7,4]	
13			BASE.PC	-7.197E-3	5.923E-3	-0.0189	-7.120E-3	4.284E-3		V.Coeff[1,1]	
14		BASE.PC x FPmg	200	-2.163E-3	5.502E-3	-0.0127	-2.283E-3	9.115E-3		V.Coeff[2,2]	
15			SD(STUDY)	0.0902	0.0296	0.0266	0.0910	0.1483		sigma.Z[1]	
16			SD(STUDY x FPmg)	0.0613	0.0254	0.0116	0.0606	0.1125		sigma.Z[2]	
17	Note: CONSTANT & Factor effects are determined at the mean of the covariate(s).										
18	Interpret these cautiously when Factor x Covariate terms have been fitted.										
19											
20	**Model**	['ICS Meta'!$A$1:$I$50]									
21	Distribution	Normal									
22	Link	Identity									
23	Response	CFB.mn; CFB.se; CFB.df									
24	Fixed	FPmg*(ICS.USE + YR2K + FORM)									
25	Covariates	BASE.PC + BASE.PC:FPmg									
26	Random	STUDY + STUDY:FPmg									
27											
28	**Priors**										
29	CONSTANT	N(mu=0.176, sigma=23.6)									
30	FPmg	N(mu=0, sigma=23.6)									
31	ICS.USE	N(mu=0, sigma=23.6)									
32	YR2K	N(mu=0, sigma=23.6)									
33	FORM	N(mu=0, sigma=23.6)									
34	FPmg x ICS.USE	N(mu=0, sigma=23.6)									
35	FPmg x YR2K	N(mu=0, sigma=23.6)									
36	FPmg x FORM	N(mu=0, sigma=23.6)									
37	BASE.PC	N(mu=0, sigma=4.69)									
38	BASE.PC x FPmg	N(mu=0, sigma=4.69)									
39	STUDY	N(0,tau^2); tau ~ Half-N(sigma=1.18)									
40	STUDY x FPmg	N(0,tau^2); tau ~ Half-N(sigma=1.18)									
41	V(obs. mean)	Inv-Gamma(0.001, 0.001)									
42											
43	**WinBUGS MCMC Settings**										
44	Burn-In: 10000 Samples: 10000 (Thin:6; Chains:1)										
45	Run took 2.5 minutes										

FIGURE 5.10
Summaries of posterior distributions, model, priors, and MCMC settings for FP meta-analysis.

settings were chosen (a burn-in of 10,000 followed by 10,000 samples after one-fifth thinning). The default priors were unchanged (see cells A29:B41 in Figure 5.10), resulting in the output shown in Figures 5.10 and 5.11.

The last term in the list of summarized priors is labeled 'V(obs mean)', and is the only one without a summarized posterior. This term exists because degrees of freedom were specified for each of the standard errors, implying that the true variance of each observed mean response is unknown (ω_{ijklm}^2 in the algebraic representation of the model given earlier). BugsXLA models this by including these variance parameters in the model, one for each observed mean, and giving them independent inverse-gamma priors. The posteriors for these parameters are not summarized as they are not considered to be of interest. One would need to work with the WinBUGS code directly if these were of interest (see Chapter 11 and Appendix C).

Inspection of the posterior summaries shows that most of the terms in the model are having some effect on the FEV_1 response. Of particular interest are the two relatively large interactions: treatment by prior ICS usage, and treatment by year of study. These can be better examined by looking at the predicted means.

N	O	P	Q	R	S	T	U	V	W
Label	Mean	St.Dev.	2.5%	Median	97.5%		WinBUGS Name		
Predicted Mean Response									
Following all at: ICS.USE(ALL) YR2K(post2K) STUDY(dist. mean) BASE.PC(mean)									
FPmg(0) FORM(MDI)	-0.130	0.055	-0.240	-0.130	-0.021		Pred.Ave[1]		
FPmg(0) FORM(DPI)	-0.043	0.071	-0.181	-0.043	0.096		Pred.Ave[2]		
FPmg(200) FORM(MDI)	0.254	0.047	0.163	0.254	0.347		Pred.Ave[3]		
FPmg(200) FORM(DPI)	0.243	0.063	0.122	0.242	0.366		Pred.Ave[4]		
Following all at: YR2K(post2K) FORM(DPI) STUDY(dist. mean) BASE.PC(mean)									
FPmg(0) ICS.USE(NONE)	0.255	0.073	0.113	0.254	0.397		Pred.Ave[5]		
FPmg(0) ICS.USE(MIX)	0.145	0.078	-0.010	0.145	0.301		Pred.Ave[6]		
FPmg(0) ICS.USE(ALL)	-0.043	0.071	-0.181	-0.043	0.096		Pred.Ave[7]		
FPmg(200) ICS.USE(NONE)	0.402	0.063	0.277	0.402	0.529		Pred.Ave[8]		
FPmg(200) ICS.USE(MIX)	0.250	0.072	0.106	0.250	0.390		Pred.Ave[9]		
FPmg(200) ICS.USE(ALL)	0.243	0.063	0.122	0.242	0.366		Pred.Ave[10]		
Following all at: ICS.USE(ALL) FORM(DPI) STUDY(dist. mean) BASE.PC(mean)									
FPmg(0) YR2K(til2K)	-0.155	0.057	-0.266	-0.156	-0.042		Pred.Ave[11]		
FPmg(0) YR2K(post2K)	-0.043	0.071	-0.181	-0.043	0.096		Pred.Ave[12]		
FPmg(200) YR2K(til2K)	0.271	0.056	0.160	0.271	0.382		Pred.Ave[13]		
FPmg(200) YR2K(post2K)	0.243	0.063	0.122	0.242	0.366		Pred.Ave[14]		
FPmg(0) ICS.USE(ALL) YR2K(post2K) FORM(DPI) STUDY(new) BASE.PC(mean)									
	-0.042	0.135	-0.313	-0.041	0.224		Pred.Ave[15]		

FIGURE 5.11
Predictions from FP meta-analysis.

The second set of predicted means in Figure 5.11 clearly shows the nature of the treatment by prior ICS usage interaction. ICS naïve patients (NONE) gave a much larger placebo response than ICS experienced patients (ALL). Although these naïve patients also gave a high response on treatment, the effect of treatment, defined as the difference from placebo, is much smaller for these patients. This is important information for the design of future studies, because if this interaction effect holds true for future treatments of asthma, it implies we would need much larger studies if we recruited naïve rather than experienced users of similar treatments. The third set of predicted means show that although there has been little change in the mean response to treatment over time, the mean placebo response has increased in recent years. This increase in placebo response is a common phenomenon in studies of treatments that have been on the market for a long period of time. The final row of the predictions is the predictive distribution for the placebo mean response in a future study with other settings as specified in the header for this prediction. As previously discussed, this could be used as the prior for the placebo mean response in a future study. For convenience of use, this prior distribution could be adequately approximated by a Normal with mean −0.04 L and standard deviation 0.135 L. The relatively large standard deviation for this prior is due to the variation between studies in the placebo response, implying that our knowledge of the placebo response for a future study is fairly imprecise. As a result, this informative prior would not allow us to greatly reduce the number of subjects required on the placebo arm. If it were possible to identify other features of these studies that could explain the variation between studies, it might be possible to obtain a more precise prior distribution for a future study. An excellent discussion of the broader issues that need to be considered when undertaking a meta-analysis is provided by Higgins et al. (2009).

Case Study 7.3 provides another example where the approach discussed here is used to analyze data in the form of summary statistics. The objective in this case is to obtain a pooled estimate of some quantity that has been repeatedly measured in numerous experiments.

Case Study 5.4 Random Coefficients Modeling of Rat Growth Curves
The WinBUGS 'Birats' example is taken from Section 6 of Gelfand et al. (1990), and involves modeling the growth curves of 30 young rats whose weights were measured weekly for 5 weeks. A simple linear relationship between weight and time over the 5 weeks was initially considered adequate. Although each rat will have a different growth rate, it was believed that the rates should be sufficiently similar to warrant modeling them as "random effects." Also, it was believed that the weights of the rats at a common time point should be similar, and that the model should be formulated to reflect this belief. Figure 5.12 shows some of the data and the BugsXLA model specification form.

Rat is defined as a factor with 30 levels, and Day is a variate that contains the day at which each weight was taken. In the classical terminology the model specified is often referred to as a random coefficients model, which can be represented algebraically:

$$y_{ij} = \mu + \alpha_i + \beta_i x_{ij} + e_{ij} \quad i = 1, 2, \ldots, 30 \quad j = 1, 2, \ldots, 5$$

$$(\alpha_i, \beta_i) \sim \text{MVN}\big((0, \beta_\mu), s\big)$$

$$e_{ij} \sim N(0, \sigma^2)$$

where
  y is the response variable Weight
  x is the covariate Day with associated regression coefficients β_i
  α_i are the effects of the factor Rat on the mean weight

The belief that rats will have both similar growth rates and similar weights at each time point is formally modeled by the multivariate Normal Distribution with mean vector $(0, \beta_\mu)$ and variance–covariance matrix Σ; the prior mean of the α_i is fixed at zero as these effects are defined relative to the model constant μ. It is important to be aware of the specific type of "similarity" that this represents, and believe that this assumption is appropriate. It is recommended that a good text on longitudinal models be consulted for advice on how to use the data to help assess the appropriateness of such an assumption, for example, Fitzmaurice et al. (2004) or Diggle et al. (2002). Here we take as given that this assumption is adequate and focus on other aspects of the analysis.

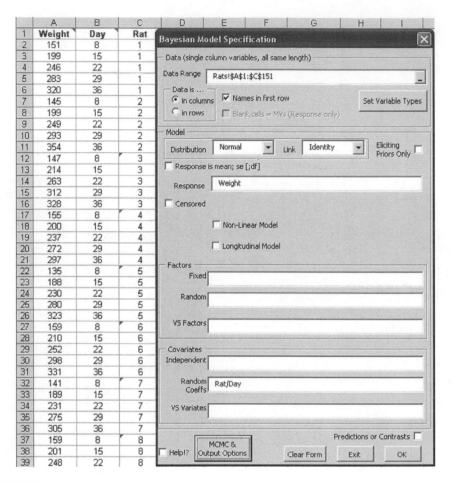

FIGURE 5.12
Data and model specification for rat growth curve study.

The simple model defaults for the MCMC settings were chosen (a burn-in of 5,000 followed by 10,000 samples). Figure 5.13 shows the BugsXLA form where the prior for the random coefficients can be altered, referred to as "Exchangeable Covariates" on the form. The mean of the regression coefficients, or 'slopes', β_μ (labeled 'mu' on the form shown in Figure 5.13), is given a Normal prior distribution with mean and standard deviation parameters that can be set by the user. The default provides a very flat prior relative to the information in the data. The prior for the variance–covariance matrix, Σ (labeled 'VCov' on the form shown in Figure 5.13), is specified via separate priors for the two standard deviations and the correlation parameter. The default prior for each standard deviation parameter is a much dispersed Half-Normal Distribution. BugsXLA offers the Half-Cauchy and the Uniform Distributions as alternatives. The same elicitation process exists

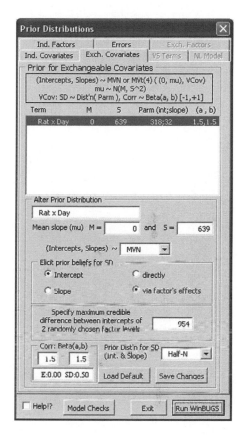

FIGURE 5.13
Prior for random coefficients in rat growth curve study.

for these parameters as exists on the random factors page (see Case Study 5.1). The correlation parameter is given a Beta prior distribution scaled on the range [−1, +1]. A uniform prior, Beta(1,1), is not taken as the default to avoid giving too much prior weight to the extremes. However, the default is likely to be very flat relative to the likelihood for most cases. It is important to note that BugsXLA automatically centers covariates, and the 'intercepts' are therefore the mean weight of each rat at a time point equal to the mean of the Day variate. Centering the covariates affects the interpretation of the correlation parameter as well as the intercept parameters. The default priors were unchanged (see worksheet cells A19:B24 in Figure 5.14), and the model checking functions were requested as explained in Section 2.9. Figure 5.14 shows the results of the analysis as imported back into Excel.

The summarized posterior distributions in Figure 5.14 are very similar to those reported in Spiegelhalter et al. (1996), despite the different prior distributions used in each analysis. This is because both sets of priors are very weak relative to the information in the likelihood. Inspection of the BugsXLA produced plots of the model checking functions cast doubt on the assumption of a simple linear relationship between weight and time.

	A	B	C	D	E	F	G	H	I	J	K
1			Label	Mean	St.Dev.	2.5%	Median	97.5%		WinBUGS Name	
2			CONSTANT	243.2000	2.7140	237.9000	243.3000	248.6000		Beta0	
3			Intercept at 0	106.8000	2.5700	101.7000	106.8000	112.0000		alpha	
4		(Slope mean)	Rat x Day	6.2010	0.1079	5.9870	6.2010	6.4140		W.MuCoeff[1]	
5		(Corr)	Rat x Day	0.6025	0.1484	0.2691	0.6206	0.8429		Crnvn.Beta[1]	
6		(Intercept)	SD(Rat x Day)	14.8300	2.0340	11.4600	14.6100	19.2800		sigma.WZ[1]	
7		(Slope)	SD(Rat x Day)	0.5274	0.0935	0.3772	0.5167	0.7424		sigma.Wcoeff[1]	
8			SD(residual)	6.0840	0.4541	5.2770	6.0550	7.0480		sigma	
9	Note: CONSTANT & Factor effects are determined at the mean of the covariate(s).										
10	Var-Cov(slope-intercept) terms are also for centred data.										
11											
12	Model	[Rats!$A$1:$C$151]									
13	Distribution	Normal									
14	Link	Identity									
15	Response	Weight									
16	Random Coeffs	Rat/Day									
17											
18	Priors										
19	CONSTANT	N(mu=243, sigma=6000)									
20	Rat x Day	MVN((0,mu), V); mu ~ N(mean=0, sigma=639)									
21		V: sd(int) ~ Half-N(sigma=318)									
22		V: sd(slope) ~ Half-N(sigma=32)									
23		V: Correlation ~ Beta(1.5,1.5) [-1,+1]									
24	V(residual)	Inv-Gamma(0.001, 0.001)									
25											
26	WinBUGS MCMC Settings										
27	Burn-In: 5000 Samples: 10000 (Thin:1; Chains:1)										
28	Run took 90 seconds										

FIGURE 5.14
Summaries of posterior distributions, model, priors, and MCMC settings for rat growth curve study.

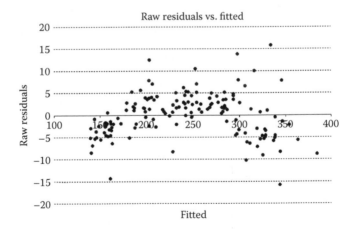

FIGURE 5.15
Plot of raw residuals versus fitted values for rat growth curve study.

Figure 5.15 shows the plot of the raw residuals against the fitted values. As this is a Bayesian analysis, it is the posterior means of the raw residuals and the posterior means of the fitted values that are actually plotted, as both of these statistics are functions of the parameters in the model, and hence, are themselves estimated quantities.

There is clear evidence of curvature in these residuals, suggesting that some form of curvilinear relationship between weight and time should be modeled. An obvious extension of the current model would be to consider a second order polynomial relationship. BugsXLA v5.0 is only able to model a factor jointly with one variate as random coefficients, and so it is not possible to extend this part of the model to include both a linear and quadratic Day term. Although not completely satisfactory, if one wanted to use BugsXLA to fit additional terms that modeled this curvature, one could enter either 'Day^2' or 'Rat:Day^2' in the independent covariates text box on the Bayesian model specification form (see Figure 5.12). Entering 'Day^2' forces the quadratic parameter to be identical for all rats. While entering 'Rat:Day^2' fits different quadratic parameters for each rat, but under the assumption of complete independence, that is, no correlation between the quadratic parameters, or between these parameters and the linear or intercept parameters specified as random coefficients. These data are reanalyzed in Case Study 9.2, where models involving quadratic terms are compared with the one fitted here, as well as with models that accommodate serial correlation between the residuals.

6

Generalized Linear Mixed Models

Generalized linear mixed models (GLMMs) extend normal linear mixed models (NLMMs) in the same way that generalized linear models (GLMs) extend normal linear models (NLMs), that is, they permit a wide range of non-Normal responses to be modeled using random effects as flexibly as in the Normal case. GLMMs still consist of three components as discussed in Chapter 4 (linear predictor, link function, and error distribution), but the linear predictor contains random effect terms as discussed in Chapter 5. Most modern books explaining mixed models discuss both NLMMs and GLMMs. So the same books referenced at the beginning of Chapter 5 provide useful coverage of the models discussed in this chapter: Brown and Prescott (2006) or McCulloch and Searle (2001) from a classical perspective, and Gelman et al. (2004) or Ntzoufras (2009) from a Bayesian one. In the Bayesian literature, both NLMMs and GLMMs are often referred to as hierarchical models due to the hierarchy in the model parameters that the random effects assumption imposes. The term "hierarchical linear model" is usually reserved for the NLMM because of the non-linearity caused by the link functions typically used in GLMMs as discussed in Chapter 4. When random effects are included in the model, it is important to be aware of how a non-linear link function affects how the parameters can be interpreted. Care is needed because, unlike in linear models, averaging over the random effects on the link scale and back-transforming is not equivalent to averaging on the original scale. So, for example, when random effects are used to model differences between subjects in a clinical trial, the parameters representing the effects of treatments are subject specific, that is, they represent the expected difference that would occur for an individual who was given one treatment instead of another. These should not be interpreted as population effects. The differences between these two types of inferences are often covered in more detail in books discussing longitudinal data, where marginal models are typically recommended when population effects are of interest; see Fitzmaurice et al. (2004) or Diggle et al. (2002), for example. Marginal models cannot be fitted using BugsXLA.

It is in no small part due to the ease with which relatively complex models, such as the more complex GLMMs, can be fitted using MCMC methods that the popularity of the Bayesian approach has grown over the last 20 years. Classical approaches rely on a variety of different approximate methods to handle the much more difficult numerical issues that arise when deriving maximum likelihood estimates in GLMMs. In many cases, consensus has not yet been reached on the best estimation methods to use, and this is still an active area of research. It would not be unusual for different software packages to give quite different estimates for the more complex GLMMs. However, it should not be assumed that software implementing Bayesian methods are guaranteed to always give reliable estimates in these more complex models. When building complex models, it is strongly recommended that they are built slowly, starting with smaller models that are expected to fit well. As more complexity is added, MCMC convergence should be carefully checked as discussed in Section 1.2. Whenever informative priors can be justified, even if they are only strong enough to rule out totally incredible parameter values, their use should be explored. The prior distribution, often spoken of as a weakness of the Bayesian approach, can sometimes provide the additional information needed to obtain a sensible model fit to the data. It is in the more complex models where the advantages of the Bayesian approach goes beyond legitimizing the natural interpretation of the interval estimates and allowing prior knowledge to be formally incorporated, to the more fundamental matter of being able to obtain a good model fit at all.

BugsXLA follows the same general approach to defining priors as discussed in Chapter 5 (refer to Section 2.7 for more details).

Case Study 6.1 Respiratory Tract Infections
Case Study 4.1 discussed a fixed effects meta-analysis of a binomial response obtained from a series of clinical studies. The response from each study was the number of respiratory tract infections and total number of patients in each treatment group. Here we reanalyze these data using a random effects model, comparing the output and discussing how one might choose between the two types of model. Figure 6.1 shows the data and completed BugsXLA model specification form.

The variables are the same as before. The model differs from before in a couple of important ways. The only factor that is fitted as a fixed effect is trt, since there is no reason to believe that the infection proportions should be similar in those groups receiving the active treatment and those receiving the control. The factor study is fitted as a random effect since it is expected that the infection proportions will be similar in the different studies for groups receiving the same treatment. One way of imagining the nature of their similarity is to think of the study effects as if they had been independently drawn at random from a common Normal Distribution. The standard deviation parameter associated with this distribution determines the degree of similarity between the effects. It will be shown later how information external to the study regarding the effects' similarity can be usefully incorporated into the Bayesian analysis. As well as the main effects of both treatment and study, this model also includes a study-by-treatment interaction term as a random effect. It is this term that differentiates the random effects meta-analysis from the fixed effects approach. By including this interaction term, the model is representing the belief that the treatment effect differs between

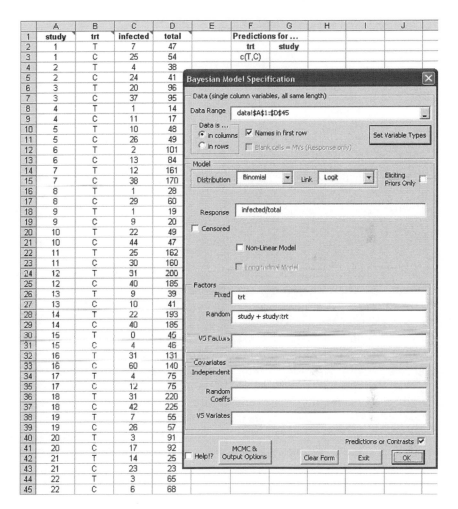

	A	B	C	D	E	F	G	H	I	J
1	study	trt	infected	total		Predictions for ...				
2	1	T	7	47		trt	study			
3	1	C	25	54		c(T,C)				
4	2	T	4	38						
5	2	C	24	41						
6	3	T	20	96						
7	3	C	37	95						
8	4	T	1	14						
9	4	C	11	17						
10	5	T	10	48						
11	5	C	26	49						
12	6	T	2	101						
13	6	C	13	84						
14	7	T	12	161						
15	7	C	38	170						
16	8	T	1	28						
17	8	C	29	60						
18	9	T	1	19						
19	9	C	9	20						
20	10	T	22	49						
21	10	C	44	47						
22	11	T	25	162						
23	11	C	30	160						
24	12	T	31	200						
25	12	C	40	185						
26	13	T	9	39						
27	13	C	10	41						
28	14	T	22	193						
29	14	C	40	185						
30	15	T	0	45						
31	15	C	4	46						
32	16	T	31	131						
33	16	C	60	140						
34	17	T	4	75						
35	17	C	12	75						
36	18	T	31	220						
37	18	C	42	225						
38	19	T	7	55						
39	19	C	26	57						
40	20	T	3	91						
41	20	C	17	92						
42	21	T	14	25						
43	21	C	23	23						
44	22	T	3	65						
45	22	C	6	68						

Bayesian Model Specification ☒

Data (single column variables, all same length)

Data Range: data!$A$1:$D$45

Data is ...
⦿ in columns ☑ Names in first row
◯ in rows ☐ Blank cells = MVs (Response only)

Set Variable Types

Model

Distribution: Binomial ▾ Link: Logit ▾ ☐ Eliciting Priors Only

Response: infected/total

☐ Censored

☐ Non-Linear Model

☐ Longitudinal Model

Factors
Fixed: trt
Random: study + study:trt
VS Factors:

Covariates
Independent:
Random Coeffs:
VS Variates:

Predictions or Contrasts ☑

☐ Help!? MCMC & Output Options Clear Form Exit OK

FIGURE 6.1
Data and model specification for respiratory tract infections random effects meta-analysis.

studies, but that these effects are similar. The main effect of treatment, fitted as a fixed effect, now represents the "population mean" treatment effect. It can be thought of as the treatment effect one would expect if an infinitely large number of studies could be run. The model specified is a mixed effects analysis of variance, and can be represented algebraically:

$$\eta_{ij} = \mu + s_i + \gamma_j + (s\gamma)_{ij} \quad i = 1, 2, \ldots, 22 \quad j = 1, 2$$

$$\text{logit}(\pi_{ij}) = \eta_{ij}$$

$$r_{ij} \sim \text{Bin}(\pi_{ij}, m_{ij})$$

$$s_i \sim N(0, \sigma_1^2)$$

$$(s\gamma)_{ij} \sim N(0, \sigma_2^2)$$

where

r and m are the response variables infected and total, respectively

s_i are the random effects of the factor study

γ_j are the fixed effects of the factor trt

$(s\gamma)_{ij}$ are the random effects of the study-by-treatment interaction

σ_1^2 and σ_2^2 are the variance components associated with the two random effects

If this was the first time one had run a model and data set with this level of complexity using BugsXLA, it would be very important to assess convergence. With experience it becomes possible to identify when it is likely to be safe to skip this activity, and then how to look for signs in the output that point to the need to go back and check convergence. In order to facilitate convergence checking, the novel model defaults for the MCMC settings were chosen (a burn-in of 5,000 followed by 1,000 samples on each of 3 chains, with the auto quit set to 'No'). The number of samples following burn-in was increased to 5000 and the 'history' and 'bgr diag' plots produced from within WinBUGS as explained in Section 1.2. All of the plots indicated that convergence had occurred, as illustrated by Figures 6.2 and 6.3, which show the history and Brooks–Gelman–Rubin ('bgr') plots for the model constant parameter.

Since convergence was rapid, the analysis was rerun using the simple model MCMC settings (a burn-in of 5,000 followed by 10,000 samples) with

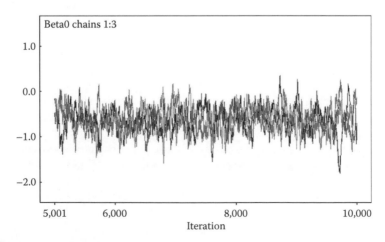

FIGURE 6.2

WinBUGS history plot for the model constant parameter in the respiratory tract infections random effects meta-analysis.

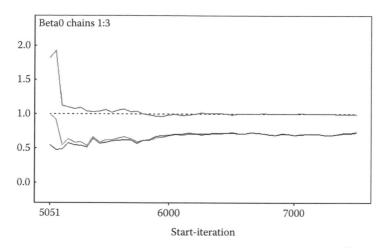

FIGURE 6.3
WinBUGS Brooks–Gelman–Rubin plot for the model constant parameter in the respiratory tract infections random effects meta-analysis.

the default priors (see cells A15:B18 in Figure 6.4). The contrast between the treated and control groups' means was requested (see cells F2:F3 in Figure 6.1). The model checking functions and DIC statistic were also requested (see Section 2.9). Figure 6.4 shows the parameters in the model and a summary of their marginal posterior distributions as imported back into Excel.

Comparison of the treatment effect estimate obtained here with that obtained when the effect was assumed to be constant for all studies (see Figure 4.2)

	A	B	C	D	E	F	G	H	I	J	K
1			Label	Mean	St.Dev.	2.5%	Median	97.5%		WinBUGS Name	
2			CONSTANT	-0.6269	0.2920	-1.2290	-0.6273	-0.0507		Beta0	
3		trt	T	-1.3970	0.2281	-1.8770	-1.3850	-0.9773		X.Eff[1,2]	
4			SD(study)	1.0660	0.2334	0.6724	1.0410	1.5860		sigma.Z[1]	
5			SD(study x trt)	0.5987	0.1556	0.3445	0.5798	0.9589		sigma.Z[2]	
6											
7	**Model**	[data!$A$1:$D$45]									
8	Distribution	Binomial									
9	Link	Logit									
10	Response	infected/total									
11	Fixed	trt									
12	Random	study + study:trt									
13											
14	**Priors**										
15	CONSTANT	N(mu=0, sigma=100)									
16	trt	N(mu=0, sigma=100)									
17	study	Norm(0,tau^2); tau ~ Half-N(sigma=5)									
18	study x trt	Norm(0,tau^2); tau ~ Half-N(sigma=5)									
19											
20	**WinBUGS MCMC Settings**										
21	Burn-In: 5000 Samples: 10000 (Thin:1; Chains:1)										
22	Run took 19 seconds										

FIGURE 6.4
Summaries of posterior distributions, model, priors, and MCMC settings for respiratory tract infections random effects meta-analysis.

TABLE 6.1

Summary of Treatment Odds-Ratio (T/C) Estimates
for the Respiratory Tract Infections Meta-Analyses

Model	Mean	St. Dev	2.5%ile	97.5%ile
Fixed effects	0.34	0.031	0.29	0.41
Random effects	0.25	0.057	0.15	0.38

shows that the effect is expected to be more extreme in this random effects analysis, although it is estimated with much less precision. It is easier to interpret the estimates when back-transformed to the odds scale (Table 6.1).

This random effects estimate is very similar to that given in Smith et al. (1995). One way to compare the goodness of fit of the two models is to use the deviance information criteria (DIC) as discussed in Case Study 4.2; Table 6.2 shows the DIC and pD values. The difference of 40 between the DIC statistics for the two models clearly indicates that the random effects model should be preferred, even though the pD values indicate that this model effectively includes an additional 14 parameters. The pD value for the fixed effects analysis almost exactly matches the actual number of parameters in this model, and probably only differs due to MCMC error. The model checking statistics and plots (not shown) do not suggest any major issues with the random effects model. Whereas with the fixed effects model when a relatively large number of moderately extreme observations were observed, here the most extreme scaled residual is about 1.5. Note that the residuals provided by BugsXLA are conditional on the random effects and should not be used as the only way to assess goodness of fit.

This case study will now be used to illustrate how alternative assumptions regarding the nature of the random effects can be assessed using BugsXLA. In Chapter 5 of Spiegelhalter et al. (2004), they discuss how one might construct an informative prior for a random effect parameter representing a log(odds-ratio). In our model the treatment-by-study interaction term, which has the prior

$$(s\gamma)_{ij} \sim N(0, \ \sigma_2^2)$$

TABLE 6.2

DIC Statistics for the Respiratory
Tract Infections Meta-Analyses

	Fixed Effects	Random Effects
DIC	294	254
pD	22.9	36.6

measures the degree with which the treatment effect in each study deviates from the population mean effect. Whereas the population mean log(odds-ratio) effect is

$$\theta_0 = \gamma_2 - \gamma_1$$

the log(odds-ratio) effect for the kth study is

$$\theta_k = \gamma_2 - \gamma_1 + (s\gamma)_{k2} - (s\gamma)_{k1}$$

which has the inferred prior distribution, conditional on θ_0 and σ_2, of

$$\theta_k \sim N(\theta_0, 2\sigma_2^2).$$

Following the general approach outlined by Spiegelhalter et al. (2004) we note that, if we take plus or minus two standard deviations as being at the extremes of the Normal Distribution, then this prior infers that it is very unlikely that two studies will differ in their θ_k values by more than $4\sqrt{2}\sigma_2 = 5.7\sigma_2$. Since this value relates to a difference in log(odds-ratio)s, it follows that the ratio of two studies' odds-ratios is very unlikely to be greater than $\exp(5.7\sigma_2)$. In the context of meta-analyses, Smith et al. (1995) suggest that odds-ratios between studies are unlikely to vary by more than an order of magnitude. This implies that a high value for σ_2 could be derived by equating $\exp(5.7\sigma_2) = 10$, giving $\sigma_2 = 0.4$. Figure 6.5 shows how this belief can be used to derive an informative prior distribution in BugsXLA. On the 'Prior Distributions' form, the hierarchical standard deviation parameter is labeled 'tau', which in the case of the 'study × trt' term represents the parameter σ_2 here. BugsXLA gives the option to elicit prior beliefs for tau directly, which has been done here by entering the value 0.4 as its maximum credible value. BugsXLA then determines the value of the parameter for the Half-Normal prior distribution for tau, 0.2, which corresponds with this belief. Section 2.7 provides more details of how this parameter is derived from the elicited value; this information is also summarized in the help obtained by selecting the 'Help!?' check box on the form.

Eliciting an informative prior for the study random effect is more difficult since the parameterization implies that the between studies log(odds-ratio) effect is

$$s_i - s_{i'} + (s\gamma)_{ij} - (s\gamma)_{i'j}$$

which has an inferred Normal prior distribution with mean zero and variance $2(\sigma_1^2 + \sigma_2^2)$. However, it is probably reasonable to assume that the main effect of study will dominate the interaction effect, and making this assumption

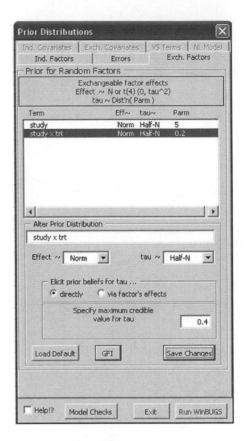

FIGURE 6.5
Informative prior on hierarchical standard
deviation parameter directly for the respiratory tract infections meta-analysis.

will lead to a more conservative dispersed prior distribution for σ_1. Again
following the suggestion given in Smith et al. (1995), a conservative assumption is that the overall response rate is very likely to lie within a range of
0.02–0.98, which implies the log(odds-ratio) between two studies should lie
between –7.8 and 7.8. This implies that, if we again take two standard deviations as being at the extreme, a high value for σ_1 could be derived by equating $2\sqrt{2}\sigma_1 = 7.8$, giving $\sigma_1 = 2.8$. This can be set using BugsXLA as was done for
the interaction term, giving an implied Half-Normal parameter of 1.4. Smith
et al. (1995) also suggest that a weakly informative prior for the population
log-odds ratio would be a Normal Distribution with mean zero and variance
of 10. Figure 6.6 shows how to specify this using BugsXLA (standard deviation for trt factor of $\sqrt{10} = 3.2$), as well as specifying an extremely conservative
prior for the model constant, representing the population mean placebo log
odds response.

The parameter estimates obtained using these informative priors, shown
in Table 6.3, are mostly quite similar to those obtained using the defaults
generated by BugsXLA. The largest difference is between the estimates of
the study-by-treatment random effects, which have been shrunk by the

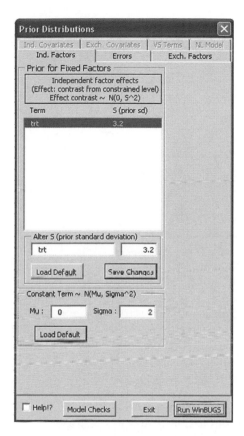

FIGURE 6.6

Informative priors for independent fixed effect parameters in the respiratory tract infections meta-analysis.

TABLE 6.3

Comparison of Parameter Estimates Obtained Using Different Priors for the Respiratory Tract Infections Meta-Analyses

Parameter	Default Vague Priors Post. Mean (St. Dev.)	Weakly Informative Priors Post. Mean (St. Dev.)
Constant	−0.63 (0.29)	−0.63 (0.23)
trt T	−1.40 (0.23)	−1.32 (0.18)
SD(study)	1.07 (0.23)	1.05 (0.20)
SD(study × trt)	0.60 (0.16)	0.42 (0.09)
Odds-ratio[T/C]	0.25 (0.06)	0.27 (0.05)

informative prior as indicated by the much smaller standard deviation parameter for this term. This reduction in the estimated between study variation in the treatment effect is also the major cause of the improved precision in the population mean treatment effect. This illustrates that even weakly informative prior distributions can sometimes provide substantial improvement in the precision with which quantities of interest are estimated. It is interesting

TABLE 6.4

Contrasts Needed to Create a Forest
Plot for a Meta-Analysis

Row	F	G
1	Predictions for ...	
2	trt	study
3	c(T,C)	*
4	c(T,C)	

to note that the DIC statistic increased to 257, which is only slightly larger than that obtained when the default priors were used.

A common way to visualize the results of a meta-analysis is to create a Forest Plot. BugsXLA can be used to calculate the values to plot, and then it is a simple matter to graph these using Excel's standard charting options. In order to obtain the separate treatment effects for each study, as well as the population mean effect, back-transformed onto the odds-ratio scale, the contrasts requested in cells F2:F3 in Figure 6.1 need to be changed to those shown in Table 6.4. The range F2:G4 would then be specified on the 'Predictions and Contrasts' form (see Section 2.6). Figure 6.7 shows the estimated contrasts after they have been imported back into Excel, and slightly edited in preparation for graphing. Also, two additional columns of data have been added, U and V, which are the differences between columns O and Q (mean and 2.5%ile) and columns S and O (97.5%ile and mean) respectively. If preferred, the median could be plotted instead of the mean as the best point estimate, in which case column R would be used instead of column O in this plotting procedure.

The first step is to select the range of cells N5:O29 (from cell containing "study(1)" to cell containing "0.2703"), which includes the labels and the posterior means to be plotted. Note this includes a couple of empty rows that will separate the individual study estimates from the pooled estimate on the plot. From the Excel charting options, select the line chart with markers, removing the default legend and formatting the plot to your preferred style. The data series should be formatted to remove the line joining the points, and adding custom 'Y Error Bars' specifying the range V5:V29 for the positive errors and U5:U29 for the negative errors. It is recommended that the scale for the Y-axis be changed to logarithmic and the crossing point for the X-axis be moved so that the labels are at the bottom of the plot. Figure 6.8 shows the Forest Plot produced. As with all good graphics, the Forest Plot makes interpretation of the results much easier. If one wished to compare the pooled estimates obtained using different models or priors, it should be clear how the procedure described could be modified to include the additional interval estimates on the graph.

Baker et al. (2009) used BugsXLA to facilitate a Bayesian mixed treatment comparison meta-analysis. They analyzed over 30,000 patients from 43 trials of various treatments for chronic obstructive pulmonary disease. Their

N	O	P	Q	R	S	T	U	V
Label	Mean	St.Dev.	2.5%	Median	97.5%			
	Predicted Probability							
	Predicted Odds							
	Following all at: contr[trt T / C]							
study(1)	0.2378	0.0939	0.1024	0.2231	0.4674		0.1354	0.2296
study(2)	0.1565	0.0685	0.0577	0.1456	0.3238		0.0988	0.1673
study(3)	0.3798	0.1115	0.2043	0.3676	0.6431		0.1755	0.2633
study(4)	0.1767	0.0951	0.0485	0.1577	0.4038		0.1282	0.2271
study(5)	0.2553	0.0973	0.1122	0.2396	0.4926		0.1431	0.2373
study(6)	0.2144	0.0963	0.0752	0.1995	0.4450		0.1392	0.2306
study(7)	0.2892	0.0895	0.1505	0.2758	0.4996		0.1387	0.2104
study(8)	0.1497	0.0728	0.0462	0.1371	0.3264		0.1035	0.1767
study(9)	0.2043	0.1083	0.0596	0.1833	0.4814		0.1447	0.2771
study(10)	0.1401	0.0611	0.0521	0.1293	0.2879		0.0880	0.1478
study(11)	0.6530	0.1897	0.3648	0.6261	1.0990		0.2882	0.4460
study(12)	0.5816	0.1473	0.3465	0.5641	0.9156		0.2351	0.3340
study(13)	0.5711	0.2563	0.2338	0.5185	1.2280		0.3373	0.6669
study(14)	0.4274	0.1142	0.2440	0.4125	0.6912		0.1834	0.2638
study(15)	0.2681	0.1535	0.0730	0.2354	0.6540		0.1951	0.3859
study(16)	0.3918	0.0988	0.2318	0.3804	0.6159		0.1600	0.2241
study(17)	0.3119	0.1397	0.1215	0.2872	0.6535		0.1904	0.3416
study(18)	0.6208	0.1529	0.3705	0.6038	0.9703		0.2603	0.3495
study(19)	0.2167	0.0844	0.0921	0.2038	0.4148		0.1246	0.1981
study(20)	0.2265	0.0974	0.0835	0.2112	0.4560		0.1430	0.2295
study(21)	0.1804	0.0932	0.0546	0.1624	0.4095		0.1258	0.2291
study(22)	0.3959	0.2029	0.1368	0.3515	0.8911		0.2591	0.4952
pooled	0.2703	0.0478	0.1880	0.2671	0.3732		0.0823	0.1029
	contr[trt T / C] study(dist. mean)							

FIGURE 6.7
Estimated contrasts for the respiratory tract infections meta-analysis, edited in preparation for the Forest Plot.

primary response variables were exacerbation and mortality rates, and they used a GLMM similar to that discussed above to model their data. The paper includes a comparison with a more traditional meta-analysis model, and they conclude that the Bayesian mixed treatment comparison approach increased the precision enabling them to draw firmer conclusions from their study.

Case Study 6.2 Clinical Trial of a Treatment to Prevent Epileptic Seizures
The WinBUGS examples include a data set initially provided by Thall and Vail (1990) concerning seizure counts in a study designed to assess the efficacy of an anti-convulsant therapy in epilepsy. In this study, 59 patients were randomized to either the active treatment or a control, and the number of seizures that occurred in each of four time periods was reported by the patient

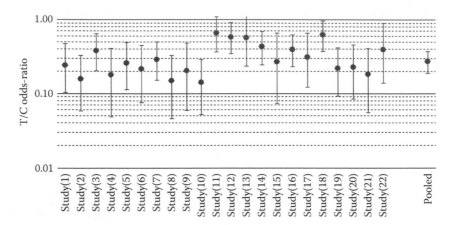

FIGURE 6.8
Forest Plot for the respiratory tract infections random effects meta-analysis using weakly informative priors.

at each return visit to the clinic. Both the patient's age and disease severity, as indicated by the number of seizures that occurred during the prerandomization baseline period, were considered to be potentially important covariates. It was considered more appropriate to transform these covariates to log of age and log of one-quarter of the baseline count, respectively. An interaction term between treatment and this transformed baseline count covariate was included in the model. There was also some concern that the fourth reporting period may have a different seizure rate to the first three periods, and so an indicator variable was added to the model to accommodate this effect if it occurred. Although a Poisson error distribution for the seizure data is a reasonable first choice, a simple fixed effects model for data such as these nearly always exhibit over dispersion. Such over dispersion can be explained by additional variation in the seizure rates both between patients and also between reporting periods within the same patient. One way to accommodate these additional sources of variation is to include them as random effects in the model. Figure 6.9 shows some of the data and the completed BugsXLA model specification form.

The log-transformed variables were calculated in Excel using the built-in functions provided, for example, cell H2 contains the formula '=LN(D2/4)' which was then simply copied down the column creating the new variable, noting that by default Excel uses relative referencing when copying formulae. The variables patient, trt, visit, and V4 were all defined as factors. Both trt and V4 have two levels labeled '0' and '1', with the latter level representing the active treatment and the fourth visit respectively. The two random effects were specified by entering visit as nested within patient, which adds both a main effect of patient as well as an interaction term that will model any additional variability in the seizure rate, on top of the natural Poisson variation, that occurs between patients and between visits

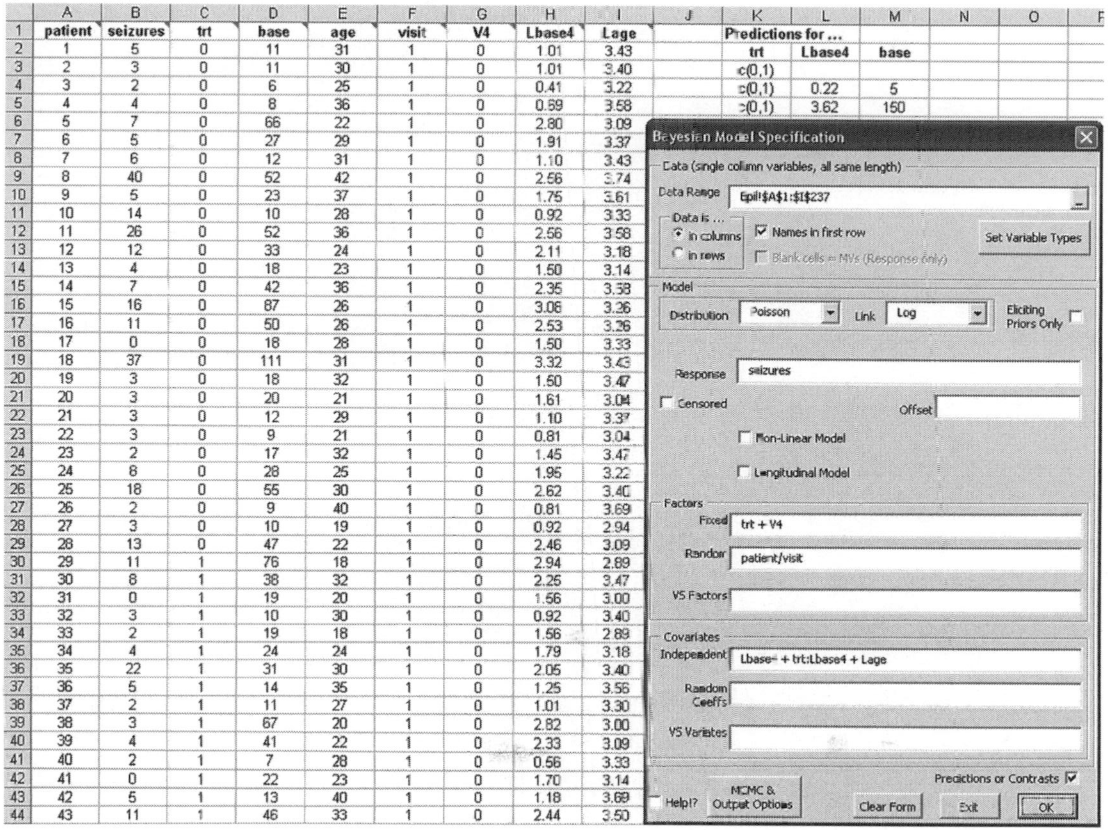

patient	seizures	trt	base	age	visit	V4	Lbase4	Lage
1	5	0	11	31	1	0	1.01	3.43
2	3	0	11	30	1	0	1.01	3.40
3	2	0	6	25	1	0	0.41	3.22
4	4	0	8	36	1	0	0.89	3.58
5	7	0	66	22	1	0	2.80	3.09
6	5	0	27	29	1	0	1.91	3.37
7	6	0	12	31	1	0	1.10	3.43
8	40	0	52	42	1	0	2.56	3.74
9	5	0	23	37	1	0	1.75	3.61
10	14	0	10	28	1	0	0.92	3.33
11	26	0	52	36	1	0	2.56	3.58
12	12	0	33	24	1	0	2.11	3.18
13	4	0	18	23	1	0	1.50	3.14
14	7	0	42	36	1	0	2.35	3.38
15	16	0	87	26	1	0	3.06	3.36
16	11	0	50	26	1	0	2.53	3.26
17	0	0	18	28	1	0	1.50	3.33
18	37	0	111	31	1	0	3.32	3.43
19	3	0	18	32	1	0	1.50	3.47
20	3	0	20	21	1	0	1.61	3.04
21	3	0	12	29	1	0	1.10	3.37
22	3	0	9	21	1	0	0.81	3.04
23	2	0	17	32	1	0	1.45	3.47
24	8	0	28	25	1	0	1.95	3.22
25	18	0	55	30	1	0	2.62	3.40
26	2	0	9	40	1	0	0.81	3.69
27	3	0	10	19	1	0	0.92	2.94
28	13	0	47	22	1	0	2.46	3.09
29	11	1	76	18	1	0	2.94	2.89
30	8	1	38	32	1	0	2.25	3.47
31	0	1	19	20	1	0	1.56	3.00
32	3	1	10	30	1	0	0.92	3.40
33	2	1	19	18	1	0	1.56	2.89
34	4	1	24	24	1	0	1.79	3.18
35	22	1	31	30	1	0	2.05	3.40
36	5	1	14	35	1	0	1.25	3.56
37	2	1	11	27	1	0	1.01	3.30
38	3	1	67	20	1	0	2.82	3.00
39	4	1	41	22	1	0	2.33	3.09
40	2	1	7	28	1	0	0.66	3.33
41	2	1	22	23	1	0	1.70	3.14
42	5	1	13	40	1	0	1.18	3.69
43	11	1	46	33	1	0	2.44	3.50

Predictions for...

	trt	Lbase4	base
	c(0,1)	0.22	5
	c(0,1)	3.62	150

Bayesian Model Specification

Data (single column variables, all same length)

Data Range: Epil!$A$1:$I$237

Data is...
- In columns
- In rows

☑ Names in first row
☐ Blank cells = M/s (Response only)

Set Variable Types

Model

Distribution: Poisson   Link: Log

Response: seizures

☐ Censored

☐ Non-Linear Model
☐ Longitudinal Model

Offset:

☐ Eliciting Priors Only

Factors

Fixed: trt + V4

Random: patient/visit

VS Factors:

Covariates

Independent: Lbase4 + trt:Lbase4 + Lage

Random Coeffs:

VS Variables:

Help?   MCMC & Output Options

Predictions or Contrasts ☑

Clear Form   Exit   OK

FIGURE 6.9

Data and model specification for epilepsy trial.

for each patient. The two covariates and the factor-by-covariate interaction were specified as independent fixed effects. The model specified can be represented algebraically:

$$\eta_{ijkl} = \mu + \delta_i + \alpha_j + \beta x_k + (\delta\beta)_i x_k + \gamma z_k + p_k + (pv)_{kl} \quad i = 1,2 \quad j = 1,2$$

$$k = 1,2,\ldots,59 \quad l = 1,2,3,4$$

$$\log(\lambda_{ijkl}) = \eta_{ijkl}$$

$$y_{ijkl} \sim \text{Pois}(\lambda_{ijkl})$$

$$p_k \sim N(0, \sigma_1^2)$$

$$(pv)_{kl} \sim N(0, \sigma_2^2)$$

where
  y_{ijkl} are the seizure counts reported by the kth patient at their lth visit
  δ_i are the effects of the factor trt
  α_j are the effects of the factor V4
  β is the regression coefficient associated with the linear effect of Lbase4
  x_k is the value of Lbase4 for the kth patient
  $(\delta\beta)_i$ are the treatment-by-baseline interaction effects
  γ is the regression coefficient associated with the linear effect of Lage
  z_k is the log(age) of the kth patient
  p_k are the effects of the factor patient
  $(pv)_{kl}$ are the effects of the factor visit nested within patient
  σ_1^2 and σ_2^2 are the variance components associated with the random effects

Although there are only two levels for the fixed effects treatment factor, and hence, the single parameter estimated will represent the effect of the active treatment over placebo, this parameter estimate will be difficult to interpret due to it being on the log-link scale. By requesting the contrast explicitly, the estimate will be back-transformed such that the effect is defined as a ratio of the seizure rates, making interpretation easier. Since a treatment-by-covariate interaction is included in the model, the treatment effect cannot be considered a single quantity to be estimated as it will vary depending upon the value of the covariate. So that the treatment effect can be fully characterized, it is necessary to request a range of contrasts to be estimated for different values of the covariate, which is Lbase4 in this case. Cells K2:L5 in Figure 6.9 show how this can be specified using BugsXLA. The first row below the header, cells K3:L3, only specifies the treatment contrast explicitly, and so, the value of Lbase4 will be set to its mean value in the data set. In the next two rows, cells K4:L5, a value for the covariate is specified, which in this case covers the whole range observed in the study. Note that for clarity, the actual baseline seizure counts were entered in column M and an Excel formula used in column L to calculate the transformed covariate value. Only cells K2:L5 need to be referenced on the 'Predictions & Contrasts' form when instructing BugsXLA to estimate these contrasts. Although not shown in Figure 6.9, in order to provide a more precise understanding of the treatment-by-baseline interaction, a total of 10 contrasts were specified at different, approximately log-spaced, settings of the baseline seizure count: 5, 7.5, 10, 15, 22, 33, 50, 75, 110, and 160.

After confirming there were no convergence issues, the simple model MCMC settings were chosen (a burn-in of 5,000 followed by 10,000 samples), and the default priors were unchanged (see cells A23:B30 in Figure 6.10). Figure 6.10 shows the parameters in the model and a summary of their marginal posterior distributions as imported back into Excel. Although the prior distributions are not the same as used in the WinBUGS example, in both cases they are very vague relative to the likelihood and hence result in very similar fitted models. Some care is needed in comparing the parameter estimates directly as the two fixed effects factors here are parameterized as covariates with regression coefficients in the WinBUGS model. Note the warning provided by BugsXLA regards interpretation of some of the parameter estimates when interactions are in the model, which is pertinent in this case as previously discussed.

	A	B	C	D	E	F	G	H	I	J	K
1			Label	Mean	St.Dev.	2.5%	Median	97.5%		WinBUGS Name	
2			CONSTANT	1.7520	0.1165	1.5190	1.7540	1.9770		Beta0	
3			Intercept at 0	-1.5040	1.2720	-4.1170	-1.4870	0.9626		alpha	
4		trt	1	-0.3176	0.1592	-0.6364	-0.3138	-0.0156		X.Eff[1,2]	
5		V4	1	-0.1002	0.0897	-0.2745	-0.1000	0.0735		X.Eff[2,2]	
6			Lbase4	0.8695	0.1341	0.6001	0.8723	1.1220		V.Coeff[1,1]	
7			Lage	0.5173	0.3745	-0.2106	0.5133	1.2760		V.Coeff[2,1]	
8		trt x Lbase4	1	0.3879	0.2080	-0.0379	0.3888	0.7857		V.Coeff[3,2]	
9			SD(patient)	0.5079	0.0726	0.3796	0.5029	0.6636		sigma.Z[1]	
10			SD(patient x visit)	0.3672	0.0438	0.2856	0.3657	0.4562		sigma.Z[2]	
11	Note: CONSTANT & Factor effects are determined at the mean of the covariate(s).										
12	Interpret these cautiously when Factor x Covariate terms have been fitted.										
13											
14	Model	[Epil!$A$1:$I$237]									
15	Distribution	Poisson									
16	Link	Log									
17	Response	seizures									
18	Fixed	trt + V4									
19	Covariates	Lbase4 + trt:Lbase4 + Lage									
20	Random	patient/visit									
21											
22	Priors										
23	CONSTANT	N(mu=1.59, sigma=111)									
24	trt	N(mu=0, sigma=111)									
25	V4	N(mu=0, sigma=111)									
26	Lbase4	N(mu=0, sigma=149)									
27	Lage	N(mu=0, sigma=498)									
28	trt x Lbase4	N(mu=0, sigma=149)									
29	patient	Norm(0,tau^2); tau ~ Half-N(sigma=5.54)									
30	patient x visit	Norm(0,tau^2); tau ~ Half-N(sigma=5.54)									
31											
32	WinBUGS MCMC Settings										
33	Burn-In: 5000 Samples: 10000 (Thin:1; Chains:1)										
34	Run took 5.1 minutes										

FIGURE 6.10
Summaries of posterior distributions, model, priors, and MCMC settings for epilepsy trial.

'no effect'	base	Mean	St.Dev.	2.5%	Median	97.5%	WinBUGS Name		
1	5	2.6690	0.9644	1.1950	2.5140	4.9780	Pred.Ave[2]		
1	7.5	2.2280	0.6391	1.1830	2.1480	3.7140	Pred.Ave[3]		
1	10	1.9690	0.4711	1.1700	1.9220	3.0160	Pred.Ave[4]		
1	15	1.6630	0.3087	1.1250	1.6360	2.3150	Pred.Ave[5]		
1	22	1.4270	0.2303	1.0360	1.4040	1.9340	Pred.Ave[6]		
1	33	1.2220	0.2158	0.8763	1.1940	1.7130	Pred.Ave[7]		
1	50	1.0510	0.2402	0.6807	1.0180	1.6010	Pred.Ave[8]		
1	75	0.9130	0.2735	0.5077	0.8704	1.5740	Pred.Ave[9]		
1	110	0.8050	0.3040	0.3815	0.7468	1.5750	Pred.Ave[10]		
1	160	0.7164	0.3308	0.2870	0.6474	1.5830	Pred.Ave[11]		

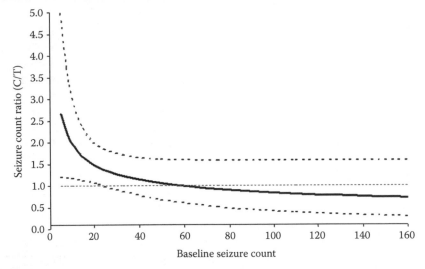

FIGURE 6.11
Contrasts and treatment-by-baseline interaction plot for epilepsy trial.

Figure 6.11 shows a plot produced from the estimated contrasts imported back into Excel. Before producing the plot, the worksheet had to be edited so that the 10 estimated contrasts were in consecutive rows. Two additional columns were added: one labeled "no effect" containing the value one in every cell, and another labeled "base" with the baseline seizure counts associated with the contrast estimates. The plot is created by first selecting the two columns of data labeled "base" and "Mean," and from the Excel charting options select the XY (scatter) plot with points connected by smoothed lines without markers, removing the default legend and formatting the plot to your preferred style. It is then a simple matter to add three data series to the scatter plot, using columns labeled "2.5%," "97.5%," and "no effect," and format them as dashed lines.

Note that the contrast is the difference, first level of treatment factor subtract second level, back-transformed as a ratio. Hence, since the active treatment is the second level, as defined when the factor levels were set up initially (see Section 2.4), the ratio is control over active treatment, with values larger than one indicating that the treatment is reducing the number of seizures. The plot shown in Figure 6.11 suggests that the treatment was only clearly

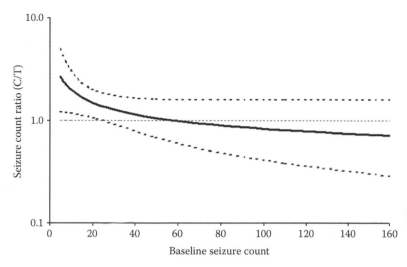

FIGURE 6.12
Treatment-by-baseline interaction plot with log-scale Y-axis for epilepsy trial.

superior to the control for the less severe sufferers of epilepsy as measured by their baseline seizure count. It is probably more appropriate to plot the seizure count ratio on the log-scale, as a doubling in one direction represents an effect size that is equal in magnitude to a halving in the other direction. Figure 6.12 shows such a plot, which also better represents the relative precision of the estimated effect, which is highest at the average value of the covariate, approximately 23 seizures.

Case Study 6.3 Cold Feet Incidence in Hypertension Trial
Case Study 4.6 discussed a fixed effects ordinal logistic regression analysis of an ordered categorical response. The response was on a scale of 1–5 measuring the incidence of cold feet reported in a clinical trial designed to compare three treatments for hypertension. Here we reanalyze these data using random effects to model the additional variation between subjects that is likely to exist. This is an attempt to accommodate any differences between individuals' natural susceptibility to cold feet that are unexplained by their pre-treatment baseline measure. The only change to the BugsXLA model specification form shown in Figure 4.23 is the inclusion of the factor Patient as a random effect. The algebraic representation of the model is as before with the addition of a set of parameters representing the random effects of the 283 patients. These random effects are assumed to be independently and identically distributed as Normal variates with mean zero and a common variance parameter to be estimated from the data. The same default priors were used as for the fixed effects analysis, with the addition of the Half-Normal Distribution with a scale parameter of five as the default prior for the random effects standard deviation parameter. As before, the simple model

TABLE 6.5

Comparison of Fixed and Random Effects
Models for Cold Feet Analysis

	Model	
	Fixed Effects	**Random Effects**
DIC	1243	950
pD	13.0	126
	Posterior mean (St. Dev.)	
Odds-ratio (A/B)	0.5 (0.1)	0.3 (0.2)
Odds-ratio (A/C)	1.8 (0.5)	3.7 (2.8)
Odds-ratio (B/C)	3.7 (0.9)	13.8 (11)

defaults for the MCMC settings were chosen (a burn-in of 5,000 followed
by 10,000 samples) and all pair wise contrasts (odds ratios) as well as predicted means (probabilities) for all levels of the factor Treat were requested.
As should be expected, this model was even slower to run than that without
the random effects, taking about 20% longer. The MCMC convergence diagnostics were fine, although there was high autocorrelation and the occasional
large spike in some of the history plots. Table 6.5 provides a comparison of
the fixed and random effects models.

The DIC statistic clearly indicates that the random effects term is a useful
addition to the model, even though it effectively adds over 100 parameters
to the 13 in the fixed effects model. There are very large differences in the
estimated odds-ratios obtained from the two models, with the most striking
difference being between the posterior means of the odds-ratios comparing
treatments B and C. However, one should also note the vastly reduced precision with which the odds-ratios are being estimated in the random effects
model. When such remarkable differences such as these are seen it should
alert the user to interrogate the data thoroughly, discussing any findings
with the subject matter experts, before confidently reporting the results. After
checking for any obvious errors in the data or model specification, the next
thing should be to reassess the MCMC convergence. A much longer run was
undertaken, using the regular model MCMC settings (a burn-in of 10,000 followed by 10,000 samples after one fifth thinning), and this gave very similar
results. The effect of the starting values was also assessed by running multiple chains, and this also gave compatible results. Since each of these longer
analyses took just over 2 h to run, these settings were not used for any further
analyses of these data; although the output from the first of these settings
was used in the assessment of the random effects now described.

Once we are content that the model's parameters have been estimated
appropriately, it is advisable to try and understand the nature of the random effects more thoroughly. The standard deviation associated with these
effects has a posterior mean (standard deviation) of 3.0 (0.4), which, given
this is on the logit scale, is both very large in an absolute sense as well as

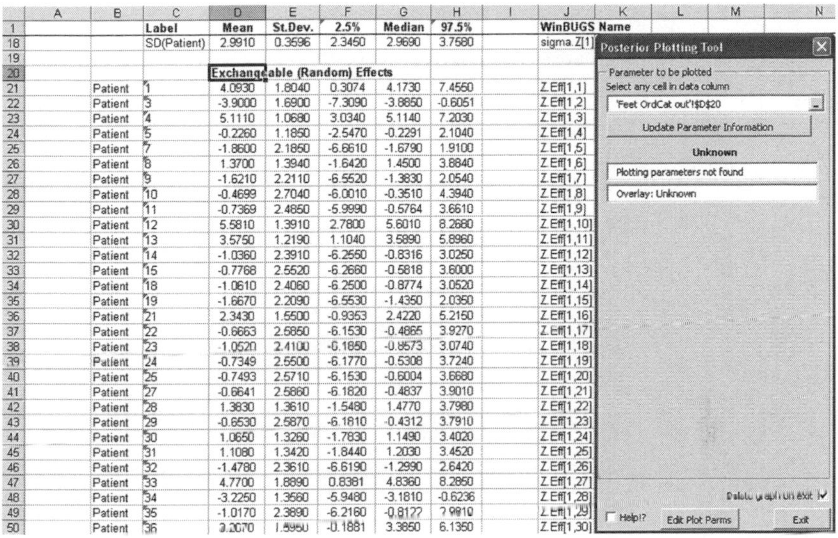

	Label	Mean	St.Dev.	2.5%	Median	97.5%		WinBUGS Name
18	SD(Patient)	2.9910	0.3596	2.3460	2.9690	3.7580		sigma.Z[1]
19								
20		**Exchangeable (Random) Effects**						
21	Patient 1	4.0930	1.8040	0.3074	4.1730	7.4550		Z.Eff[1,1]
22	Patient 3	-3.9000	1.6900	-7.3090	-3.8850	-0.6051		Z.Eff[1,2]
23	Patient 4	5.1110	1.0680	3.0340	5.1140	7.2030		Z.Eff[1,3]
24	Patient 5	-0.2260	1.1850	-2.5470	-0.2291	2.1040		Z.Eff[1,4]
25	Patient 7	-1.8600	2.1850	-6.6610	-1.6790	1.9100		Z.Eff[1,5]
26	Patient 8	1.3700	1.3940	-1.6420	1.4500	3.8840		Z.Eff[1,6]
27	Patient 9	-1.6210	2.2110	-6.5520	-1.3830	2.0540		Z.Eff[1,7]
28	Patient 10	-0.4699	2.7040	-6.0010	-0.3510	4.3940		Z.Eff[1,8]
29	Patient 11	-0.7369	2.4850	-5.9990	-0.5764	3.6610		Z.Eff[1,9]
30	Patient 12	5.5810	1.3910	2.7800	5.6010	8.2680		Z.Eff[1,10]
31	Patient 13	3.5750	1.2190	1.1040	3.5890	5.8960		Z.Eff[1,11]
32	Patient 14	-1.0360	2.3910	-6.2560	-0.8316	3.0250		Z.Eff[1,12]
33	Patient 15	-0.7768	2.5520	-6.2660	-0.5818	3.6000		Z.Eff[1,13]
34	Patient 18	-1.0610	2.4060	-6.2500	-0.8774	3.0520		Z.Eff[1,14]
35	Patient 19	-1.6670	2.2090	-6.5530	-1.4350	2.0350		Z.Eff[1,15]
36	Patient 21	2.3430	1.5500	-0.9353	2.4220	5.2150		Z.Eff[1,16]
37	Patient 22	-0.6663	2.5850	-6.1530	-0.4865	3.9270		Z.Eff[1,17]
38	Patient 23	-1.0520	2.4100	-6.1850	-0.8573	3.0740		Z.Eff[1,18]
39	Patient 24	-0.7349	2.5500	-6.1770	-0.5308	3.7240		Z.Eff[1,19]
40	Patient 25	-0.7493	2.5710	-6.1530	-0.6004	3.6680		Z.Eff[1,20]
41	Patient 27	-0.6641	2.5860	-6.1820	-0.4837	3.9010		Z.Eff[1,21]
42	Patient 28	1.3830	1.3610	-1.5480	1.4770	3.7980		Z.Eff[1,22]
43	Patient 29	-0.6530	2.5870	-6.1810	-0.4312	3.7910		Z.Eff[1,23]
44	Patient 30	1.0650	1.3260	-1.7830	1.1490	3.4020		Z.Eff[1,24]
45	Patient 31	1.1080	1.3420	-1.8440	1.2030	3.4520		Z.Eff[1,25]
46	Patient 32	-1.4780	2.3610	-6.6190	-1.2990	2.6420		Z.Eff[1,26]
47	Patient 33	4.7700	1.8890	0.8381	4.8360	8.2850		Z.Eff[1,27]
48	Patient 34	-3.2250	1.3560	-5.9480	-3.1810	-0.6236		Z.Eff[1,28]
49	Patient 35	-1.0170	2.3890	-6.2160	-0.8172	2.9910		Z.Eff[1,29]
50	Patient 36	3.2070	1.5950	-0.1881	3.3850	6.1350		Z.Eff[1,30]

Posterior Plotting Tool — Parameter to be plotted: Select any cell in data column — 'Feet OrdCat out'!$D$20 — Update Parameter Information — Unknown — Plotting parameters not found — Overlay: Unknown — Help!? Edit Plot Parms Exit

FIGURE 6.13
Initial state of posterior plotting tool form when plotting random effects in the cold feet analysis.

being "statistically significant" relative to its posterior standard deviation. The posterior summaries of the random effects can be imported back into Excel by requesting the statistics for the Random Terms be imported on the 'MCMC & Output Options' form prior to running WinBUGS (see Section 2.5). It is then possible to plot these using the Posterior Plotting tool provided by BugsXLA as discussed in Section 2.11. Note that before using this plotting tool, it is necessary to add a blank row to separate the header row for the 'Exchangeable (Random) Effects' from the last row of the parameter estimates above it. To plot the posterior means of the random effects, select a cell in this column and click on the 'Post Plots' icon to bring up the form as shown in Figure 6.13.

Since the header for this column of data, assumed to be in the first non-blank cell (cell D:20 in this case), does not contain a comment specifying how to plot these data, the 'Posterior Plotting Tool' does not offer any default settings. The user must click on the button labeled 'Edit Plot Parms' and complete the 'Edit Plotting Parameters' form as shown in Figure 6.14. The name has been altered, and the estimated Normal Distribution for the population of random effects specified as an overlay. Figure 6.15 shows the resultant plot after manually altering the default settings for the histogram.

It is clear from this plot that the assumed Normal Distribution for the random effects is not supported by the data. There appears to be at least two groups of patients, with the bulk having very similar effects (as indicated by the group centered about –1), and another group being less homogeneous and having larger positive effects. There may also be a small group of

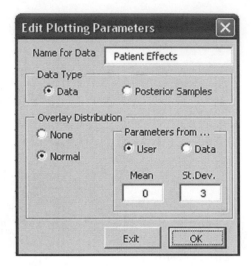

FIGURE 6.14
Edit plotting parameters form when plotting random effects in the cold feet analysis.

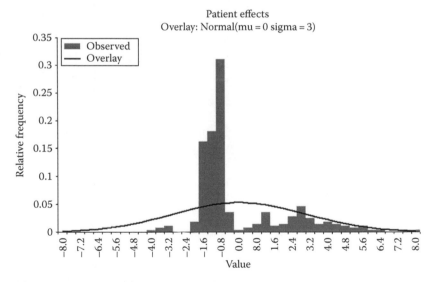

FIGURE 6.15
Histogram of the estimated random effects with estimated population distribution overlaid in the cold feet analysis.

patients who are outliers at the negative end of this range. Without having access to other information concerning these patients, or to the subject matter experts, it is difficult to take this analysis much further. This analysis does suggest how any further investigation into the nature of the additional variation between patients might proceed, for example, one might assess whether there were similarities between patients in each group identified that could explain the differences between groups, or one could try to build a more sophisticated hierarchical model that might accommodate the variation observed. It might be that these data, despite being numerous, are insufficient to fit this more complex model very precisely. A simple tabulation of the end of study and baseline cold feet data shows that about 80% and 85%, respectively, are in category 1, that is, no cold feet occurred. This doesn't leave a lot of variation in the responses with which to estimate the effects of treatment, as well as the other factors. Further exploratory analysis reveals that virtually all of the patients in the main homogenous group shown in Figure 6.15 were in category 1 at both baseline and end of treatment for all visits. These findings suggest that there is a relatively small minority of patients who are susceptible to cold feet, and who can provide information with which to differentiate between treatments. Models analogous to the Zero Inflated Binomial (see, e.g., Chapter 8 of Ntzoufras (2009)) might be useful for data such as these, although fitting such models is well beyond the scope of BugsXLA.

It is interesting to note that Brown and Prescott (2006) accommodate the within patient correlation by fitting a variety of covariance pattern models, rather than a random effects model directly as here, using a quasi-likelihood approach in the statistical package SAS. They report the estimated odds-ratios from the compound symmetry model, which should be the closest equivalent to the Bayesian model fitted here. Their estimates are very similar to those obtained from our Bayesian fixed effects analysis, and hence, for reasons I cannot explain, are quite different to these Bayesian random effects estimates.

Case Study 6.4 Split-Plot Experiment with Mixture Variables to Identify Optimal Film Manufacturing Conditions

Robinson et al. (2009) describe a Bayesian analysis of a film manufacturing experiment first presented by Robinson et al. (2004). This experiment was designed to assess the effects on film quality of three mixture variables defining the film's formulation, as well as three process variables. A total of 5 formulations were replicated in 13 batches. Four pieces of film were cut from each batch and randomly assigned to levels of the three process variables according to a half-fractional factorial design. In the split-plot terminology the batches are considered the "whole plots" with the effects of the mixture variables being estimated at this whole plot level (see Cochran and Cox (1992) for a thorough discussion of split-plot designs). The effects of the process variables should be more precisely estimated, since they are being studied at the split-plot level, which eliminates the additional variation between batches from their effect estimates. A mixture variable defines the proportion of a composite that is made up of the variable's associated substance. Hence, the values of these variables are constrained to sum to one for every formulation. This constraint has led to the development of special methods for the design and analysis of mixture experiments; Cornell (2002) is the classic text covering this topic. Figure 6.16 shows some of the data and the completed BugsXLA model specification form.

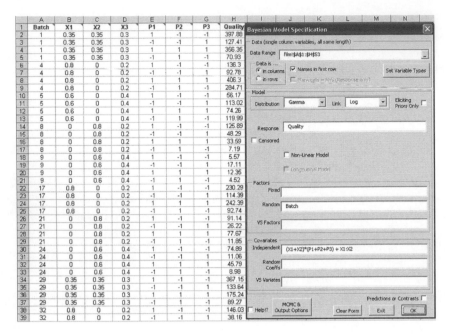

FIGURE 6.16
Data and model specification for film experiment.

The variable Batch is defined as a factor, and is included in the model as a random effect. The variables X1, X2, and X3 define the mixture components of the formulation, while variables P1, P2, and P3 are the process variables with coded levels. The response, Quality, is assumed to have a Gamma error distribution. Since the three mixture variables are constrained to sum to one, their effects cannot be determined independently. Robinson et al. fitted a Scheffé model in which the intercept is not included. The intercept is the predicted mean response when all of the mixture variables are at zero; this is purely a theoretical quantity since this mixture combination is impossible. This model cannot be fitted using BugsXLA because the model constant parameter is defined as the mean response when all the covariates are at their mean value in the data, and so, excluding this parameter by giving it a prior mean of zero with zero variance does not constrain the intercept to zero. Instead a reparameterized version of the Scheffé model was fitted by excluding one of the linear mixture terms from the model, which was chosen to be X3 in this case. The model specified can be represented algebraically:

$$\eta_{ij} = \beta_0 + \beta_1 x_{1i} + \beta_2 x_{2i} + \beta_{12} x_{1i} x_{2i} + \gamma_1 p_{1ij} + \gamma_2 p_{2ij} + \gamma_3 p_{3ij} + (\beta\gamma)_{11} x_{1i} p_{1ij}$$

$$+ (\beta\gamma)_{12} x_{1i} p_{2ij} + (\beta\gamma)_{13} x_{1i} p_{3ij} + (\beta\gamma)_{21} x_{2i} p_{1ij} + (\beta\gamma)_{22} x_{2i} p_{2ij} + (\beta\gamma)_{23} x_{2i} p_{3ij} + \delta_i$$

$$i = 1, 2, \ldots, 13 \quad j = 1, 2, 3, 4$$

$$\log(\theta_{ij}) = \eta_{ij}$$

$$\mu_{ij} = r_{ij}/\theta_{ij}$$

$$y_{ij} \sim Ga(r_{ij}, \mu_{ij})$$

$$\delta_i \sim N(0, \sigma_1^2)$$

where
- y_{ij} is the quality of the jth film taken from the ith batch, which has expected value θ_{ij}
- β's are the regression coefficients associated with the mixture variables x that, importantly here, are not centered
- γ's are the regression coefficients associated with the process variables p
- $(\beta\gamma)$'s are the regression coefficients associated with the mixture-by-process interactions
- δ_i is the effect of the ith batch
- σ_1^2 is the variance component associated with the random batch effects

Leaving the default priors unchanged, a preliminary run suggested that the MCMC chains converged relatively quickly and were mixing fairly well, albeit with high autocorrelation. The regular model MCMC settings were then chosen (a burn-in of 10,000 followed by 10,000 samples after one-fifth thinning), but further convergence checking was undertaken within WinBUGS due to lack of experience with fitting mixture models in a GLMM setting. The history plots looked reasonably well behaved with the exception of that for the hierarchical batch sigma parameter; for the last few hundred iterations of the chain, it appeared to have got stuck at values very close to zero. Another 10,000 samples after one-fifth thinning were run, and Figure 6.17 shows the history plot of all 20,000 samples for this parameter. The history plot shows that for about 1,500 iterations, centered on the 20,000th iteration, the chain was stuck as described. Inspection of the history plots for the regression coefficients did not reveal any noticeable change in their behavior during this time. Given the apparent lack of sensitivity of the other parameters to the behavior of the batch parameter, and the fact that it mixes reasonably well both before and after the region in which it got stuck, we decided to base our final analysis on a burn-in of 10,000 followed by 10,000 samples after a 1/20th thinning; this also gave MCMC errors no greater than about 5% of the posterior standard deviations (a good rule of thumb). If there had been evidence of sensitivity to the batch sigma parameter, further investigation would be appropriate. A more informative prior for the variation between batches, based on information external to the experimental data, might help. A similar exercise for all the other prior distributions should also be considered. In the event that no legitimate alterations to the model improved the mixing, one is left with the only other option, which

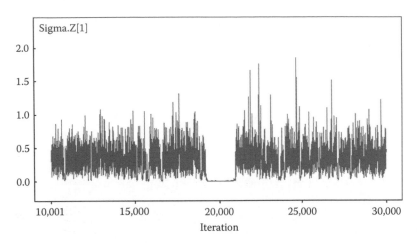

FIGURE 6.17
WinBUGS history plot for the batch sigma parameter in the film experiment.

is to choose a very aggressive thinning strategy in order to reduce the number of samples in any consecutive region that are included in the summarized posterior. Interestingly, Robinson et al. (2009) based their results on 10,000 samples after a 1/100th thinning.

Figure 6.18 shows the parameters in the model and a summary of their marginal posterior distributions, as well as the specified model and default priors (see cells A30:B44). When fitting a mixture model, the parameter CONSTANT is not meaningful and should be ignored. It is easily shown (see Cornell, 2002) that the only difference between the parameters in the model fitted here and those in the standard Scheffé model is in the linear mixture variable coefficients: the intercept at zero here is equivalent to the Scheffé coefficient for X3, while X1 and X2 here are equivalent to the differences between X1 and X3 or X2 and X3, respectively, in the standard Scheffé parameterization. Taking these facts into account, the estimates shown in Figure 6.18 are only slightly different to those presented by Robinson et al. (2009); the difference between the estimated posterior means of each of the regression coefficients is less than 10% of their posterior standard deviations. Only the two dispersion parameters differ by more than this margin, probably reflecting the effect that the degree of thinning has on the mixing issue discussed earlier. Note that as well as providing a posterior summary for the scale parameter, Gamma r, BugsXLA also summarizes the posterior for the residual coefficient of variation, Gamma CV. The CV is often stated as a percentage, in which case the values shown here should be multiplied by 100.

Once we are satisfied that the MCMC process has converged, and the sampling scheme is giving adequate precision, the model fit itself can be assessed using the checking functions provided by BugsXLA as discussed in Section 2.9. The only difference of note when fitting a Gamma error distribution is that the scaled residuals are Bayesian versions of deviance

	A	B	C	D	E	F	G	H	I	J	K
1			Label	Mean	St.Dev.	2.5%	Median	97.5%		WinBUGS Name	
2			CONSTANT	5.2460	0.3063	4.6440	5.2470	5.8440		Beta0	
3			Intercept at 0	1.1870	0.8964	-0.7163	1.2170	2.8700		alpha	
4			X1	5.0100	1.2600	2.6330	4.9720	7.6220		V.Coeff[1,1]	
5			X2	2.8870	1.2780	0.4758	2.8460	5.5940		V.Coeff[2,1]	
6			P1	-0.6717	0.5559	-1.7590	-0.6803	0.4429		V.Coeff[3,1]	
7			P2	0.3586	0.5543	-0.7326	0.3634	1.4610		V.Coeff[4,1]	
8			P3	0.2774	0.5436	-0.8053	0.2815	1.3480		V.Coeff[5,1]	
9			X1 x X2	10.5000	2.6410	5.3660	10.5000	15.5700		V.Coeff[6,1]	
10			X1 x P1	1.1590	0.7878	-0.4182	1.1700	2.7140		V.Coeff[7,1]	
11			X1 x P2	-0.2164	0.7843	-1.7610	-0.2255	1.3300		V.Coeff[8,1]	
12			X2 x P1	1.9450	0.7828	0.3649	1.9560	3.4980		V.Coeff[9,1]	
13			X1 x P3	-0.4482	0.7691	-1.9690	-0.4579	1.0800		V.Coeff[10,1]	
14			X2 x P2	-1.0190	0.7833	-2.5520	-1.0240	0.5259		V.Coeff[11,1]	
15			X2 x P3	-0.1099	0.7675	-1.6170	-0.1205	1.3930		V.Coeff[12,1]	
16			SD(Batch)	0.2971	0.1741	3.947E-3	0.2901	0.6772		sigma.Z[1]	
17			Gamma CV	0.4889	0.0627	0.3828	0.4841	0.6261		CV.gamma	
18			Gamma r	4.3890	1.1080	2.5520	4.2670	6.8250		r.gamma	
19	Note: CONSTANT & Factor effects are determined at the mean of the covariate(s).										
20		Interpret these cautiously when Factor x Covariate terms have been fitted.									
21											
22	Model	[film!$A$1:$H$53]									
23	Distribution	Gamma									
24	Link	Log									
25	Response	Quality									
26	Covariates	(X1+X2)*(P1+P2+P3) + X1:X2									
27	Random	Batch									
28											
29	Priors										
30	CONSTANT	N(mu=4.03, sigma=123)									
31	X1	N(mu=0, sigma=363)									
32	X2	N(mu=0, sigma=371)									
33	P1	N(mu=0, sigma=122)									
34	P2	N(mu=0, sigma=122)									
35	P3	N(mu=0, sigma=122)									
36	X1 x X2	N(mu=0, sigma=2750)									
37	X1 x P1	N(mu=0, sigma=258)									
38	X1 x P2	N(mu=0, sigma=258)									
39	X2 x P1	N(mu=0, sigma=244)									
40	X1 x P3	N(mu=0, sigma=258)									
41	X2 x P2	N(mu=0, sigma=244)									
42	X2 x P3	N(mu=0, sigma=244)									
43	Batch	Norm(0,tau^2); tau ~ Half-N(sigma=6.15)									
44	Gamma r	Gamma(0.1, 0.1)									

FIGURE 6.18
Summaries of posterior distributions, model, and priors for film experiment.

residuals. These should be approximately Normal distributed, and we now show how a Normal probability plot of these residuals can be created in Excel. After importing the results back from WinBUGS as usual, the deviance residuals will be found on the 'Mdl Chks(#)' sheet in the column with header 'ScRes'. A copy of this column of data should be made in a new sheet and additional columns created as shown in Figure 6.19. The 52 deviance residuals have been copied to column D, and sorted into ascending order using Excel's built-in sorting routine. Column A simply contains the integers 1–52. Column B contains the plotting positions, generated by entering the Excel formula '=A2/53' in cell B2 and copying this down to B53. The Normal score against which the residuals will be plotted is generated by entering the Excel formula '=NORMSINV(B2)' into cell C2 and copying this down to cell C53. A Normal probability plot of the residuals can then be created by using

	A	B	C	D
1		Index	NormScore	ScRes
2	1	0.02	-2.08	-1.153
3	2	0.04	-1.78	-0.6604
4	3	0.06	-1.58	-0.6408
5	4	0.08	-1.44	-0.6126
6	5	0.09	-1.31	-0.6037
7	6	0.11	-1.21	-0.5683
8	7	0.13	-1.12	-0.5362
9	8	0.15	-1.03	-0.5042
10	9	0.17	-0.95	-0.4984

FIGURE 6.19
Deviance residuals for film experiment with additional columns in preparation for creation of
Normal probability plot.

Excel's XY (scatter) chart to plot columns C and D. After a little bit of formatting the plot shown in Figure 6.20 can be obtained.

As was noted by Robinson et al. (2009), the smallest observation is perhaps an outlier, being noticeably off the straight line formed by the rest of
the residuals in this plot. The probability of a replicated value being more
extreme for this observation (PrExt on the 'MdlChks' sheet) was about 0.03,
which although quite small, does not trip the default limit set in BugsXLA of
0.025. Although not shown, the posterior plotting tool discussed in Section
2.11 could also be used to produce a histogram of these residuals with the
best fitting Normal Distribution overlaid. In this case, this plot does not
reveal anything more than can be gleaned from the plot in Figure 6.20.

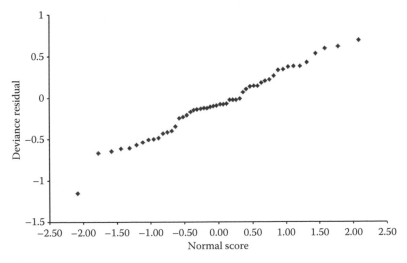

FIGURE 6.20
Normal probability plot of the deviance residuals for film experiment.

Given the complexity of this model and data set, it would be appropriate to begin the analysis by using BugsXLA's novel model MCMC settings in which three chains are run from dispersed starting points. If this is done, WinBUGS fails to complete the analysis and displays the following error message in its on-screen log:

cannot bracket slice for node r.gamma

This cannot be a problem with the code created by BugsXLA, otherwise the previous analysis using a single chain would not have completed successfully. Since the only difference between using a single chain and multiple chains are the initial starting values, these must be the cause of the problem. The automatically generated initial values are one of the few features that cannot be altered in BugsXLA v5.0. Future versions of the program should provide this facility, as well as improvements to the automatically generated initial values. In the meantime, users experienced with the WinBUGS program can use the approach described in Section 11.2 to fix this problem.

A more complicated example of a nested design is provided by Miller et al. (2010) who show how BugsXLA was used to estimate the trend in vegetation cover in the Southwest Alaska Network. This involved fitting a hierarchical random effects model to account for the different sampling strata, with a covariate for year and a log-normal error distribution.

7

Emax or Four-Parameter Logistic Non-Linear Models

Although the linear models discussed in Chapters 3 through 6 will be adequate for the majority of data analyses that are undertaken, there are situations when it is clearly necessary to use a more complex non-linear model. A common case is when the relationship between a covariate and the response is "sigmoidal." That is, initially the response changes very slowly as the covariate increases, then a stronger covariate–response relationship occurs, and finally this slows again as some sort of plateau is reached. In the biological sciences, when dose or concentration–response relationships are being assessed, a commonly used model to describe this is the Emax model. This model has four parameters: the response when the dose is zero, E0, the maximum possible effect, Emax, the dose that produces 50% of this maximum, ED_{50}, and a shape parameter, Hill, which determines the steepness of the dose response. The Hill parameter is named after the scientist who first showed the relevance of this equation to pharmacology, Hill (1910). The Emax model formula is

$$E0 + (Emax\ X^{Hill})/(ED_{50}^{Hill} + X^{Hill})$$

where X is the dose, or more generally the value of the covariate. This is a non-linear model since the formula given above implies that the mean response is not a weighted linear combination of the unknown parameters. Figure 7.1 shows the shape of the Emax model, and how it is affected by the Hill parameter. The E0 and Emax parameters simply shift and stretch the curves vertically, respectively, while the ED_{50} parameter stretches the curves horizontally, in an obvious manner. Although the curves plotted all show an increasing response with increasing dose, a decreasing response is implied by a negative Emax parameter; the parameterization adopted by BugsXLA forces Hill to be positive in both cases. Note that between the doses that give 20% and 80% of the effect, the relationship is very close to being linear on the log-dose scale. For this reason, it is usually much easier to interpret plots of the data using the log-scale for the dose. Also, if it is known that the experiment is being run in this central region, a much simpler linear model, with log-dose as the covariate, should be adequate. In this case, the models discussed in the earlier chapters could be used.

The Emax model is the only option that BugsXLA provides in the Non-Linear Model part of the Model Specification form. However, as will be shown

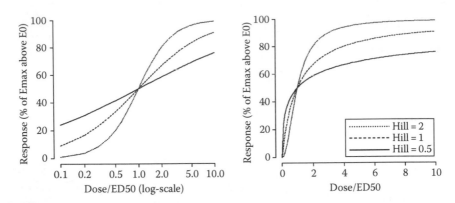

FIGURE 7.1
Response as a function of a covariate, dose, under the Emax model.

in the case studies, multiple Emax models can be fitted simultaneously, and in the case where the parameters can be considered exchangeable and "similar," a Bayesian random effects non-linear model can be specified.

In some application areas, notably biochemistry, an equivalent model that is frequently used is the four-parameter logistic that, using notation that makes the link with the Emax model clear, is

$$E0 + Emax/(1 + (ED_{50}/X)^{Hill})$$

or

$$E0 + Emax/(1 + 10^{-Hill(logX - logED50)})$$

where logX and logED50 are $\log_{10}$ of X and ED_{50}, respectively. It is also common for the Emax parameter to be replaced by the difference between a parameter representing the response as X tends to infinity and the E0 parameter. Sometimes a simpler Emax model, also known as the three-parameter logistic, is specified, in which the Hill parameter is assumed equal to one. This constraint is implemented in BugsXLA by specifying a prior distribution for Hill that has zero variance about a mean of one.

Since BugsXLA allows the Emax relationship between the response and a covariate to be included as part of a bigger model that includes other terms, the E0 parameter is not explicitly included. In the simplest case when there are no other terms in the model, the E0 parameter is the constant term.

Case Study 7.1 Pharmacology Biomarker
An essential part of the drug discovery process is to demonstrate that a novel compound affects human pharmacology in a way that is likely to have a beneficial impact on a particular disease. One such experiment involved

taking human blood and spiking it with a challenge agent that is known to induce an inflammatory response. A concentration–response relationship between a pharmacology biomarker of inflammation and a novel compound was assessed by spiking separate aliquots of this challenged blood with different amounts of the compound. The response used was a derived measure that represented the N-fold change relative to an aliquot without any spiked compound; an increase implies the compound affects the pharmacology as hypothesized. Figure 7.2 shows some of the data and BugsXLA model specification form. Although the blood was obtained from five different donors, the initial model fitted will assume an identical Emax relationship for all donors with no correlation between the measurements obtained from the same donor. It will be shown later how to fit a model that can take account of any additional variation that exists due to differences between the donors' blood.

The Emax non-linear model is specified by selecting the "Non-Linear Model" check box, and entering the name of the covariate between curly

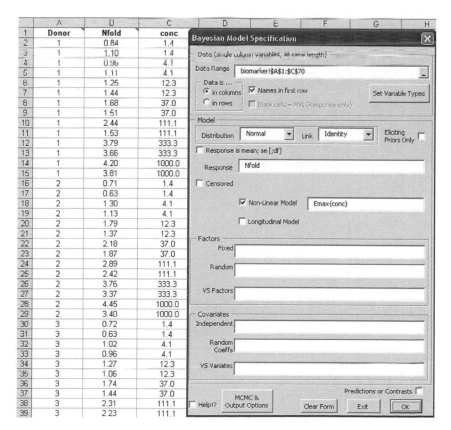

FIGURE 7.2
Data and model specification for biomarker experiment.

brackets, "{ }", after specifying the non-linear model "Emax", as shown in Figure 7.2. This model can be represented algebraically:

$$y_i = \alpha + (\delta\, x_i^\beta)/(\gamma^\beta + x_i^\beta) + e_i \quad i = 1, 2, \ldots, 69$$

$$e_i \sim N(0, \sigma^2)$$

where
    y is the response variable N-fold
    x is the covariate conc
    α is the E0 parameter
    δ is the Emax parameter
    γ is the ED_{50} parameter
    β is the Hill parameter
    σ^2 is the residual variance

Although non-linear models can be difficult to fit, a simple Excel XY (scatter) plot of the data, shown in Figure 7.3, indicates that it should not be difficult to obtain a reasonably good fit in this case. For this reason, the simple model defaults for the MCMC settings were chosen (a burn-in of 5,000 followed by 10,000 samples). So that the posterior plotting tool can be used to help assess the influence of the priors (see Section 2.11 for details), samples were requested for the "Fixed Terms" on the "MCMC and Output Options" form.

When fitting an Emax model it is particularly important to scrutinize the default priors for the parameters. Noting that the E0 parameter is the model constant, the default prior is a highly dispersed Normal Distribution

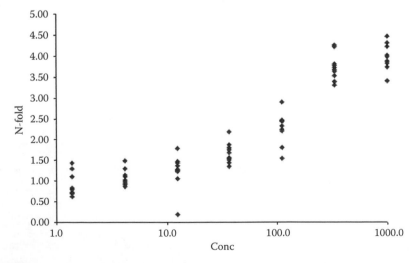

FIGURE 7.3
Scatter plot of N-fold against conc (log-scale) for the biomarker experiment.

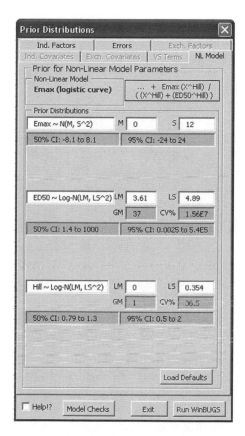

FIGURE 7.4
Default priors for Emax parameters in bio-marker experiment.

centered on the mean of all the data, 2.09. Although it is known that E0 will be at one extreme of the response, the standard deviation (SD) of the prior is sufficiently large, 120, to make this prior approximately uniform over all values remotely close to the observed responses. Since it is expected that the lower concentrations will have virtually no effect and hence give a mean response very close to E0, this prior should not be unduly informative. Figure 7.4 shows the BugsXLA form where the priors for the remaining three Emax parameters can be altered.

The default prior for the Emax parameter is centered on zero, with high dispersion relative to the variation in the data. If the data are such that both the upper and lower plateaus can be reliably estimated, this prior for Emax should not be influential. To help the user in setting priors for these parameters, the implied 50% and 95% credible intervals are shown. The default prior for the ED_{50} parameter is a Log-Normal Distribution with a median (geometric mean [GM]) equal to the GM of the smallest and largest non-zero concentration values (1.4 and 1000), and a scale parameter such that the ED_{50} is believed to lie between these two values with 50% probability. Again, if the data are such that both the upper and lower plateaus can be reliably estimated, this prior for ED_{50} should not be influential. Unlike the other parameters, the Hill parameter is deliberately given a reasonably informative default prior: a Log-Normal Distribution centered on one, with an implied 95% probability of being between 0.5 and 2. This is based on personal experience in its application in the discovery and development of medicines, where it is very common for the experimental data to provide little information with which to estimate this parameter, and where it is often expected that this value will be close to one. All of these defaults can be altered as explained in Section 2.12. In this particular case, it was expected that the data would be sufficient to give a good estimate of the Hill parameter, and so the prior was significantly widened by increasing the log-scale SD parameter, LS, from 0.354 to 2 on the form shown in Figure 7.4. Note that the form shows the GM and coefficient of variation (CV%) implied by the specified parameters for the two Log-Normal prior distributions as an additional aid to setting values for the parameters.

Model checking functions as well as the DIC statistics were requested from the "Model Checking Options" form. Figure 7.5 shows the results of the analysis as imported back into Excel. The primary objective of this experiment was to determine whether there was any evidence of a concentration–response relationship between the test compound and the biomarker. The Emax parameter defines the maximum possible effect of the compound on the response, and in the case when the compound has no effect, this parameter equals zero. Hence, since the credible interval for Emax does not include zero and is positive, there is clear evidence of an increasing concentration response. The best way to interpret the other parameter estimates is by overlaying the fitted Emax curve on the scatter plot shown in Figure 7.3. BugsXLA does not offer this as an option, but it is simple to produce this

	A	B	C	D	E	F	G	H	I	J	K
1			Label	Mean	St.Dev.	2.5%	Median	97.5%		WinBUGS Name	
2			CONSTANT	0.9901	0.0979	0.7733	0.9967	1.1630		Beta0	
3			Emax	3.3820	0.3236	2.8900	3.3320	4.1660		NL.Emax	
4			ED50	140.7000	25.2300	105.6000	136.0000	203.4000		NL.ED50	
5			Hill	1.2600	0.2524	0.8293	1.2380	1.8160		NL.Hill	
6			SD(residual)	0.3413	0.0299	0.2886	0.3394	0.4054		sigma	
7											
8	**Model**	[biomarker!$A$1:$C$70]									
9	Distribution	Normal									
10	Link	Identity									
11	Response	Nfold									
12	NL model	Emax{conc}									
13											
14	**Priors**										
15	CONSTANT	N(mu=2.09, sigma=120)									
16	Emax	N(mean=0, sd=12)									
17	ED50	LogN(mu=3.61, sigma=4.89)									
18	Hill	LogN(mu=0, sigma=2)									
19	V(residual)	Inv-Gamma(0.001, 0.001)									
20											
21	**WinBUGS MCMC Settings**										
22	Burn-In: 5000 Samples: 10000 (Thin:1; Chains:1)										
23	Run took 31 seconds										

FIGURE 7.5
Summaries of posterior distributions, model, priors, and MCMC settings for biomarker experiment.

W	X	Y	Z	AA	AB	AC	AD
Mean	St. Dev.	2.5%	Median	97.5%		WinBUGS Name	
Fitted							
1.0080	0.0841	0.8351	1.0100	1.1660		fit[1]	
1.0080	0.0841	0.8351	1.0100	1.1660		fit[2]	
1.0460	0.0679	0.9117	1.0460	1.1790		fit[3]	
1.0460	0.0679	0.9117	1.0460	1.1790		fit[4]	
1.1710	0.0584	1.0590	1.1700	1.2890		fit[5]	
1.1710	0.0584	1.0590	1.1700	1.2890		fit[6]	
1.5570	0.0973	1.3670	1.5570	1.7500		fit[7]	
1.5570	0.0973	1.3670	1.5570	1.7500		fit[8]	
2.4580	0.0865	2.2820	2.4590	2.6200		fit[9]	
2.4580	0.0865	2.2820	2.4590	2.6200		fit[10]	
3.5050	0.0910	3.3270	3.5050	3.6850		fit[11]	
3.5050	0.0910	3.3270	3.5050	3.6850		fit[12]	
4.0680	0.1053	3.8640	4.0670	4.2760		fit[13]	
4.0680	0.1053	3.8640	4.0670	4.2760		fit[14]	

FIGURE 7.6
Fitted values for biomarker experiment.

using the output provided and Excel's plotting facility. The "Mdl Chks" sheet created to display the requested model checking functions also contains the fitted values with credible intervals. Figure 7.6 shows the fitted values for the first donor. Since the only covariate in the model is the concentration, there are only seven unique fitted values. These seven values of the fitted mean, as well as their lower and upper credible limits can be copied to a convenient location on the spreadsheet and aligned with their respective concentration values. It is then a simple matter to add three data series to the scatter plot, and format them as lines, in order to produce a plot as shown in Figure 7.7.

It is important to note that the credible interval shown is for the fitted mean, not a prediction interval for individual observations. A prediction interval could be obtained, but this would involve specifying predicted values and requesting predictions for individual responses as explained in Section 2.6. Inspection of the plot in Figure 7.7 makes it easier to interpret the estimated parameters. The plot also suggests some aberrant responses, and the more formal model checking diagnostics on the "Mdl Chks" sheet (not shown) identifies between two and four responses that have suspiciously large residuals. For now, the influence of these values will not be assessed, but this could be done using the robust methods discussed in Chapter 10. Apart from a few potential outliers, the model appears to be a reasonably good fit to the data. A good fit suggests that the priors have not been unduly influential over the range of the data, as well as supporting the initial belief that convergence of the MCMC run would occur with the settings chosen. Even so, for non-linear models convergence should be checked and although

1.4	1.0080	0.8351	1.1660					
4.1	1.0460	0.9117	1.1790					
12.3	1.1710	1.0590	1.2890					
37.0	1.5570	1.3670	1.7500					
111.1	2.4580	2.2820	2.6200					
333.3	3.5050	3.3270	3.6850					
1000.0	4.0680	3.8640	4.2760					

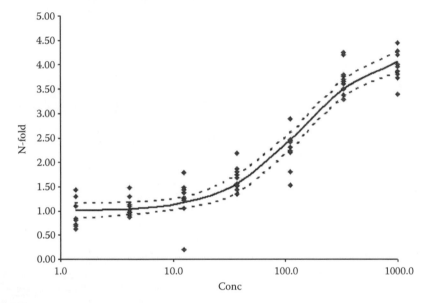

FIGURE 7.7
Scatter plot of N-fold against conc (log-scale) with fitted Emax curve (solid line) and 95% CI (dashed lines) overlaid for the biomarker experiment. Values used to create the Emax curve shown above the plot.

not shown, the diagnostic plots were fine for this analysis. The final assessment that is recommended when fitting an Emax model is to inspect the plots of the prior and posterior overlaid, as explained in Section 2.11. None of the parameters showed any signs of being dominated by their prior, that is, the prior was approximately flat over the credible range of the posterior. Figure 7.8 shows the plot produced for the Hill parameter, which had the largest gradient in the prior density over the posterior credible range.

Since the blood used in this experiment came from five separate donors, it is reasonable to believe that the concentration–response relationship is different for each donor. It is possible to model such grouped data by altering the "Non-Linear Model" field on the "Model Specification" form shown in Figure 7.2 to

Emax{Donor/conc}

This can be interpreted as an Emax model for the variate conc, nested within the grouping factor Donor, that is, a different concentration response for

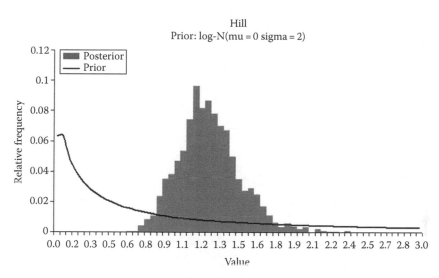

FIGURE 7.8
Posterior and prior overlaid for the Hill parameter in the biomarker experiment.

each donor. Since this is likely to greatly increase the complexity of the model, it is strongly recommended that the novel model MCMC settings are selected (a burn-in of 5,000 followed by 1,000 samples on each of three chains, with auto quit set to "No"), enabling convergence checking from within WinBUGS using the guidance provided in Section 1.2. The objective when using these settings is to determine an appropriate MCMC sampling scheme before running the analysis proper.

When including a grouping factor in the Emax model, each parameter can be assumed to vary between the levels of this factor in one of three ways:

1. Constant: the parameter is constrained to be identical for each level
2. Fixed Effect: its value is estimated independently for each level
3. Random Effect: its values are assumed to be "similar" for each level

Although it is not believed to be the best way to model these data, it will be shown how to fit completely independent Emax curves to each of the donors, with a discussion of the issues this raises. It is known that even when the data cover more than 80% of the range of the Emax effect, the parameters of the Emax model can be difficult to estimate using the data alone (i.e., using classical or likelihood methods), particularly when the sample size is small relative to the residual variation. Since the default priors of BugsXLA are designed to be very flat relative to the likelihood, a naïve analysis using these defaults will often suffer the same difficulties. A common way to deal with this problem when classical statistical methods are being used is to constrain one or more of the parameters to fixed values, for example, adopt the three-parameter logistic by fixing the Hill parameter to one. The Bayesian approach allows a part-way solution between "complete ignorance" and "absolute certainty" regarding the parameters' values via the use of informative priors.

The way that the E0 parameter is modeled is determined by how the factor Donor is specified in other parts of the "Model Specification" form:

1. Constant: not specified in any other parts of the model
2. Fixed Effect: Donor included in the "Fixed Factors" part of the model
3. Random Effect: Donor included in the "Random Factors" part of the model

In this example, Donor was included as a fixed effects factor. Figure 7.9 shows the BugsXLA form where the priors for the other Emax parameters can be altered. By default, both the Emax and ED_{50} parameters are assumed to be "similar" between levels of the grouping factor, while the Hill parameter is assumed to be identical for each level. The reason for the tighter constraint on the Hill parameters is for the same reasons as given earlier: experience

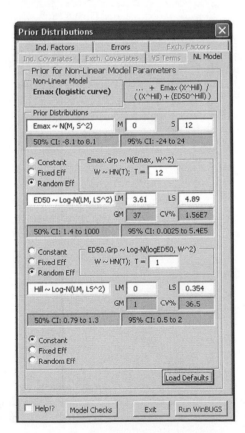

FIGURE 7.9
Default priors for grouped Emax parameters in biomarker experiment.

has shown that there is often insufficient information in the data to precisely estimate this parameter. In order to fit completely independent Emax curves, all of the parameters are given fixed effects priors by selecting the "Fixed Eff" radio buttons. The prior for the Hill parameter was altered as before, that is, the log-scale SD was increased to two.

This model can be represented algebraically:

$$y_{ij} = \alpha_i + (\delta_i x_{ij}^{\beta i}) / (\gamma_i^{\beta i} + x_{ij}^{\beta i}) + e_{ij} \quad i = 1, 2, \ldots, 5 \quad j = 1, 2, \ldots, n_i$$

$$e_{ij} \sim N(0, \sigma^2)$$

with the only change from before being the additional donor level subscript on each of the parameters.

After clicking on the "Run WinBUGS" button, all seems to be going fine until after a few thousand iterations WinBUGS crashes showing the "Trap" window in Figure 7.10. These error trap messages are only decipherable by those with intimate knowledge of the program that underlies WinBUGS. The most likely cause when using BugsXLA is that there is insufficient information in the data with which to estimate all the parameters in the model. It is also possible that the initial values chosen by BugsXLA were so poor that the MCMC process ran into numerical problems (refer to Appendix B for more information on the initial values). Version 5.0 of BugsXLA does not provide any user control on the initial values. However, they are chosen to be credible under the prior, so they can be affected indirectly via the choice of prior distributions. Even if the initial values are reasonable, if the prior is very vague then the Bayesian MCMC approach can suffer analogous problems to conventional statistical software, and fail to find a good model fit to the data, which will often lead to an error trap message.

A reasonable approach when this happens is to review the priors and assess whether more informative priors can be justified, being careful not to allow the observed data to have undue influence. One could start by reviewing all the parameters, tightening all of their priors, hopefully leading to a reasonable model fit. This should be followed up by relaxing each of the priors in turn to discover which parameters needed the additional information provided by its prior distribution. Alternatively, one could make an informed judgment on which parameter was most likely to need a tighter prior, and review each parameter in turn until a reasonable model fit was obtained. The post-hoc nature of this process makes the assessment of sensitivity of inferences to prior beliefs absolutely critical. The latter of these two approaches was used to determine priors (see cells A32:B37 in Figure 7.11) that were weakly informative, leading to MCMC convergence. As expected, the Emax model parameters mixed very slowly, and so for the final BugsXLA run the regular model MCMC settings were chosen (a burn-in of 10,000 followed by 10,000 samples after a one-fifth thinning). As for the simpler Emax

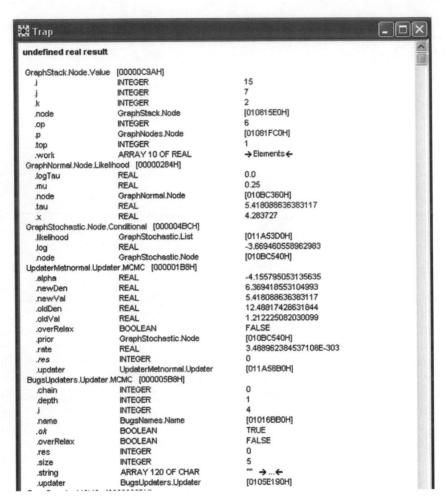

FIGURE 7.10
The dreaded WinBUGS error trap message.

model, samples were requested to be imported for the "Fixed Terms," to aid assessment of the influence of the priors. Figure 7.11 shows the results of the analysis as imported back into Excel.

As should be expected, there is considerably greater uncertainty in the estimates shown in Figure 7.11 than for the single Emax model (refer back to Figure 7.5). Again, a plot of the fitted curves greatly helps to interpret the parameter estimates. Figure 7.12 shows the data plotted using symbols to denote donor, with the five fitted curves. One of the curves, shown as a dotted light grey line, appears noticeably different from the other four; its associated data are the light grey triangles. This is the curve for donor 2. Referring back to the estimates shown in Figure 7.11, it is the Hill parameter for this curve that is most noticeably different.

	A	B	C	D	E	F	G	H	I	J	K
1			Label	Mean	St.Dev.	2.5%	Median	97.5%		WinBUGS Name	
2			CONSTANT	0.9920	0.1860	0.5758	1.0090	1.3110		Beta0	
3			Emax 1	3.9850	1.1140	2.7810	3.6900	7.1370		NL.EmaxGrp[1]	
4			Emax 2	4.3020	1.1110	2.8720	4.0720	7.1650		NL.EmaxGrp[2]	
5			Emax 3	4.2700	1.0560	3.0180	4.0370	6.8910		NL.EmaxGrp[3]	
6			Emax 4	4.3290	1.2350	3.0260	3.9910	7.7150		NL.EmaxGrp[4]	
7			Emax 5	3.2650	0.5699	2.5360	3.1550	4.7200		NL.EmaxGrp[5]	
8			ED50 1	285.4000	341.6000	110.7000	198.0000	1057.0000		NL.ED50Grp[1]	
9			ED50 2	151.8000	396.7000	25.1100	85.1100	604.3000		NL.ED50Grp[2]	
10			ED50 3	200.8000	251.2000	76.9000	150.4000	601.8000		NL.ED50Grp[3]	
11			ED50 4	298.7000	437.9000	101.1000	192.7000	1186.0000		NL.ED50Grp[4]	
12			ED50 5	163.8000	95.7000	90.7900	144.8000	351.5000		NL.ED50Grp[5]	
13			Hill 1	1.1400	0.3780	0.5685	1.0860	2.0280		NL.HillGrp[1]	
14			Hill 2	0.6883	0.2030	0.3706	0.6617	1.1590		NL.HillGrp[2]	
15			Hill 3	1.0230	0.2964	0.5366	0.9914	1.7200		NL.HillGrp[3]	
16			Hill 4	1.0290	0.2975	0.5387	0.9990	1.6980		NL.HillGrp[4]	
17			Hill 5	1.4100	0.4111	0.7345	1.3700	2.3420		NL.HillGrp[5]	
18			Donor 2	-0.4695	0.4749	-1.5330	-0.4173	0.2982		X.Eff[1,2]	
19			Donor 3	-0.2521	0.2822	-0.8353	-0.2471	0.2973		X.Eff[1,3]	
20			Donor 4	-0.3194	0.2637	-0.8506	-0.3153	0.2091		X.Eff[1,4]	
21			Donor 5	0.2493	0.2402	-0.2035	0.2412	0.7501		X.Eff[1,5]	
22			SD(residual)	0.3130	0.0323	0.2570	0.3104	0.3825		sigma	
23											
24	Model	[biomarker!$A$1:$C$70]									
25	Distribution	Normal									
26	Link	Identity									
27	Response	Nfold									
28	NL model	Emax{Donor/conc}									
29	Fixed	Donor									
30											
31	Priors										
32	CONSTANT	N(mu=1, sigma=5)									
33	Emax	N(mean=6, sd=5)									
34	ED50	LogN(mu=3.61, sigma=4.09)									
35	Hill	LogN(mu=0, sigma=0.354)									
36	Donor	N(mu=0, sigma=10)									
37	V(residual)	Inv-Gamma(0.001, 0.001)									
38											
39	WinBUGS MCMC Settings										
40	Burn-In: 10000 Samples: 10000 (Thin:5; Chains:1)										
41	Run took 3.1 minutes										

FIGURE 7.11
Summaries of posterior distributions, model, priors, and MCMC settings for grouped Emax fixed effects fit to biomarker experiment.

Inspection of the plots of the prior and posterior overlaid did not show any evidence that the E0, ED_{50}, or Emax parameters were dominated by their priors. However, it was very clear that the Hill parameters were heavily influenced. Figure 7.13 shows these plots for donors 2 and 4. In the latter case, which is very similar to that for the other three donors, the plot indicates that there is very little information in the data alone with which to reliably estimate the Hill parameter, since the posterior is almost identical to the prior. For donor 2, the data are such that the posterior has deviated slightly from the prior. It is important to be aware that for complicated models, and in particular non-linear models such as the Emax, an informative prior on one parameter might not only influence its own posterior, but also that of other parameters. If it is not possible to justify a reasonably informative prior for parameters that need to be constrained in order for the model to fit at all, it is necessary to assess the sensitivity by undertaking a series of model fits using a range of different prior assumptions. Although in this case there were no

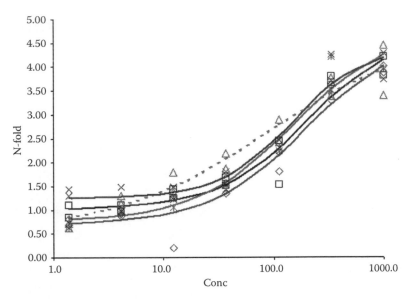

FIGURE 7.12
Scatter plot of N-fold against conc (log-scale), stratified by donor, with fitted completely independent Emax curves overlaid for the biomarker experiment.

good reasons for believing that large differences should exist in the Emax curves for the different donors, the effect of relaxing the prior for the Hill parameter was assessed. The prior log-scale SD was set to one, and this had its most noticeable effect on the curve for donor 2, with its Hill parameter reducing to about 0.3, and its Emax and ED_{50} parameters increasing markedly. No good reason for the "outlying" behavior of donor 2 was found, and it is possible that its fitted curve has been guided by a few influential results.

In this case, it was always believed that the Emax models should be similar for all donors, and the objective was to make inferences about human blood in general rather than specific individuals. Since it is possible to calculate the DIC statistics for both the single Emax model and the fixed effects grouped Emax models, it is interesting to compare them in Table 7.1. Based on the rule of thumb that any difference between DIC values less than 5 is unimportant, this would suggest that unless the Hill slopes were allowed to vary greatly between donors, there is little to be gained in fitting separate curves to each donor. The effective number of parameters, as measured by pD, virtually equals the number of distinct parameters for the single Emax model, but is less than the 21 included in the fixed effects grouped Emax models. This is probably due to the very high correlation that exists between the parameters in the latter cases.

An alternative to the fixed effects assumption that still allows the Emax curves to differ between donors is the random effects model. As discussed in earlier chapters, random effects are appropriate when we believe that the parameters should be "similar" for the different levels of the grouping factor.

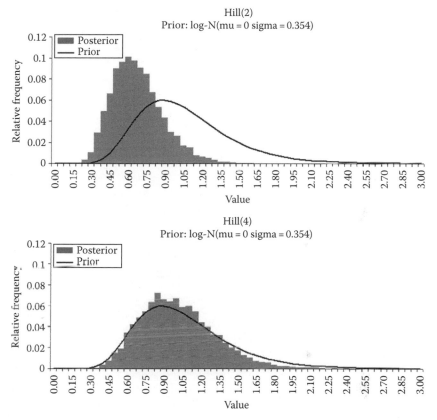

FIGURE 7.13
Posterior and prior overlaid for the Hill parameters of donors 2 and 4 in the biomarker experiment. The axes on the bottom graph have been altered from the default to match those on the top graph.

TABLE 7.1

DIC Statistics Used to Compare Different Emax Model
Fits to the Biomarker Data

	Single Emax	Fixed Effects Hill ~ log N(0, 0.354)	Fixed Effects Hill ~ log N(0, 1)
DIC	52	50	46
pD	4.9	15	15

In the Bayesian approach, any additional prior knowledge of the likely degree of similarity can be used in specifying the prior distribution; this additional information can be very useful when fitting complex models, particularly non-linear models, when the data alone are often compatible with parameter estimates that are not credible to a subject matter expert.

The E0 parameters were defined as random effects by entering Donor in the "Random Factors" field on the "Model Specification" form shown in Figure 7.2. This implies a prior of the form:

$$\alpha_i \sim N(\mu, \tau^2) \quad i = 1, 2, \ldots, 5$$

$$\tau \sim \text{Half-N}(K)$$

where
  α_i are the E0 parameters for each of the donors
  μ is the model constant, which has its prior set on the "Independent Factors" page of the "Prior Distributions" form
  τ^2 is the variance component parameter associated with these random effects
  K is the value of the Half-Normal Distribution's parameter that defines the prior distribution (alternative distributional forms for the prior can be chosen as explained in Section 2.7 and discussed in Chapter 5.)

The default value of K was altered via the "Graphical Elicitation of Prior" form (see Section 2.8 for details) obtained via the "GFI" button on the "Prior Distributions" form. It was believed to be very unlikely that the E0 parameters would differ by more than three between donors. Figure 7.14 shows the implied distribution for the difference between the E0 parameters of two randomly chosen donors that was considered a reasonable reflection of these weakly informative prior beliefs. This equated to a value of one for K.

The other Emax model parameters were given random effects priors by selecting the "Random Eff" radio buttons on the "Prior Distributions" form

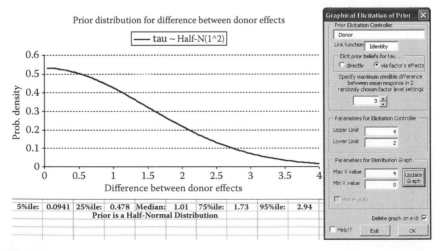

FIGURE 7.14
Informative prior for grouped random effects E0 parameters in biomarker experiment.

shown in Figure 7.9. BugsXLA does not provide the same level of assistance in setting informative random effects priors for the non-linear parameters as it does for the linear random effects. The same help is provided for the population mean parameters as is provided for the fixed effects or constant non-linear parameters, that is, credible intervals, as well as the GM and coefficient of variation for Log-Normal Distributions. But users will need to do their own calculations to determine a value for T, the Half-Normal Distribution parameter, which represents an appropriate prior belief. In this case, since it was expected that there should be a reasonable amount of information in the data to estimate these parameters, only weakly informative priors were specified (see cells A41:B46 in Figure 7.15). By default, BugsXLA only shows the population estimates of random effects parameters, plus their associated

	Label	Mean	St.Dev.	2.5%	Median	97.5%	WinBUGS Name
	CONSTANT	0.9859	0.1410	0.7064	0.9894	1.2520	Beta0
(Mean)	Emax	3.3790	0.3158	2.8620	3.3440	4.0930	NL.Emax
Emax 1		3.3580	0.3202	2.9270	3.3250	4.0940	NL.EmaxGrp[1]
Emax 2		3.4010	0.3469	2.8250	3.3680	4.2050	NL.EmaxGrp[2]
Emax 3		3.4300	0.3336	2.8890	3.3890	4.2070	NL.EmaxGrp[3]
Emax 4		3.3650	0.3125	2.8520	3.3310	4.0750	NL.EmaxGrp[4]
Emax 5		3.3460	0.3268	2.7990	3.3110	4.0800	NL.EmaxGrp[5]
(Median)	ED50	139.8000	28.0700	93.5900	136.6000	205.3000	NL.ED50
ED50 1		154.1000	32.0900	108.6000	149.5000	230.0000	NL.ED50Grp[1]
ED50 2		122.5000	30.9100	65.9800	122.3000	187.4000	NL.ED50Grp[2]
ED50 3		135.9000	25.8300	92.9800	133.5000	193.4000	NL.ED50Grp[3]
ED50 4		152.3000	31.2500	107.3000	147.2000	229.0000	NL.ED50Grp[4]
ED50 5		139.7000	28.1100	95.7100	136.0000	200.9000	NL.ED50Grp[5]
(Median)	Hill	1.2870	0.3100	0.8058	1.2490	2.0190	NL.Hill
Hill 1		1.3800	0.4278	0.8347	1.3070		NL.HillGrp[1]
Hill 2		1.0230	0.2406	0.6438	0.9917	1.5710	NL.HillGrp[2]
Hill 3		1.3840	0.3805	0.8539	1.3180	2.2380	NL.HillGrp[3]
Hill 4		1.3750	0.3370	0.8899	1.3220	2.1420	NL.HillGrp[4]
Hill 5		1.3650	0.4056	0.8032	1.3010	2.3160	NL.HillGrp[5]
	SD(Emax)	0.2271	0.2391	7.682E-3	0.1574	0.8088	NL.EmaxGrp.SD
	SD(logED50)	0.2310	0.2054	0.0132	0.1777	0.7571	NL.ED50Grp.SD
	SD(logHill)	0.2460	0.1714	0.0132	0.2144	0.6717	NL.HillGrp.SD
	SD(Donor)	0.1946	0.1516	0.0164	0.1599	0.5860	sigma.Z[1]
	SD(residual)	0.3109	0.0300	0.2585	0.3089	0.3759	sigma
Exchangeable (Random) Effects							
Donor 1		0.0220	0.1378	-0.2483	0.0136	0.3039	Z.Eff[1,1]
Donor 2		0.0237	0.1441	-0.2802	0.0184	0.3179	Z.Eff[1,2]
Donor 3		-0.0413	0.1369	-0.3400	-0.0285	0.2132	Z.Eff[1,3]
Donor 4		-0.1374	0.1439	-0.4559	-0.1216	0.0853	Z.Eff[1,4]
Donor 5		0.1345	0.1482	-0.0990	0.1166	0.4652	Z.Eff[1,5]

Model	[biomarker!$A$1:$C$70]
Distribution	Normal
Link	Identity
Response	Nfold
NL model	Emax(Donor/conc)
Random	Donor

Priors	
CONSTANT	N(mu=2.09, sigma=120)
Emax	N(mu, sigma); mu ~ N(mean=0, sd=12); sigma ~ HN(2)
ED50	LogN(mu, sigma); mu ~ N(mean=3.61, sd=4.89); sigma ~ HN(1)
Hill	LogN(mu, sigma); mu ~ N(mean=0, sd=2); sigma ~ HN(0.5)
Donor	Norm(0,tau^2); tau ~ Half-N(sigma=1)
V(residual)	Inv-Gamma(0.001, 0.001)

WinBUGS MCMC Settings	
Burn-In: 10000 Samples: 10000 (Thin:5; Chains:1)	
Run took 2.4 minutes	

FIGURE 7.15
Summaries of posterior distributions, model, priors, and MCMC settings for grouped Emax random effects fit to biomarker experiment.

variance components. In order to obtain the parameter estimates for each donor, it is necessary to request that "Random Terms" have their statistics imported on the "MCMC and Output Options" form. Figure 7.15 shows the results of the analysis as imported back into Excel.

Further exploratory analyses, similar to those discussed earlier, gave confidence in this model as a reasonable representation of the data. The DIC statistic of 47 was comparable with the best of the previous models, and the pD value of 13 indicated that the random effects assumption was more parsimonious than the fixed effects model as one should expect. Comparison of the population mean estimates with the estimates obtained from fitting a single Emax model (Figure 7.5) shows that they are very similar, suggesting that the simpler model might be adequate for characterizing the population concentration–response relationship. Note that the random effects model reflects the increased uncertainty in these parameters' values due to differences between the donors.

Case Study 7.2 Clinical Pharmacology Study

Case Study 7.1 describes a laboratory experiment, conducted with human blood in vitro, that provided confidence in the pharmacological activity of a compound under ideal conditions. In practice, the compound will be delivered to the patient in tablet form and the actual concentration that will exist in the blood will be determined by its pharmacokinetic properties. Also, the amount that can be administered will be limited by the compound's tolerability, that is, only doses below which tolerable adverse events occur can be used. A clinical study was undertaken to assess the pharmacological activity with these practical constraints. In this study, 34 subjects were randomly assigned to one of six dose levels of the compound. After sufficient time had elapsed for the compound to have reached its maximum concentration in the blood, a sample was taken and, as before, it was spiked with the challenge agent known to induce an inflammatory response. The compound's actual concentration in the blood, plus the N-fold change in inflammatory response of the blood from the pre-treatment baseline, was measured for each sample following spiking with the challenge agent ex vivo. Note that the concentration is in different units to before, and, for other reasons also, is not directly comparable to the values used in Case Study 7.1. Figure 7.16 shows a scatter plot of the data.

It is clear from Figure 7.16 that the concentration–response curve has not reached its upper plateau. In order to be confident that the next much larger study in patients could adequately test the scientific hypothesis underpinning the rationale for the treatment, it was considered important that blood concentrations at least as high as the ED_{50} could be obtained. Without undertaking any formal analysis, it is clear that the data alone would not be able to answer this question, since to estimate the ED_{50}, it is necessary to have reliable estimates of both the lower (E0) and upper plateaus (E0 + Emax). If BugsXLA is used to fit a simple ungrouped Emax model, leaving the default priors unchanged and using the regular model MCMC settings

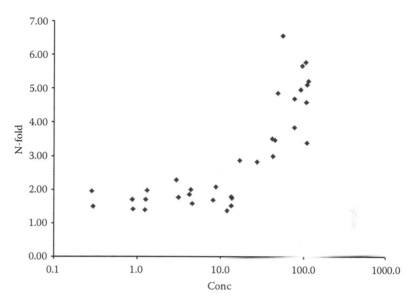

FIGURE 7.16
Scatter plot of N-fold against conc (log-scale) for the clinical pharmacology study.

(a burn-in of 10,000 followed by 10,000 samples after one-fifth thinning), the output is as shown in Figure 7.17.

Inspection of both the Emax and the ED_{50} parameters shows that neither is estimated very precisely, both having credible values that are well outside the range of the data. Results such as these should alert the user to the strong

	A	B	C	D	E	F	G	H	I	J	K
1			Label	Mean	St.Dev.	2.5%	Median	97.5%		WinBUGS Name	
2			CONSTANT	1.5070	0.2511	0.9640	1.5230	1.9480		Beta0	
3			Emax	7.2130	4.8610	3.2860	5.5760	22.5700		NL.Emax	
4			ED50	156.4000	307.2000	26.1200	64.8100	1101.0000		NL.ED50	
5			Hill	1.2660	0.4677	0.6251	1.1650	2.4060		NL.Hill	
6			SD(residual)	0.7309	0.0998	0.5642	0.7214	0.9533		sigma	
7											
8	Model	[Pharmacology!$A$1:$B$35]									
9	Distribution	Normal									
10	Link	Identity									
11	Response	Nfold									
12	NL model	Emax(conc)									
13											
14	Priors										
15	CONSTANT	N(mu=2.99, sigma=156)									
16	Emax	N(mean=0, sd=15.6)									
17	ED50	LogN(mu=1.74, sigma=4.44)									
18	Hill	LogN(mu=0, sigma=0.354)									
19	V(residual)	Inv-Gamma(0.001, 0.001)									
20											
21	WinBUGS MCMC Settings										
22	Burn-In: 10000 Samples: 10000 (Thin:5; Chains:1)										
23	Run took 34 seconds										

FIGURE 7.17
Summaries of posterior distributions, model, priors, and MCMC settings for default Emax fit to clinical pharmacology study data.

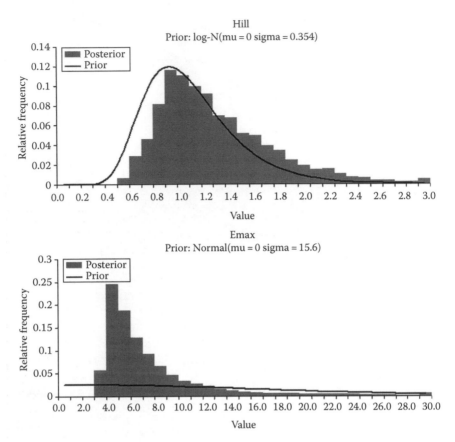

FIGURE 7.18
Posterior and prior overlaid for the Hill and Emax parameters after default Emax fit to clinical pharmacology study data.

possibility that the priors are influential, which is of concern when they are not based on justifiable prior beliefs. Figure 7.18 shows the posterior and prior plots for the Hill and Emax parameters. Although these clearly show that the Hill parameter's posterior has hardly changed from its prior, the plot for the Emax appears to suggest that its prior is not dominating over the likelihood. However, further analyses (not shown) reveal that the Emax, ED_{50}, and Hill parameter estimates are all sensitive to each others' prior distributions.

There was some reason to believe that the maximum possible effect should be the same in both the in vitro laboratory experiment and the ex vivo clinical study. If so, the posterior for the Emax parameter in Case Study 7.1 could be justified as a prior for the clinical study. After some discussion, it was decided that the additional uncertainty in the translation of the in vitro experiment should be incorporated by inflating the variance of the posterior. Specifically, the posterior for the population mean Emax shown in Figure 7.15 (summarized by a mean of 3.38 and an SD of 0.32) was used to justify a

	A	B	C	D	E	F	G	H	I	J	K
1			Label	Mean	St.Dev.	2.5%	Median	97.5%		WinBUGS Name	
2			CONSTANT	1.6010	0.2028	1.1800	1.6040	1.9810		Beta0	
3			Emax	4.0480	0.5817	3.0920	3.9780	5.3640		NL.Emax	
4			ED50	39.6200	10.8600	23.3600	37.7800	65.1600		NL.ED50	
5			Hill	1.7100	0.4518	1.0150	1.6490	2.7560		NL.Hill	
6			SD(residual)	0.7055	0.0962	0.5453	0.6963	0.9238		sigma	
7											
8	**Model**	[Pharmacology!$A$1:$B$35]									
9	Distribution	Normal									
10	Link	Identity									
11	Response	Nfold									
12	NL model	Emax{conc}									
13											
14	**Priors**										
15	CONSTANT	N(mu=2.99, sigma=156)									
16	Emax	N(mean=3.4, sd=1)									
17	ED50	LogN(mu=1.74, sigma=4.44)									
18	Hill	LogN(mu=0, sigma=0.354)									
19	V(residual)	Inv-Gamma(0.001, 0.001)									
20											
21	**WinBUGS MCMC Settings**										
22	Burn-In: 10000 Samples: 10000 (Thin:5; Chains:1)										
23	Run took 37 seconds										

FIGURE 7.19
Summaries of posterior distribution, model, priors, and MCMC settings for informative Emax fit to clinical pharmacology study data.

Normal prior distribution for the Emax in the clinical study with a mean of 3.4 and an SD of 1. The output is shown in Figure 7.19. Not surprisingly, the posterior for the Emax parameter is much more precise than before, with a mean that has been shrunk toward the prior mean. Although none of the other prior distributions were changed from the defaults, the additional information regarding the value of the Emax parameter has also greatly increased the posterior precision for the ED_{50} parameter. This should not be a surprise either, since this parameter defines the concentration that gives a response halfway between E0 and (E0 + Emax).

If the informative prior for the Emax parameter can be justified, then this study demonstrated that it was possible to achieve concentrations of the compound in the blood that were almost certainly higher than the ED_{50}. However, this conclusion relied heavily on this informative prior, and hence, the data alone were not sufficient to meet the objectives of the study. The study did clearly provide evidence of pharmacology in the ex vivo setting, since regardless of the prior distribution used the Emax parameter was conclusively greater than zero.

Case Study 7.3 Pooling Potency Estimates across Assay Runs
Once a compound has been identified as a promising candidate for testing in the clinic, its potency, as determined in pre-clinical laboratory experiments, is measured numerous times to ensure it has been estimated with adequate precision. The potency is usually defined by a value that is equivalent to the ED_{50} parameter in the Emax model discussed in this chapter. One way to pool the estimates would be to combine all the data, from all the separate

assay runs used to generate concentration response curves, and fit a grouped random effects model as discussed in Case Study 7.1. The population ED_{50} parameter would then represent the potency of the compound. However, it is more common for only the summary statistics from each assay run to be stored in a manner that easily retrievable. Here we show how these experiment level summaries can be analyzed to derive a pooled estimate of potency using a modeling approach similar to that in Case Study 5.3. Although this does not involve fitting a non-linear model, this example is included in this chapter due to the summary statistics themselves being derived from separate Emax model fits.

The data consists of five separate estimates of $\log_{10}(IC_{50})$, where IC_{50} stands for the concentration that inhibits 50% of the maximum possible effect, that is, the effect that would be seen if the compound was at its maximum concentration. The estimates are recorded on the log-scale as this makes the assumption of Normal errors more credible. Since there are only five estimates, each obtained from a separate assay run of the experiment, and it is known that there is a significant additional source of variation between assay runs, the precision of the estimated pooled potency will be sensitive to each of the individual values. Prior knowledge of the between assay run variance would be very useful in this case. As part of the quality control process for an assay, a standard compound is included in every run. It is reasonable to suppose that the variation seen between the $\log_{10}(IC_{50})$ estimates of the standard will be similar to that for most test compounds run through the same assay. Typically an assay is run more than once per day, with the assay conditions being more similar for runs on the same day than those occurring between days. Hence, the between and within "assay day" variance components need to be separately estimated. Figure 7.20 shows some of the data for the standard compound and the completed model specification form.

The variables Day and Run are both defined as factors, with the latter factor simply indexing the assay runs within each day. Each assay run produces a mean, standard error, and associated degrees of freedom for the $\log_{10}(IC_{50})$ estimate. If only the means and standard errors were available, this approach can still be used; only two variables are specified in the response, with an assumption that the standard errors are exact rather than estimates. If only the means are available, providing one is happy to assume that they are each estimated with approximately the same precision, a more conventional NLMM can be specified with a single response variable. The model specified in Figure 7.20 can be represented algebraically:

$$\eta_{ij} = \mu + d_i + (dr)_{ij} \quad i = 1, 2, \ldots, 36 \quad j = 1, 2, \ldots, n_i$$

$$y_{ij} \sim N(\eta_{ij}, \omega_{ij}^2)$$

$$se_{ij}^2 \sim Ga(0.5\ df_{ij},\ 0.5\ df_{ij}/\omega_{ij}^2)$$

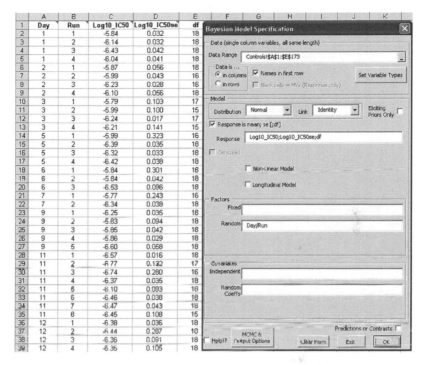

FIGURE 7.20
Data and model specification form for assay standard compound data.

$$d_i \sim N(0, \sigma_1^2)$$

$$(dr)_{ij} \sim N(0, \sigma_2^2)$$

where
  y is the response variable Log10_IC50
  se is the estimated standard error Log10_IC50se
  df is the degrees of freedom df
  d_i are the random effects of the factor Day
  $(dr)_{ij}$ are the random effects of the factor Run within Day
  σ_1^2, σ_2^2 are the variance components

The purpose of this analysis is to derive an evidence based prior distribution for the between assay run variance parameter for use when analyzing the test compound data. Since the five replicate assay runs for the test compound were each done on separate days, the between assay run variance will be equal to the sum of the two variance components being estimated in this initial analysis of the standard compound data. In order to derive the posterior distribution for this quantity, we need to request that the individual samples for the two variance parameters be imported

	A	B	C	D	E	F	G	H	I	J	K
1			Label	Mean	St.Dev.	2.5%	Median	97.5%		WinBUGS Name	
2			CONSTANT	-6.4640	0.0479	-6.5580	-6.4590	-6.3730		Beta0	
3			V(Day)	0.0853	0.0275	0.0453	0.0806	0.1498		sigma2.Z[1]	
4			V(Day x Run)	0.0777	0.0104	0.0593	0.0769	0.1005		sigma2.Z[2]	
5											
6	Model	[Controls!$A$1:$E$173]									
7	Distribution	Normal									
8	Link	Identity									
9	Response	Log10_IC50;Log10_IC50se;df									
10	Random	Day/Run									
11											
12	Priors										
13	CONSTANT	N(mu=-6.49, sigma=39.8)									
14	Day	Norm(0,tau^2); tau ~ Half-N(sigma=1.99)									
15	Day x Run	Norm(0,tau^2); tau ~ Half-N(sigma=1.99)									
16	V(obs. mean)	Inv-Gamma(0.001, 0.001)									
17											
18	WinBUGS MCMC Settings										
19	Burn-In: 5000 Samples: 10000 (Thin:1; Chains:1)										
20	Run took 27 seconds										

FIGURE 7.21
Summaries of posterior distributions, model, priors, and MCMC settings for assay standard compound data.

back into Excel. This is done on the "MCMC and Output Options" form as explained in Section 2.5 and illustrated in Case Study 3.1. Note that it is the variance not the SDs that we need to import. The simple model defaults for the MCMC settings were chosen (a burn-in of 5,000 followed by 10,000 samples), and the default prior distributions used as shown in Figure 7.21 with the other posterior summaries.

Although we may be able to derive an adequate approximation for the posterior distribution for the sum of the variances using the marginal summaries shown in Figure 7.21, it is better to work with the imported individual samples. More generally, any correlation that exists in the joint bivariate distribution will only be adequately accommodated if we work with the samples directly. Also, since an informative prior for a variance component is usually best approximated by an inverse-Gamma Distribution, it is better to fit a distribution to the inverse of the sum of these two variances, which would require additional approximate methods if we were to work with the marginal summaries shown in Figure 7.21. We fit a Gamma Distribution using the method of moments as illustrated in Figure 7.22.

	A	B	C	D	E	F
1	sigma2.Z[1]	sigma2.Z[2]	1 / VarTot		Sample Stats (1 / VarTot)	
2	0.1000	0.0786	5.5991		mean =	6.31041
3	0.0746	0.0671	7.0572		variance =	1.07217
4	0.0756	0.0699	6.8762		Fitted Gamma Distribution	
5	0.0998	0.1018	4.9598		r =	37.1
6	0.1643	0.0675	4.3135		mu =	5.9
7	0.1082	0.0647	5.7834			

FIGURE 7.22
Excel sheet showing parameters of fitted Gamma Distribution to the inverse of the sum of the two variance components in the model used to analyze the assay standard compound data.

BugsXLA imports the 10,000 samples generated by WinBUGS for each of the two variance components into the first two columns of the sheet. Comparing the WinBUGS node names in the first row of Figure 7.22 with those in column J of Figure 7.21 tells us that columns A and B of Figure 7.22 contain the samples for V(Day) and V(Day × Run) respectively. Column C in Figure 7.22 was calculated by entering the Excel formula "=1/(A2 + B2)" into cell C2 and copying this down to cell C10001, utilizing the fact that Excel uses relative referencing by default. The label "1/VarTot" was entered into the first row of Column C to denote these as being samples from the inverse of the total variance The sample mean and variance of the values in column C were calculated by entering the Excel formulae "=AVERAGE(C2:C10001)" and "=VAR(C2:C10001)" into cells F2 and F3, respectively. It is then a simple matter to calculate method of moment estimates of the parameters of the Gamma Distribution (see Appendix A) by entering the Excel formulae "=F2*F6" and "=F2/F3" into cells F5 and F6 respectively. There are more precise methods of fitting a Gamma Distribution, but the method described here should be adequate for the purposes of deriving a prior distribution. Although Normal theory, and experience, can be used to give confidence that a Gamma Distribution will provide an adequate approximation to the distribution of this variable, it is always advisable to check by overlaying the fitted distribution to a histogram of the data. This is not a simple task using Excel, so users are advised to learn how to do this using other software. Figure 7.23 shows such a plot, which confirms the adequacy of the Gamma Distribution fitted. Noting that a scaled Chi-Square Distribution is a special case of the Gamma (see Appendix A) means we can calculate the effective number of degrees of freedom associated with this estimate. This is simply two times the Gamma r parameter, which equals about 74 in this case. If the

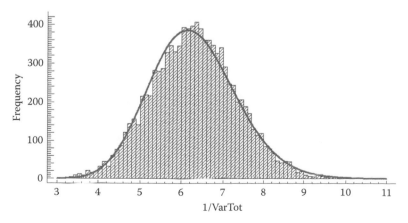

FIGURE 7.23
Plot showing fitted Gamma Distribution (solid line) and histogram of the inverse of the sum of the two variance components in the model used to analyze the assay standard compound data.

same method of moments approach is used to fit Gamma Distributions to the inverse of the individual variance components, the effective numbers of degrees of freedom are 21 and 113 for V(Day) and V(Day × Run), respectively. Since the degrees of freedom are routinely used to indicate the precision with which variance components are estimated, these statistics are a helpful aid to interpreting these estimates in relation to those obtained using classical methods.

We now show how to use this prior distribution in the analysis of the test compound data. Figure 7.24 shows the test compound data and completed model specification form. The simple model defaults for the MCMC settings were chosen (a burn-in of 5,000 followed by 10,000 samples), and the samples for the variance components again selected for importing, but this time specifying the SD. Whenever strongly informative prior distributions are specified it is recommended that BugsXLA's posterior plotting tool be used

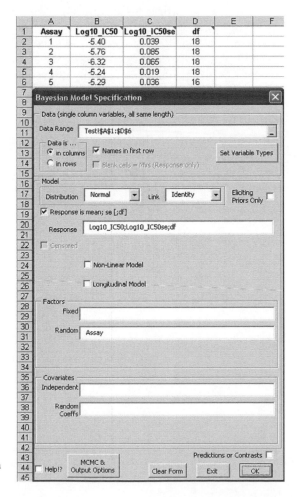

FIGURE 7.24

Data and model specification form for assay test compound data.

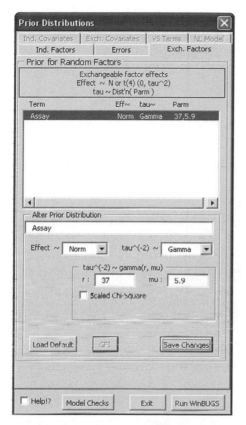

FIGURE 7.25

Specification of informative prior distribution for assay variance component in model used to analyze test compound data.

to overlay the prior onto the posterior distribution (see Section 2.11). Figure 7.25 shows the "Prior Distributions" form where the informative prior for the Assay variance component is specified.

In order to appreciate the impact of this prior distribution, the analysis was also run using the default prior distribution, which is a Half-Normal Distribution with scale parameter 2.25 for the SD of the Assay random effects. Table 7.2 compares the estimates obtained from the two analyses.

It is clear from Table 7.2 that, as one should expect, the posterior distribution for the parameter defining the random effects distribution is strongly influenced by the prior distribution derived from the data on the standard compound. This in turn has increased the precision with which the model

TABLE 7.2

Posterior Mean (SD) of Parameters When Vague or Informative Prior Distributions Are Used

Prior Distribution	Constant	SD(Assay)
Default (vague)	−5.54 (0.28)	0.64 (0.32)
Strongly informative	−5.61 (0.15)	0.40 (0.03)

constant is estimated; the model constant represents the "population" mean response, which is taken to be the test compound's true $\log_{10}(IC_{50})$.

As stated earlier, it is recommended that BugsXLA's posterior plotting tool be used to inspect the prior and posterior distributions overlaid. Figures 7.26 and 7.27 show these plots for the informative and vague prior distributions,

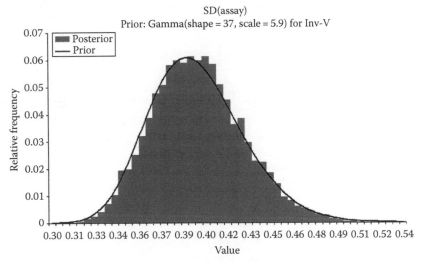

FIGURE 7.26
Posterior and informative prior distribution for assay SD component in model used to analyze test compound data.

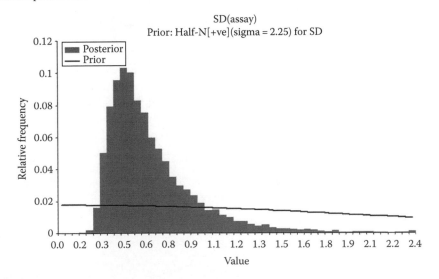

FIGURE 7.27
Posterior and default vague prior distribution for assay SD component in model used to analyze test compound data.

respectively. Notice the vastly different scales needed in the two plots in order to fairly represent each distribution. It is clear that the informative prior distribution is very strong relative to the amount of information contained in the likelihood, causing the posterior to be almost identical to the prior distribution. This means it is very important that the assumption of exchangeability between the random errors in the standard compound data and those in the test compound data is credible. If there were serious doubts about this assumption, one approach would be to elicit a prior distribution for the between assay variation from subject matter experts (see O'Hagan et al. 2006), using the posterior summaries derived from the standard compound as evidence to consider, rather than as a means to determine the prior directly. Additional sources of information could be obtained from other test compounds that had been repeatedly assayed. These data could be used to assess the heterogeneity of the variance components, perhaps building a more complex hierarchical model to adequately represent this heterogeneity. This approach could either be used to provide more information to be considered in an elicitation exercise, or to derive the prior distribution directly using an analogous approach to that outlined above.

8

Bayesian Variable Selection

An important part of any statistical analysis is the work done to check that the fitted model is fit for purpose, which often provides clues on how to improve the model. Skill in building statistical models is a key attribute of an experienced statistician. When there are numerous potential predictor variables that should be considered for inclusion, this task becomes particularly challenging. Ironically, it is in these cases when we have an abundance of potentially useful predictors that we are beset by some of the more difficult statistical issues. It is well known that indiscriminately including numerous variables in a model leads to over-fitting of the data, which results in a model with very poor predictive ability.* Hence, many statisticians have devoted large parts of their lives developing methods to help determine which variables to include in our models. Still the most common general approach is based on the concept of the p-value, in which the statistical significance of each variable is assessed, and only included if it passes some nominal threshold, typically less than 5%. Software packages such as SAS, Genstat, and R will routinely provide t and F tests for this purpose as part of their output after fitting linear models. The stepwise regression procedures are based on this approach (see Draper and Smith, 1998). A crude Bayesian version of this approach would be to simply inspect the posterior credible interval for a parameter, and decide based on whether zero was a credible value. Other uniquely Bayesian model checking functions, based on predictive distributions, also exist (see Section 2.9 for information on those provided by BugsXLA). Global measures of goodness-of-fit, such as the deviance, or information criteria such as the AIC (see Brown and Prescott, 2006), as well as the Bayesian DIC, have all been used to choose between models. Irrespective of which approach is used to identify the final model, if all the inferences are based solely on this model, then a potentially major source of uncertainty will be ignored, namely, uncertainty about the model itself. As a consequence, it is likely that uncertainty about quantities of interest will be underestimated. A vast amount of research into this topic exists, with Draper (1995) providing good coverage of the major issues and the Bayesian methods for dealing with them.

In principle, the Bayesian solution to this problem is straightforward. The model itself can be treated as another nuisance parameter, which is given a prior distribution, allowing uncertainty in its "true value" to be integrated

---

* Although Lindley, in his discussion of the paper by Draper (1995), suggests that over-fitting is not inherent but a consequence of adopting non-Bayesian methods, or inappropriate priors.

over in the usual way (see Smith, 1986). This approach comes under the general heading of "Bayesian Model Averaging" in the literature. In practice, this is difficult because it requires all the potential models to be specified in advance of seeing the data, each with their associated prior probabilities, and also due to the large computational issues it raises. Bayesian Variable Selection (BVS) refers to techniques that apply this approach to handling model uncertainty in cases where the models only differ due to the subset of a prespecified list of predictor variables that they include. A common situation is in the building of a linear regression model from a relatively large list of potential explanatory variables, and we use this to explain the approach adopted by BugsXLA; initially we also assume this is a Normal Linear Model (NLM). Given a response variable, y, and a set of candidate predictors, $x_1, x_2,..., x_k$, the conventional objective of variable selection is to find the "best" model of the form:

$$y = \alpha + \beta_1 x_1 + \beta_2 x_2 + \cdots + \beta_p x_p$$

where $x_1, x_2,..., x_p$ is a subset of $x_1, x_2,..., x_k$. Although BVS can be used as another technique to identify a single model, as its name unfortunately suggests, it could be argued that its real power comes from using it to model average as discussed above; Meyer and Wilkinson (1998) use the term "Bayesian Variable Assessment" to emphasize this point. Many BVS methods have been developed (see O'Hara and Sillanpaa (2009) for a recent review), and BugsXLA adopts an approach similar in essence to that first proposed by Kuo and Mallick (1998). The β_i parameters in the model are effectively given a mixture prior, also know as a "slab and spike" prior, in which a mass of prior probability, the spike, is assigned to zero. This reflects the belief that there is a non-negligible probability that some of the predictor variables have no effect at all on the response. Variables whose true effect size is nonzero are termed "active." This prior distribution is specified via the introduction of two auxiliary parameters:

I_i : an indicator parameter taking the value 1 if the variable is active and 0 otherwise

β_i^* : a parameter representing the "slab part" of the variable's effect size

The model parameters are then related to these auxiliary parameters via their product:

$$\beta_i = I_i \beta_i^*$$

with the slab and spike prior being induced by independent priors for the two auxiliary parameters:

$I_i \sim \text{Bern}(\pi)$

$\beta_i^* \sim N(0, \gamma^2 \sigma^2)$, where σ^2 is the usual residual variance parameter in the NLM that is, a Bernoulli and Normal Distribution, respectively. The parameter π represents the prior probability that a predictor variable is active, and this is assumed to be the same for all variables, independently of which other variables are also active. The prior for the β_i^* was chosen based on the work of Meyer and Wilkinson (1998). In the absence of informative priors for the regression coefficients based on subject matter expertise, it feels reasonable to define the likely effect size of an active variable relative to the residual variation, with the parameter γ determining the relative prior precision. This approach at least helps ensure that the prior is invariant to the scale of measurement used for the response. Note that the same γ is used for all β_i^*, and in order to make this assumption of exchangeability between these parameters more credible, all the predictor variables are scaled to unit variance. BugsXLA back-transforms the regression coefficient estimates, so that this scaling is invisible to the end user. Note that it is not possible to adopt the standard "fixed effects" vague prior for the β_i^*, that is, a Normal Distribution with extremely large variance, since this is likely to cause the phenomenon referred to as "Lindley's paradox" (see Shafer (1982) for details). The implications of this paradox are that no matter how large an effect the predictors truly have on the response, the BVS approach will always support the simplest model, that is, none of the variables will be identified as being active.

A widely recommended value for π is 0.25 (e.g., Box and Meyer 1993, Chipman et al. 1997, Meyer and Wilkinson 1998), reflecting the belief, obtained from practical experience, that the majority of predictor variables studied tend to be inactive. Meyer and Wilkinson (1998) also state that values of γ in the range of 0.5–3.0 worked well in their experience of applying this method. BugsXLA uses this information to suggest default priors for these two parameters:

$\pi \sim$ Beta(30, 90), which implies a mean of 0.25 with standard deviation 0.04

$\gamma \sim$ U(0.5, 3.0)

Typically, it is unlikely that there will be much information in the data with which to update these prior beliefs, so they should be seen as a convenient way to express uncertainty about the values for these parameters, which will then be appropriately propagated into the final inferences. As will be seen in the case studies, it is possible to alter these priors from their defaults to suit the experimental context. For example, in relatively large supersaturated designs it might be more reasonable to expect much fewer than 25% of the variables to be active, adjusting the prior for π accordingly.

As well as including the regression coefficients associated with individual predictor variables in a BVS analysis, it is possible to include cross-product terms between two variables. The coefficients associated with these cross-product terms are given the same prior as their "parent terms," with the exception that the user is given the option to put constraints on the π parameters in order to force some form of "inheritance principle" on the terms included in the model (see Chipman et al. (1997) for more details). Three additional parameters are defined representing the prior probabilities of activity for cross-product terms:

whose parents are both active	: π_{11}
with one active and one inactive parent	: π_{10}
whose parents are both inactive	: π_{00}

Each of these is set equal to some multiple of π, such that $\pi \geq \pi_{11} \geq \pi_{10} \geq \pi_{00}$. Strong heredity is forced if both π_{10} and π_{00} are zero, implying that a cross-product term can only be included in a model in which both of its parents are also included. Strict weak heredity is allowed if only π_{00} is zero, while relaxed weak heredity is allowed if all three are nonzero. Some have argued that only models satisfying the strong heredity principle should be considered, notably Nelder (1977) who repeatedly criticized models that failed to adhere to this principle. BugsXLA forces strong heredity by default.

Categorical factors can also be included in a BVS model. In fact, it is in the area of fractional factorial designed experiments that this approach has demonstrated some significant advantages over traditional approaches (see, e.g., Box and Meyer (1993), Chipman et al. (1997), and Wu and Hamada (2000)). In most cases, in order to use a method initially developed for regression coefficients, factors have been reparameterized so that their effects can be represented by coefficients associated with dummy variables (as discussed in, e.g., Dobson (1988)). Although this causes no additional issues for two-level factors, which only have one parameter to represent its effect, problems arise with multilevel factors due to the need to take account of the inevitable dependencies that exist between the parameters that define their multidimensional effects. BugsXLA circumvents this issue by specifying a prior on the factors' effects directly. The n_i effects of the ith factor, α_{ij}, are given a prior that is similar to that given to the BVS regression coefficients:

$$\alpha_{ij} = I_i \, \alpha_{ij}^*$$

$$\alpha_{ij}^* \sim N(0, \, k^2\gamma^2\sigma^2)$$

where I_i, γ, and σ are the same as before. The constant k is given the default value $\sqrt{2}$ so that the same prior is implied for a two-level factor as would be the case if it was represented by a dummy variable with coded values -1 and $+1$ for the two levels, this being the convention usually adopted in the literature. This can be seen by noting that the difference between the effects at the two levels would be

$$\alpha^*_{i2} - \alpha^*_{i1} = 2\beta^*_i$$

with implied prior variances of

$$2k^2\gamma^2\sigma^2 = 4\gamma^2\sigma^2$$

As well as the main effects of a factor, two-factor interactions can also be included in BVS models; the same inheritance principle applies as discussed for variates above.

In order to improve the MCMC performance for the general multilevel case, the actual effects fitted are constrained to sum-to-zero with an implied multivariate Normal prior distribution determined from that defined above (see Woodward and Walley (2009) for details). Note that if both BVS variates and factors are included in the same model, the same γ and π parameters are used for both types of variables. This implies that the unconstrained factor effects, scaled by a factor of k, and the regression coefficients are considered exchangeable. Although this is unlikely to be strongly believed in many cases, we do not believe this exchangeability assumption is much less realistic than those already adopted in the BVS literature. This suggests it is likely to be an adequate approximation, leading to useful results.

BugsXLA allows BVS terms to be included for most of the error distributions. The only difference between the priors in each case is how the σ parameter is defined. Section 2.7 explains how BugsXLA chooses the σ parameter for each error distribution. It is only in the Normal case that the BVS method has been extensively tested, and the adaptation to the other error distributions is ad-hoc and largely untested. Hence, it should not be surprised if in some cases the MCMC algorithm fails completely, and even when it does return results, as well as checking convergence, the sensitivity of these results to other prior assumptions should be assessed. Chapter 11 of Ntzoufras (2009) provides alternative ways of specifying BVS models using WinBUGS directly, covering both the NLM and GLM cases.

It should be clear that BugsXLA can be used to specify models that are a mixture of both BVS and conventional terms. This enables the user to ensure that some variables are always included in the model, equivalent to giving them a prior probability of being active, π, of one. For example, in designed experiments where factors are needed to represent the known additional sources of variation, for example, "blocking factors," these would be included as conventional fixed or random effects.

Case Study 8.1 Pollution
The pollution data originally published by McDonald and Schwing (1973) have been widely used to assess variable selection methods (see, e.g., Meyer and Wilkinson (1998) and O'Hara and Sillanpaa (2009)). The response variable is the 1963 age-adjusted mortality rates per 100,000 from 60 different

TABLE 8.1

Description of the Variables in the Pollution Data Set

Variable	Description
A	Average annual precipitation in inches
B	Average January temperature in °F
C	Same for July
D	% of 1960 SMSA population aged 65 or older
E	Average household size
F	Median school years completed by those over 22
G	% of housing units that are sound and with all facilities
H	Population per square mile in urbanized areas, 1960
I	% non-white population in urbanized areas, 1960
J	% employed in white collar occupations
K	% of families with income less than $3000
L	Relative hydrocarbon pollution potential
M	Same for nitric oxides
N	Same for sulfur dioxide
O	Annual average % relative humidity at 1 pm
MORT	Total age-adjusted mortality rate per 100,000 (response)

locations in the United States. Fifteen potential predictor variables were measured as described in Table 8.1, with the short coded variable names used in BugsXLA to simplify the BVS model specification statement.

Figure 8.1 shows some of the data with the completed model specification form. The model is exactly as described in the introduction to this chapter. Although BVS models are notorious for their poor MCMC mixing, the initial exploratory runs indicated that for these data the simple model defaults for the MCMC settings would be adequate (a burn-in of 5,000 followed by 10,000 samples). In order that the posterior distributions of the regression coefficients could be more fully explored, samples were requested to be imported for the 'Covariate Coeffs' on the 'MCMC & Output Options' form.

Figure 8.2 shows the BugsXLA form where the priors for the BVS terms can be altered. Refer back to the description of the prior distributions given in the introduction to this chapter for an explanation of each of the settings available. Note that since no cross-product terms were specified, the heredity principle is irrelevant in this case, that is, the values of p11, p10, and p00 in Figure 8.2 are redundant. The default priors for all the parameters were used in the analysis (see cells A28:B34 in Figure 8.3). Figures 8.3 and 8.4 show the results of the analysis as imported back into Excel.

As well as the usual summaries provided, an additional column, labeled 'Pr(Active)', is reported. This column gives the estimated posterior probability of activity for each variable, as well as the probability that none of the

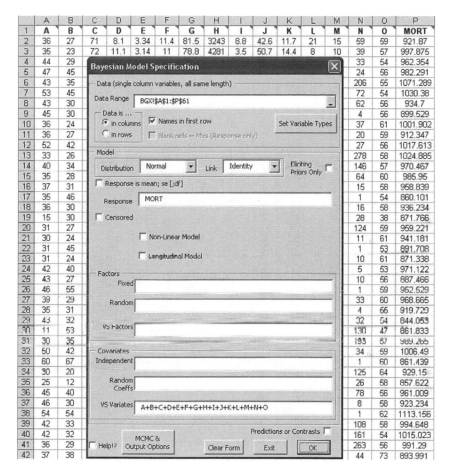

FIGURE 8.1
Data and model specification for pollution study.

terms are active, which in this case is zero. BugsXLA also provides a graphical representation of these activity probabilities, which is shown in Figure 8.4.

This analysis identifies the same important predictor variables as did Meyer and Wilkinson (1998). They used a different prior, notably a higher prior probability of activity parameter, $\pi = 0.5$, implying that all models are a-priori equally likely. This increased prior belief increased the posterior activity probabilities as one would expect. The work of O'Hara and Sillanpaa (2009) indicated that the posterior probability is approximately equal to the prior probability when there is no evidence that the variable should be included. Although in this case study the relationship was closer to a 50% reduction, their conclusion that the results should be presented in terms of a relative change from the prior probability is worth considering unless we have good reasons for the prior probability chosen.

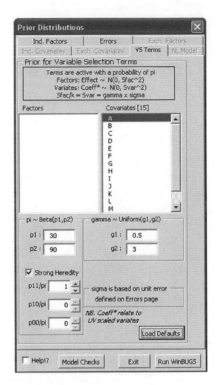

FIGURE 8.2
Default priors for pollution study.

	A	B	C	D	E	F	G	H	I	J	K
1	Pr(Active)		Label	Mean	St.Dev.	2.5%	Median	97.5%		WinBUGS Name	
2			CONSTANT	940.2000	4.6870	931.0000	940.2000	949.4000		Beta0	
3	0.75		A	1.3740	0.9797	0.0000	1.5620	3.0280		VScov.Coeff[1,1]	
4	0.78		B	-1.2890	0.8654	-2.8150	-1.4340	0.0000		VScov.Coeff[2,1]	
5	0.20		C	-0.4321	1.0630	-3.7460	0.0000	0.0000		VScov.Coeff[3,1]	
6	0.14		D	0.8402	3.0390	0.0000	0.0000	11.2300		VScov.Coeff[4,1]	
7	0.10		E	-3.5020	22.5900	-75.6300	0.0000	7.0880		VScov.Coeff[5,1]	
8	0.46		F	-8.0880	10.4000	-31.2600	0.0000	0.0000		VScov.Coeff[6,1]	
9	0.13		G	-0.1808	0.6898	-2.5800	0.0000	0.0000		VScov.Coeff[7,1]	
10	0.18		H	1.002E-3	2.696E-3	0.0000	0.0000	9.634E-3		VScov.Coeff[8,1]	
11	1.00		I	4.1750	0.9771	2.4160	4.1290	6.3330		VScov.Coeff[9,1]	
12	0.14		J	-0.2049	0.7502	-2.7860	0.0000	0.0000		VScov.Coeff[10,1]	
13	0.13		K	-0.0800	1.0380	-3.0900	0.0000	1.8560		VScov.Coeff[11,1]	
14	0.13		L	-0.0138	0.0735	-0.1937	0.0000	0.0189		VScov.Coeff[12,1]	
15	0.13		M	0.0155	0.1446	-0.1921	0.0000	0.3638		VScov.Coeff[13,1]	
16	0.87		N	0.2587	0.1311	0.0000	0.2803	0.4737		VScov.Coeff[14,1]	
17	0.08		O	0.0457	0.3131	0.0000	0.0000	1.0680		VScov.Coeff[15,1]	
18	0.00		No Active VS Terms							VS.NoneIn	
19			SD(residual)	35.9500	3.7130	29.5000	35.6100	43.8900		sigma	
20											
21	Model	[BGX!$A$1:$P$61]									
22	Distribution	Normal									
23	Link	Identity									
24	Response	MORT									
25	VS Variates	A+B+C+D+E+F+G+H+I+J+K+L+M+N+O									
26											
27	Priors										
28	CONSTANT	N(mu=940, sigma=6000)									
29	VS Variate Coeffs	N(0,Svar^2); Svar = VSgamma x VSsigma [with prob. pi, else 0]									
30	VS sigma	= SD(residual)									
31	VS gamma	Uniform(0.5, 3)									
32	VS pi	Beta(30, 90)									
33	(p11, p10, p00)/pi	(1, 0, 0)									
34	V(residual)	Inv-Gamma(0.001, 0.001)									
35											
36	WinBUGS MCMC Settings										
37	Burn-In: 5000 Samples: 10000 (Thin:1; Chains:1)										
38	Run took 3.5 minutes										

FIGURE 8.3
Summaries of posterior distributions, model, priors, and MCMC settings for the pollution study.

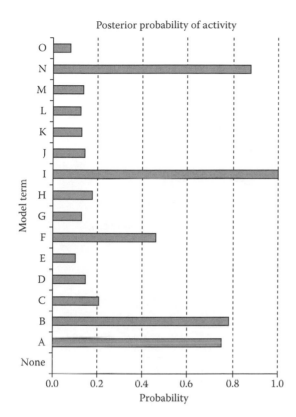

FIGURE 8.4
Plot of posterior probabilities of variable activity for the pollution study.

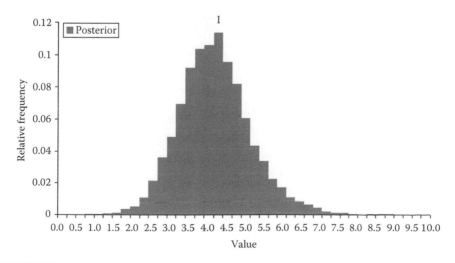

FIGURE 8.5
Posterior distribution of the regression coefficient for variable I in the pollution study.

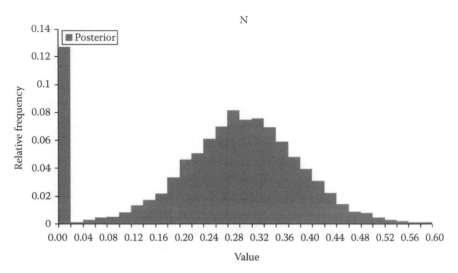

FIGURE 8.6
Posterior distribution of the regression coefficient for variable N in the pollution study.

Another summary that is useful in interpreting the results is the histogram of the posterior distribution that BugsXLA can provide via its posterior plotting tool (see Section 2.11 for details or Case Study 3.1 for an example of how to obtain). Figures 8.5 and 8.6 show the posterior distributions of the coefficients associated with variables I and N respectively. Although the posterior distribution for variable I is slightly skewed to the right, it is clearly uni-modal, reflecting the fact that this variable has a probability of activity very close to one. The same plot for variable N shows a relatively tall spike at zero, even though this variable has an activity probability of 87%. One needs to be very careful when interpreting histograms in cases such as these when there is a point mass of probability at a single value, for example, at zero here. This is because the height of the slab part of the posterior distribution is affected by the choice of bin width, where as the height of the spike at zero remains the same.

The conclusion to draw from this analysis is that variables A, B, I, and N are each likely to be important predictors of mortality, with all bar variable B being positively correlated with mortality. There is weaker evidence to suggest that variable F is negatively correlated with mortality. None of the other variables appear to be "statistically significant." In this particular example, there does not appear to be any value in using the model to make predictions (even if we were doing this analysis back in the early 1960s), so the model averaging aspect is not of value here. Presumably the main objective was to assess the impact on mortality of the pollutants measured, so that this could inform public policy on the need for increased controls

TABLE 8.2

Factors and Levels for the Blood Glucose Experiment

Variable	Description	Levels		
		1	2	3
A	Wash	Yes	No	—
B	Microvial vol. (mL)	2.0	2.5	3.0
C	Caras water level (mL)	20	28	35
D	Centrifuge (RPM)	2100	2300	2500
E	Centrifuge time (min)	1.75	3.00	4.50
F	Sensitivity absorption	0.1–2.5	0.25–2.0	0.5–1.5
G	Temperature (°C)	25	27	30
H	Dilution	1:51	1:101	1:151

on their emissions by industry. In which case, this study clearly indicated that mortality rates could have been reduced by tighter controls on sulfur dioxide emissions.

Case Study 8.2 Blood Glucose Experiment

Chipman et al. (1997) provide a BVS analysis of a data set originally presented by Henkin (1986). The data come from an 18-run mixed-level (1 two-level and 7 three-level) fractional factorial experiment designed to help improve an analytical method for the measurement of glucose in blood samples. The objective was to identify those factors that affect the mean blood glucose reading so that changes could be made in order to improve the precision of the analytical method. Table 8.2 shows the factors with their levels assessed in the experiment.

Chipman et al. treated all of the variables as continuous variates, and used polynomial terms to model the three-level factors. Here we treat all the variables as categorical factors, which should give similar results since, just like the linear and quadratic terms in a polynomial model, this will account for all the variation due to a factor with three levels. The parameterization adopted here has the theoretical advantage of modeling the dependencies between the multivariate effects of a three-level factor, as discussed previously, unlike in the polynomial approach where these were ignored. Figure 8.7 shows the data with the completed model specification form.

All the predictor variables were defined as factors. As well as assessing the main effects of each of the factors, it was hoped that the experiment might identify any large interactions between them. BugsXLA only allows second order interactions to be fitted as BVS terms. Note how the '@' operator has been used to facilitate concise specification of the model that includes all possible two-factor interactions, but no higher order terms. The list of variables separated by the '*' operator defines all terms that include these variables,

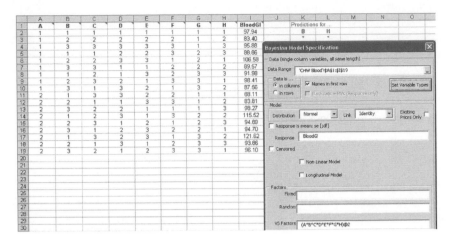

FIGURE 8.7
Data and model specification for blood glucose experiment.

from the main effects up to the eight-way interaction. By adding the '@2' at the end of the model statement, it instructs BugsXLA to exclude terms higher than the second order. Strictly speaking, the brackets are not needed in this model statement, but including them makes it explicit that the final operator acts after all of the cross-product terms have been expanded.

The model is essentially as described in the introduction to this chapter, noting that factor effects rather than the products of regression coefficients and variates define the linear predictor. A total of 28 two-factor interactions as well as the main effects of the eight factors are included in this model. Since there are only 18 experimental runs, it is clear that this model could not be fitted using conventional methods due to the high degree of aliasing that must exist between the terms. This should alert the user to the need for careful checking that the MCMC process is mixing sufficiently well and that convergence has occurred. To do this, the novel MCMC settings were chosen (a burn-in of 5,000 followed by 1,000 samples on each of three chains, with the auto quit set to 'No'). The prior distributions for the parameters were left unchanged from the defaults apart from those determining the type of heredity principle applied. Figure 8.8 shows the 'Prior Distribution' form after deselecting the 'Strong Heredity' check box.

If the strong heredity assumption is relaxed, BugsXLA suggests the settings shown in Figure 8.8, based loosely on the suggestions in Chipman et al. (1997). Since we wished to compare our results with those reported by Chipman et al., these settings were used in the analysis. Our preference is to always force the strong heredity principle, sometimes deviating from the BugsXLA default by lowering the value of p11/pi to 0.5 to reflect the belief that interactions are less likely than main effects.

After running the specified number of iterations in WinBUGS, the program's convergence checking options were used to assess the MCMC process

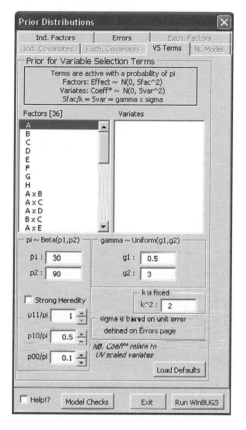

FIGURE 8.8

Prior distribution for BVS parameters in blood glucose experiment.

as discussed in Section 1.2. An initial inspection of the first 1000 samples indicated, perhaps rather surprisingly, that mixing was not too bad after all, and so only a further 4000 samples were requested. Figure 8.9 shows history plots of the first four elements of the two-dimensional array variable named 'VSfac.Eff', this being the generic name given to the WinBUGS variable that represents the BVS factor effects.

Inspection of the top two plots show that they are mirror images of each other; this is because they are the effects associated with the two-levels of factor A, which have been constrained to sum-to-zero as discussed earlier. It is also clear that much of the time the chain for these variables is at zero exactly, indicating that this factor is often not being identified as active. Although, as would be expected, there is clear evidence of autocorrelation in the chain, there is reasonably good mixing as evidenced by the relatively large number of separate occasions when the chain moves between the "active" and "inactive" states; poor mixing is seen when the chain frequently gets stuck for long periods in one relatively small region of the whole parameter space. The bottom two plots in Figure 8.9 represent the first two-levels of factor B, which are not mirror images due to this factor having three-levels. Since this factor is frequently being identified as active, resulting in the majority of the

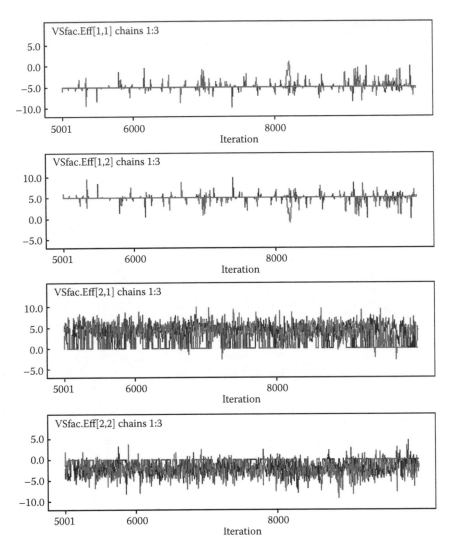

FIGURE 8.9
History plots for some of the factor effects parameters in blood glucose experiment.

sampled values being non-zero, it is easier to see that the three chains have converged and are mixing reasonably well.

If the Brooks–Gelman–Rubin (bgr diag) plot is requested for this variable, the dreaded WinBUGS error trap message appears. This is because some of the elements of this array, such as the first two shown in Figure 8.9, contain long runs of values that are identical causing the algorithm to fail. It is still possible to obtain plots for individual elements of the array, for example, entering 'VSfac.Eff[2,1]' in the 'node' field on the WinBUGS 'Sample Monitor Tool' and clicking on the 'bgr diag' button gives the plot shown in Figure 8.10.

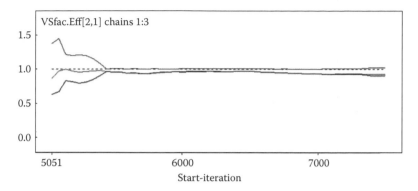

FIGURE 8.10
Brooks–Gelman–Rubin plot for effect of first level of factor B in blood glucose experiment.

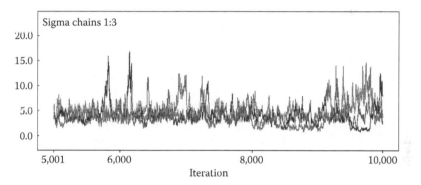

FIGURE 8.11
History plot for sigma parameter in blood glucose experiment.

Since the top line (red in WinBUGS) has converged to one, and the other two lines (green and blue in WinBUGS) have converged to the same value, this plot suggests that the MCMC chain has converged.

It is important to check that the other variables have converged, and Figure 8.11 shows the history plot for the sigma parameter. The mixing does not look as good as for the factor effects parameter with occasions where the three chains are clearly separated for relatively long periods. However, the three chains do appear to be randomly fluctuating about the same general region of the parameter space.

The chain for the sigma parameter exhibits extremely high autocorrelation, which even after applying 1/10th thinning is still evident as shown in Figure 8.12.

After exiting from WinBUGS, the MCMC settings were changed back to a single chain with a burn-in of 10,000 followed by 10,000 samples after a 1/50th thinning to greatly reduce their autocorrelation. Since we were expecting that the only active terms would include the factors B and H, we also requested that the predicted means for all combinations of the levels of these two factors

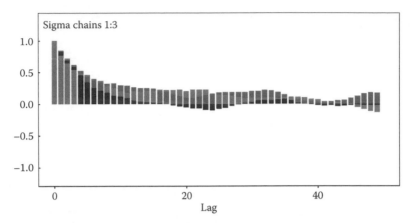

FIGURE 8.12
Autocorrelation plot after 1-10th thinning for sigma parameter in blood glucose experiment.

be estimated (see cells K2:L3 in Figure 8.7). This analysis took 2.9 hours to run. Clearly if one wished to explore model sensitivity using these MCMC settings, it would be a very slow and laborious process. In these circumstances, it is sensible to first assess the sensitivity of the results to these settings. Table 8.3 compares the output obtained from this initial run, labeled "complex" here, with that obtained from using BugsXLA's simple and regular MCMC settings.

Inspection of the output reveals that we were unnecessarily cautious in adopting a 1/50th thinning, since BugsXLA's regular model settings, with a one-fifth thinning, give almost identical results. Although one cannot generalize from a single example, we have found the regular model settings

TABLE 8.3

Comparison of Output Obtained Using Three Different MCMC Settings for the Blood Glucose Experiment

	MCMC Settings		
	Simple	Regular	Complex
Burn-in	5,000	10,000	10,000
Samples	10,000	10,000	10,000
Thinning ratio	1	5th	50th
Run time (min)	5	24	175
Parameter			
Pr(Active) B	0.64	0.71	0.69
Pr(Active) BH	0.93	0.95	0.94
Pr(Active) F	0.08	0.10	0.10
sigma mean (sd)	4.7 (2.0)	4.3 (1.7)	4.2 (1.8)
B(1)H(2) mean (sd)	112 (6.2)	113 (5.7)	113 (5.9)

B(1)H(2) refers to the predicted mean response when factors B and H are at their first and second levels, respectively.

to be adequate for data sets and models with this level of complexity, even when there is very high autocorrelation as indicated in this case (refer back to Figure 8.12). The MCMC errors reported when using the regular model settings were all less than 5% of their associated parameters' posterior standard deviations, meeting the rule of thumb for adequate MCMC precision. It is perhaps surprising that even the simple model MCMC settings gave similar results to the run with 1/50th thinning. This suggests that a reasonable approach in this case would be to do much of the initial model development and assessment with the faster settings, and only resort to thinning to improve the precision of the final inferences.

The results of this analysis were compatible with those reported by Chipman et al. (1997) whose best-fitting models contained the linear term for B plus many of the BH cross-product linear and quadratic terms. In our analysis, the only two terms with posterior activity probabilities greater than 10% were those for B and the interaction BH. The best way to try and understand the effects of these two factors is to visualize their interaction. This can be done by producing a line plot using Excel's built-in charting features as described in Case Study 5.2. Figure 8.13 shows the predicted means (using the regular MCMC settings), with manually inserted blank rows, and the interaction plot produced using the values in columns R, S, and T. The plot clearly shows that the cause of the interaction is the surprisingly high response, when compared to the general pattern, for the combination B(1)H(2). This suggests that the subject matter experts should investigate why this particular combination might behave so differently, as this could generate new insights into the analytical method and ideas for how to improve its precision.

Note that this BVS approach to analyzing fractional factorial experiments not only indicates which factors are likely to be active, but also provides interval estimates, which take account of the uncertainty in the best-fitting model. In the traditional approach to analyzing such experiments, uncertainty in the effect estimates are either not quantified or derived using ad-hoc rules for pooling the smaller effects into an estimate of residual variance, with the usual concerns regarding underestimating the true uncertainty. The posterior distribution obtained from this Bayesian approach provide a coherent and efficient way to both quantify and visualize the uncertainty in the size of the effects. Woodward and Walley (2009) also analyze these data using BugsXLA, forcing the strong heredity principle. They include a plot of the posterior distributions of the sum of the main and interaction effects of the factors B and H, which provides a useful addition to the plot shown in Figure 8.13.

Case Study 8.3 Transmission Shaft Experiment

Davis (1995) describes a fractional factorial experiment aimed at improving the reliability of a transmission centre shaft. This experiment had censored reliability data as the response, which allows us to illustrate how the BVS method can be used on complex non-Normal data. Table 8.4 shows the factors with their levels assessed in the experiment.

R	S	T	U	V	W	X	Y	Z
Label	Mean	St. Dev.	2.5%	Median	97.5%		WinBUGS Name	
Predicted Mean Response								
Following all at: A(zero) C(zero) D(zero) E(zero) F(zero) G(zero)								
B(1) H(1)	92.1400	3.6850	84.5400	92.2300	99.5500		Pred.Ave[1]	
B(1) H(2)	113.0000	5.7010	96.3800	114.3000	120.6000		Pred.Ave[2]	
B(1) H(3)	92.2600	3.6710	84.7800	92.4300	99.5200		Pred.Ave[3]	
B(2) H(1)	100.6000	3.5360	92.7700	100.9000	107.1000		Pred.Ave[4]	
B(2) H(2)	85.6000	4.2170	79.1800	84.8600	96.4700		Pred.Ave[5]	
B(2) H(3)	96.4700	3.1120	90.2200	96.4400	102.8000		Pred.Ave[6]	
B(3) H(1)	95.2700	3.1350	89.2400	95.1700	101.7000		Pred.Ave[7]	
B(3) H(2)	89.4700	3.5620	83.2600	89.1000	97.5000		Pred.Ave[8]	
B(3) H(3)	98.6800	3.2380	92.1600	98.7100	105.2000		Pred.Ave[9]	

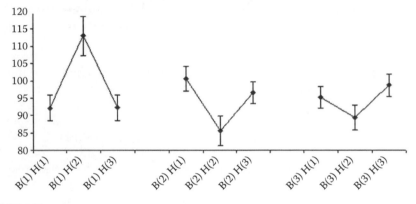

FIGURE 8.13

Plots of predicted means with error bars (posterior standard deviations) depicting the BH interaction in the blood glucose experiment.

TABLE 8.4

Factors and Levels for the Shaft Experiment

		Levels		
Variable	Description	1	2	3
A	Spline end	Spherical	Grooved	—
B	Annealing amount (h)	Nil	1	4
C	Shaft diameter (mm)	16.1	17.7	18.8
D	Shot intensity (Almen)	3	6	9
E	Shot coverage (%)	200	400	600
F	Tempering temp. (°C)	140	160	180
G	Shot blasting	Without	With	—

Davis, like Chipman et al. in Case Study 8.2, treated all of the variables as continuous variates, and used polynomial terms to model the three-level factors. His analysis was more conventional in only considering the main effects in the initial analysis, in which a linear model with a log-normal error distribution was fitted. This was followed by a further analysis in which "insignificant" factors were dropped, allowing interactions between those still in the model to be estimated. Here we treat all the variables as categorical factors, and use the BVS approach, avoiding the need to resort to an ad-hoc iterative model fitting strategy. Figure 8.14 shows the data with the completed model specification form.

The response, FailTime, is the survival time in units of 1000 cycles on the test rig, with blank cells where the time is right censored at the value shown in variable TimeCen. All of the predictor variables were defined as factors, with their levels sorted into ascending order, and the variable TimeCen was defined as type censor as described in Case Study 4.5. The model is essentially as in Case Study 8.2, but with a log-normal error distribution and the linear predictor being linked to the location parameter of the log-normal; this is effectively equivalent to a log-link to the median of the response.

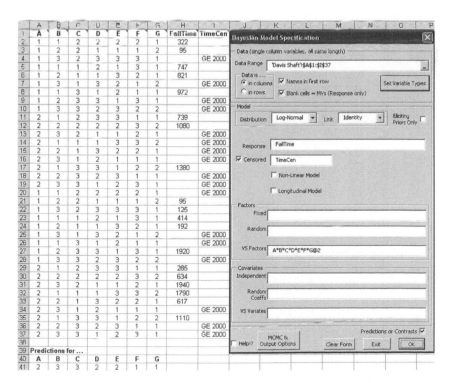

FIGURE 8.14
Data and model specification for shaft experiment.

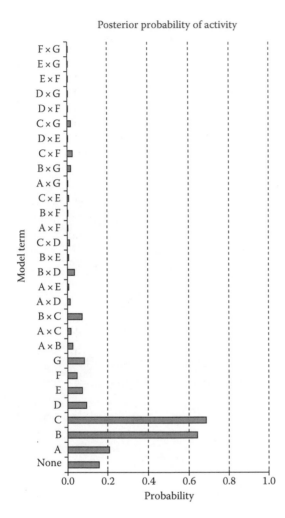

Posterior probability of activity

FIGURE 8.15
Activity probability plot for shaft experiment.

The regular MCMC settings (a burn-in of 10,000 followed by 10,000 samples after a one-fifth thinning) were used, and the default priors left unchanged. WinBUGS took 23.1 min to complete the calculations. Figure 8.15 shows the plot produced by BugsXLA from the WinBUGS output depicting the posterior activity probabilities for each term in the model.

Only the main effects of factors B and C are clearly identified as active, with factor A having the next largest probability, but this being smaller than the 0.25 prior probability assigned to each term. Davis concluded from his more conventional analysis that factors A, B, and C all have active main effects, with little evidence of any interactions. He also fitted a proportional hazards model and concluded that factors B, C, and E were active. It is interesting to note that B and C are the only factors considered active in both of the models fitted by Davis. Davis proceeded to make predictions of the lifetime for shafts made with the optimal settings, which, using the log-normal model, were a mean log-lifetime of 9.74 with a conventional 95% confidence interval of 8.68–10.80. This illustrates one of the criticisms with the conventional approach to model selection, that once a model is selected, none of the uncertainty in the model itself is accounted for in any inferences that are subsequently made. To obtain this prediction using BugsXLA, the analysis was run with the predictions requested by cells A40:G41 in Figure 8.14; the optimal settings were determined simply by noting the level for each factor associated with the largest positive effect, as estimated from the initial analysis. As well as requesting the 'Summary Stats' for the 'Mean Response', both the 'Posterior Samples' and 'Individual Response' check boxes were also selected on the 'Predictions & Contrasts' form. By importing the predictions for an individual response, this will allow us to make probabilistic statements about the lifetime of an individual transmission shaft, arguably of more interest to the end user of this product than a prediction for the population mean. Table 8.5 shows the summary statistics for the predictions requested.

BugsXLA only reports the predictions in the original units of 1000 cycles. The reason that the population level prediction is of the median lifetime is that the location parameter of the log-normal, although representing the mean log-lifetime, when exponentiated represents the median on the natural scale. The estimate quoted by Davis can be transformed to the natural scale, noting that the mean will transform to the median, giving the estimates shown in Table 8.5. The BVS predicted lifetime is shorter than that obtained by Davis, and the prediction interval is wider. This reflects the additional uncertainty associated with the model-selection process that is more appropriately reflected in the Bayesian approach. The prediction interval for an individual shaft is very wide, which is typical for such studies, which are designed to identify the important factors leading to reliability improvement rather than to estimate lifetimes. An alternative prediction that could be of value is the probability that a centre shaft will survive longer than the product's guaranteed lifetime; this value would be useful in estimating the expected number of warranty claims. If, as is common in such reliability tests, the censoring time of 2×10^6 cycles was chosen to represent this period, then Figure 8.16 shows how BugsXLA's posterior plotting tool can be used to calculate this probability (refer to Case Study 3.1 for an explanation of how

TABLE 8.5

Summary Statistics for the Predicted Lifetime of Shafts Made with Optimal Settings

	Mean	St. Dev.	2.5%	Median	97.5%
Shaft median lifetime	18,000	40,000	1300	8,400	93,000
Davis (transformed)			5900	17,000	49,000
Individual shaft lifetime	42,000	160,000	250	8,900	300,000

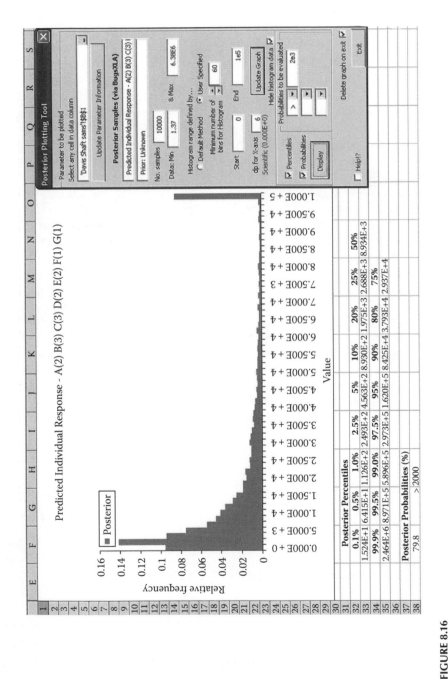

FIGURE 8.16
Predictive distribution for lifetime of an individual transmission shaft.

to use this feature). The output in cells F37:H38 show that there is approximately 80% probability that a centre shaft made with the optimal settings will survive past its guaranteed lifetime. Note that the right most bin on the histogram shows a distinct spike, indicating that that there is still a sizeable tail to the predictive distribution outside the plotting range specified. The percentiles of this distribution, shown in the cells beneath the plot, provide more information on the length of this tail.

If the same analysis is repeated using a Weibull Distribution the same two factors, B and C, are the only ones clearly active. Using this Weibull model, the probability that a centre shaft will survive longer than 2×10^6 cycles is about 90%. This illustrates the sensitivity of inferences involving the tails of a distribution when there is insufficient data to model these tails. If this analysis is repeated using the Gamma Distribution, WinBUGS fails to complete the analysis. Inspection of the log produced by WinBUGS reveals the following error message:

cannot sample from interval censored gamma full conditional Y[3]

This is probably related to the same problem that occurred in Case Study 4.5, providing more evidence to suggest that WinBUGS has a problem when modeling censored data using the Gamma Distribution.

9

Longitudinal and Repeated Measures Models

Data are defined as being longitudinal if each individual, or other experimental unit, has more than one measurement taken through time. This contrasts with what are sometimes called cross-sectional data, which consist of a single outcome on each experimental unit. It is only when a study is designed to obtain longitudinal data that an assessment of within-unit changes in the response over time can be made. In the context of population studies, longitudinal data are required in order to be able to separate out what are termed "cohort" and "age" effects. Cohort effects are caused by differences between the groups of individuals in the study, typically with each group being a different age cohort. Age effects are caused by changes that occur within an individual over time. In a purely cross-sectional study, when only one measurement is taken from each individual, even when the individuals span a wide range with respect to their age, the cohort and age effects are completely confounded.

Repeated measures data also consist of multiple measurements taken on each experimental unit, typically with each measurement being under different experimental conditions. The distinction between repeated measures and longitudinal data is that in the former case the multiple measurements on an experimental unit are typically taken at the same time point, although these commensurate measurements are sometimes also repeated at different times. Repeated measures experimental designs are used, not primarily to assess changes in the response over time, but to improve the precision with which the treatment effects are estimated and/or to assess the magnitude and causes of variation within an experimental unit. Due to the similarity between these two types of data, the distinction between longitudinal and repeated measures data is sometimes blurred in the literature. Due to the methods for analyzing repeated measures data having their origins in agricultural research, the term "repeated measures analysis" is often taken as being synonymous with analysis of variance (ANOVA) techniques for analyzing split-plot experiments (see, e.g., Cochran and Cox (1992)). Even though much research has occurred, and is still ongoing, ANOVA models are still a useful starting place for analyzing either repeated measures or longitudinal data.

With the advent of better models for describing data with multiple sources of variation, complex correlation structures, and non-normal error distributions, along with the computing power and software to fit these models, there are now numerous methods for analyzing longitudinal and repeated measures data. Much can be done simply by utilizing the additional flexibility provided by the mixed models discussed in Chapters 5 and 6. There are now excellent texts that both describe the underlying theory, as well as providing incredibly valuable practical advice, for example, Fitzmaurice et al. (2004) or Diggle et al. (2002). The only additional feature of specific relevance to longitudinal data that BugsXLA provides is the option to model serial correlation between measurements taken on the same experimental unit over time. Specifically, a first-order autoregressive process can be specified for the within unit residual errors, also known as the exponential correlation model. This serial correlation can be supplemented by additional measurement error, which is appropriate when the assumption of perfect correlation between measurements taken at the same time point is likely to be a poor approximation. The theoretical details underpinning these modeling options, as well as the broader approach of using parametric models to impose a covariance structure, are provided in Chapter 5 of Diggle et al. (2002). The practical details of how to implement this using BugsXLA are provided in Case Study 9.2. These serial correlation models can only be applied when the error distribution is Normal, Log-Normal, or t-distribution; strictly speaking the theory does not hold for the t-distribution, but it is included as it is likely to provide an adequate approximation in most cases. When the error distribution is Poisson an analogous approach provided by BugsXLA is the transition model introduced by Zeger and Qaqish (1988), and discussed in Section 10.4 of Diggle et al. (2002). The practical implementation of this approach using BugsXLA is illustrated in Case Study 9.3.

It should be recognized that some of the case studies discussed in earlier chapters of this book involve longitudinal or repeated measures data. Table 9.1 lists these case studies, describing how they relate to the types of data being discussed in this chapter. Also, Case Study 6.2 has been analyzed by others in a longitudinal context. Instead of treating the baseline epilepsy count as a covariate in the model, it is possible to include it as part of the response variate, which then makes these longitudinal data with responses at two different time points for each patient. Both Fitzmaurice et al. (2004) and Diggle et al. (2002) analyze this case study using this explicit longitudinal interpretation.

TABLE 9.1

Case Studies Containing Longitudinal or Repeated Measures Data

Case Study	Data Type	Experimental Units	Objective
5.2	Repeated measures and longitudinal	One device	Sources of variation Time trend (stability)
5.4	Longitudinal	Rats	Growth curve
6.3	Longitudinal	Patients	Treatment effects
6.4	Repeated measures	Batches	Optimal process settings
7.1	Repeated measures	Donors	Exposure response curve

Before discussing how to include a serial correlation component to a longitudinal model, we first illustrate how one of the most widely used experimental designs that generates longitudinal data can be analyzed using BugsXLA.

Case Study 9.1 Pain Treatment Cross-Over Trial
The cross-over design has been extensively used to investigate the safety and efficacy of potential medicines in the early stages of clinical development. In their book dedicated solely to the design and analysis of such experiments, Jones and Kenward (2003) describe a three-period cross-over trial of an analgesic medicine for the treatment of pain in patients suffering from dysmenorrhoea. The response analyzed here is the patients' reported outcome at the end of each period, which indicates whether or not they felt relief from their pain symptoms. Two doses of the compound plus a placebo control were included in the trial, with each patient receiving each of the three treatments once. A total of 86 women were randomized to one of the six possible treatment sequences, with constraints to minimize the amount of imbalance in order to ensure the treatment and period factors would be almost completely unconfounded. This type of three period cross-over design also allows the carry-over effect to be assessed; carry-over is defined as the effect that a treatment in one period has on the response in the following period. Although in practice we agree with the skepticism raised in Section 17.2.4 of Senn (2007) regarding the pharmacological incredibility of the standard statistical approach to modeling carry-over in drug trials, we illustrate here how one can fit such a term for the benefit of those who wish to use the technique in circumstances where it does make scientific sense. Figure 9.1 shows some of the data and the completed model specification form.

All of the five variables, including the response, relief, were defined as factors. The two levels of the response represent the answers "Yes" and "No" to whether the patient obtained pain relief in that period. In order that the probability of obtaining relief was modeled, Y was set as the first level for the factor relief. The levels for the treatment factor, trt, are P for placebo, L for low dose, and H for high dose. The carry-over factor, carry, was parameterized such that the levels L and H indicate that this treatment was administered in the previous period. Level 0 of carry indicates that either this is the first period or placebo was administered in the previous period. Parameterizing carry in this way does not imply that there is no placebo carry-over effect. The difference between the mean responses in period one when there is no carry-over and that in period two or three when placebo was administered in the previous period is confounded with the period effect. The variables pat and per represent the patient number and period, respectively. This model can be represented algebraically:

$$\eta_{ijkl} = \mu + p_i + \alpha_j + \gamma_k + \zeta_l \quad i = 1, 2, \ldots, 86 \quad j = 1, 2, 3 \quad k = 1, 2, 3 \quad l = 1, 2, 3$$

$$\text{logit}(\pi_{ijkl}) = \eta_{ijkl}$$

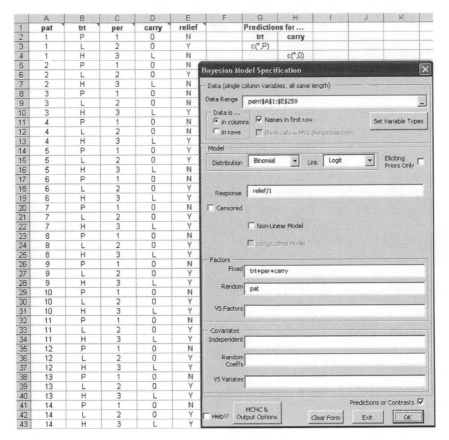

FIGURE 9.1
Data and model specification for pain cross-over trial.

$$r_{ijkl} \sim \text{Bern}(\pi_{ijkl})$$

$$p_i \sim N(0, \sigma_1^2)$$

where
     r is the Bernoulli response variable relief
     p_i are the random effects of the factor pat
     α_j are the fixed effects of the factor trt
     γ_k are the fixed effects of the factor per
     ζ_l are the fixed effects of the factor carry
     σ_1^2 is the variance component associated with the pat random effect

A preliminary MCMC run indicated that all the chains converged relatively quickly, and that all bar the variance component were mixing well; the chain for the σ_1 parameter appeared to have converged, but a relatively long run was needed to obtain adequate MCMC precision. It was found that the regular model defaults for the MCMC settings (a burn-in of 10,000 followed by 10,000 samples after one-fifth thinning) were adequate to reduce the MCMC error to about 5% of this parameter's posterior standard deviation. To aid interpretation of the treatment and carry-over effects, contrasts between the test compound and placebo were requested (see cells G2:H4 in Figure 9.1). In addition to the usual summary statistics, the "Posterior Samples" check box was selected on the "Predictions and Contrasts" form to allow more refined inferences to be made later. Figure 9.2 shows the usual results summary, including the prior distributions (cells A19:B23) that were unchanged from the defaults.

Inspection of the credible intervals indicates that there is clear evidence of an effect due to the analgesic. There does not appear to be any large effects due to periods, but there is some evidence of a carry-over effect for the highest dose. Figure 9.3 shows the treatment and carry-over contrasts requested. Since these have been back-transformed from the logit-scale, they represent the odds-ratios of obtaining pain relief. Although the credible intervals for the treatment odds-ratio effects are fairly wide, this study has very conclusively shown that the drug is superior to placebo, with a best estimate of about a 10-fold increase in the odds of pain relief when receiving the drug compared to placebo. The odds-ratio contrast between the high and low doses is not shown, but the difference is small relative to

	A	B	C	D	E	F	G	H	I	J	K
1			Label	Mean	St.Dev.	2.5%	Median	97.5%		WinBUGS Name	
2			CONSTANT	-1.1650	0.3487	-1.8690	-1.1590	-0.4984		Beta0	
3		trt	L	2.0940	0.4153	1.3080	2.0890	2.9280		X.Eff[1,2]	
4		trt	H	2.3890	0.4212	1.5880	2.3800	3.2430		X.Eff[1,3]	
5		per	2	0.4538	0.4822	-0.4671	0.4451	1.4200		X.Eff[2,2]	
6		per	3	0.6344	0.4951	-0.3167	0.6265	1.6380		X.Eff[2,3]	
7		carry	L	-0.2064	0.5271	-1.2510	-0.2023	0.8139		X.Eff[3,2]	
8		carry	H	-0.9054	0.5062	-1.9480	-0.8944	0.0640		X.Eff[3,3]	
9			SD(pat)	0.3827	0.2478	0.0191	0.3464	0.9462		sigma.Z[1]	
10											
11	Model	[pain!$A$1:$E$259]									
12	Distribution	Bernoulli									
13	Link	Logit									
14	Response	relief/1 : Pr[relief = Y]									
15	Fixed	trt+per+carry									
16	Random	pat									
17											
18	Priors										
19	CONSTANT	N(mu=0, sigma=100)									
20	trt	N(mu=0, sigma=100)									
21	per	N(mu=0, sigma=100)									
22	carry	N(mu=0, sigma=100)									
23	pat	Norm(0,tau^2); tau ~ Half-N(sigma=5)									
24											
25	WinBUGS MCMC Settings										
26	Burn-In: 10000 Samples: 10000 (Thin:5; Chains:1)										
27	Run took 7.7 minutes										

FIGURE 9.2
Summaries of posterior distributions, model, priors, and MCMC settings for pain cross-over trial.

N	O	P	Q	R	S	T	U	V
Label	Mean	St.Dev.	2.5%	Median	97.5%		WinBUGS Name	
	Predicted Probability							
	Predicted Odds							
	Following all at: per(1) carry(0) pat(dist. mean)							
contr[trt L / P]	8.8580	3.9390	3.6980	8.0780	18.6900		Pred.Odds[1]	
contr[trt H / P]	11.9400	5.4580	4.8930	10.8100	25.6000		Pred.Odds[2]	
	Following all at: trt(P) per(1) pat(dist. mean)							
contr[carry L / 0]	0.9341	0.5291	0.2862	0.8169	2.2570		Pred.Odds[3]	
contr[carry H / 0]	0.4583	0.2397	0.1425	0.4088	1.0660		Pred.Odds[4]	

FIGURE 9.3
Treatment and carry-over odds-ratio effects for pain cross-over trial.

the precision with which it has been estimated. The median estimate of the carry-over effect for the highest dose is equivalent to a 60% (0.6 = 1 − 0.4) reduction in the odds of pain relief in periods two or three among patients who received this dose in the previous period rather than placebo, regardless of the currently assigned treatment. If this carry-over effect is real, it is an interesting phenomenon because, contrary to the usual concern of a pharmacological effect due to residual amounts of the drug still being active in the body, the carry-over effect of the highest dose has reduced the observed efficacy, not increased it. Such a result should prompt further investigation after discussions with the subject matter experts. For example, could this effect be due to some delayed adverse events caused by the higher dose, or were there other symptoms associated only with the higher dose in the previous period that induced a negative psychological effect in the following period?

It is possible to make more refined inferences about the odds-ratio effects summarized in Figure 9.3 using BugsXLA's posterior plotting tool (refer to Case Study 3.1 for more details). Figure 9.4 shows a histogram of the samples imported back from WinBUGS representing the posterior distribution of the high dose to placebo carry-over odds-ratio effect, that is, the variable labeled "contr[carry H/0]" in Figure 9.3. Percentiles of this distribution are provided beneath the plot, enabling credible intervals other than the 95% default to be determined. Finally, the probability that this odds-ratio is greater than one has been estimated, which is about 3%, providing a Bayesian analogy to the conventional p-value obtained when testing a one-sided hypothesis of "no effect." One might be tempted to double this probability in order to make it more like a p-value associated with the two-sided hypothesis that would almost certainly have been specified in a conventional analysis, but this is not recommended as there is no obvious Bayesian interpretation of what this probability represents.

Case Study 9.2 Modeling Serial Correlation in Rat Growth Measurements
Case Study 5.4 showed how random regression coefficients could be used to model the growth curves of the weights of 30 young rats. Here we reanalyze

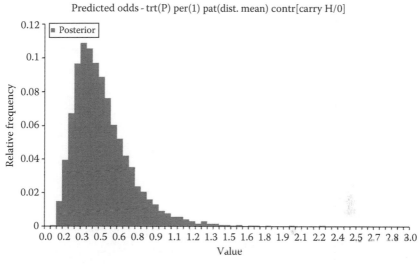

FIGURE 9.4
Posterior distribution of high dose/placebo carry-over odds-ratio effect for pain cross-over trial.

these data and assess whether there is any benefit in adding terms that model any serial correlation that might exist between measurements taken on the same rat. Initially, we fit a model that assumes a common intercept and slope for all rats, with all the within rat correlation being explained by serial correlation between residual errors. Figure 9.5 shows some of the data and the model specification form.

All the variables are as defined before. A first-order autoregressive process is specified for the within rat residual errors by first selecting the "Longitudinal Model" check box. In the text box that appears, the name of the factor, Rat, that defines the experimental unit is entered between curly brackets, "{ }," after specifying the covariance structure "AR1," as shown in Figure 9.5. This model can be represented algebraically:

$$y_{ij} = \mu + \alpha_i + \beta x_{ij} + e_{ij} \quad i = 1, 2, \ldots, 30 \quad j = 1, 2, \ldots, 5$$

$$e_{ij} \sim N(\varphi e_{ij-1}, (1 - \varphi^2)\sigma^2) \quad j = 2, 3, 4, 5$$

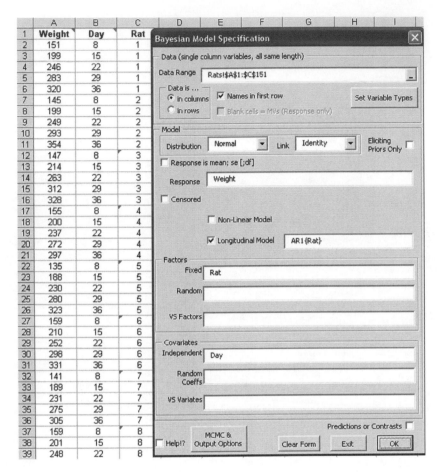

FIGURE 9.5
Data and initial model specification for serial correlation in rat growth measurements.

$$e_{i1} \sim N(0, \sigma^2)$$

$$-1 \le \varphi \le 1$$

where

y, x, α and β are essentially as before

φ determines the degree of autocorrelation between the residual errors on the same rat

σ^2 is the unconditional variance of the e_{ij}

When there is no autocorrelation, φ equals zero and the model reduces to the simple Normal linear model. If the residuals exhibit high positive correlation, then φ would be close to one, and all the e_{ij} on the same rat would be virtually identical. At the other extreme, when the residuals exhibit high negative correlation, φ would be close to -1 and the e_{ij} on the same rat would alternate between positive and negative errors. Note that it is assumed that the measurements on each rat are in time order and that the time interval is the same throughout the study for all rats. This is true for this study, but if it was not then it is possible to specify the time associated with each measurement. In this case, this would be done by replacing the longitudinal model statement with

$$AR1\{Rat/Day\}$$

The only change to the model defined is

$$e_{ij} \sim N(\varphi^{(T_{ij}-T_{ij-1})}e_{ij-1}, (1-\varphi^{2(T_{ij}-T_{ij-1})})\sigma^2)$$

where $(T_{ij} - T_{ij-1})$ is the time difference between the jth and (j–1)th measurement on the ith rat. Note that φ now represents the correlation between residuals one unit of time apart. If any of the time differences are non-integer, then φ must be constrained non-negative in order for this parameter to be estimable. As the time difference tends to zero then the correlation between residual errors tends to one, and as the time difference tends to infinity this correlation tends to zero for all φ less than one. When the individual measurements either involve some kind of sampling within units or are obtained using a method that is itself subject to random errors, the assumption that measurements taken at the same time point will be identical is not credible. BugsXLA allows a serial correlation plus measurement error covariance structure to be specified using the following terminology:

$$AR1e\{Rat\} \quad or \quad AR1e\{Rat/Day\}$$

The changes to the model are

$$y_{ij} = \mu + \alpha_i + \beta x_{ij} + e_{ij} + w_{ij}$$

$$w_{ij} \sim N(0, \omega^2)$$

where ω^2 is the additional measurement error variance. If the data contains replicate measurements at the same time, then it is possible to estimate ω^2 directly. Otherwise, estimation of the measurement error variance will be confounded with model lack-of-fit and may be strongly dependent on the assumed parametric model. In many cases, when there is no true replication, unless an informative prior is specified for ω^2 WinBUGS will crash and fail to return any estimates. The priors for the additional longitudinal model parameters can be altered on the "Errors" page as shown in Figure 9.6.

Note that the sigma parameter refers to the unconditional standard deviation of the residual errors. The measurement error variance is labeled

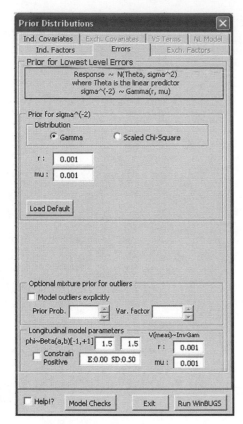

FIGURE 9.6
Default priors for longitudinal model with first-order autocorrelation plus measurement error.

"V(meas)" and is given the standard inverse-gamma prior distribution with very small parameter values by default. The autocorrelation parameter phi is given a highly dispersed Beta prior distribution scaled to the range [−1, +1]. The mean (E) and standard deviation (SD) of this Beta Distribution are displayed to help interpret this prior. A uniform prior distribution, Beta(1, 1), is not recommended as this can cause numerical problems due to it giving relatively high prior credence to correlations very close to ±1. The autocorrelation can be constrained to be positive, in which case the range for the Beta prior distribution is [0, 1].

Since this rat study did not contain any replicate measurements, the additional measurement error term was not included in the model. In addition to the model specified in Figure 9.5, five other longitudinal models were used to analyze these data. Table 9.2 lists these models using the exact terminology required on the BugsXLA model specification form. In each case the default priors and the regular model MCMC settings (a burn-in of 10,000 followed by 10,000 samples after one-fifth thinning) were used.

Table 9.3 provides summary statistics that can be used to compare the six models. In terms of speed, none took a great deal of time to complete the 60,000 iterations, although as should be expected the most complex model, number

TABLE 9.2

Longitudinal Models Compared for Analysis of Rat Growth Measurements

Model No.	1	2	3	4	5	6
Longitudinal model	AR1(Rat)	AR1(Rat)	AR1(Rat)	AR1(Rat)		
Fixed factor	Rat					
Random factor		Rat				
Independent covariate	Day	Day		Rat:Day^2		Rat:Day^2
Random coefficients			Rat/Day	Rat/Day	Rat/Day	Rat/Day

TABLE 9.3

Summaries Used to Compare Different Longitudinal Models for Analysis of Rat Growth Measurements

Model No.	1	2	3	4	5	6
Run time (min)	3.4	2.8	6.5	7.6	4.4	4.7
Constant	238	240	241	247	243	247
	(21)	(2.7)	(2.8)	(2.9)	(2.8)	(2.8)
Day or slope mean	6.17	6.17	6.17	7.96	6.19	7.95
	(0.10)	(0.10)	(0.11)	(0.27)	(0.11)	(0.26)
Correlation (slope – intercept)			0.67	0.67	0.60	0.70
			(0.25)	(0.19)	(0.15)	(0.15)
SD (residual)	19	14	10	4.4	6.1	3.7
	(11)	(2.2)	(2.2)	(1.8)	(0.5)	(0.3)
Phi(AR1)	0.86	0.82	0.68	0.08		
	(0.09)	(0.06)	(0.12)	(0.35)		
pD	39.9	11.6	32.2	73.5	51.8	76.2
DIC	1137	1093	1071	906	1018	894

Most are posterior means with posterior standard deviations in brackets.

four, took the longest time to run. The model constant is only comparable for models 2–6 where it represents the mean weight over the population of rats at the mean time point. In model one it represents the weight for the first rat, since the factor Rat is entered as a fixed effect in this model, which explains the much larger posterior standard deviation for this parameter here. The model constant is very similar for all the comparable models, with a slight increase in those that include quadratic terms for the growth curve. Both the posterior means and standard deviations for the linear regression coefficient for Day are very similar for all the models that do not include quadratic terms, even between models that treat this term as constant for all rats or random across rats. When quadratic terms for Day are included in the model the linear regression coefficient has a different interpretation and so we should not be surprised that it changes. In those cases where the linear coefficients for Day and Rat intercepts were modeled as correlated parameters, their correlation remained relatively high irrespective of the other terms included in the model. The biggest difference between the models is in the amount of unexplained residual variation they have, which also correlates here with the DIC statistic. The two that include quadratic terms for Day clearly provide the best fits to the data, with the next best model having a DIC value more than 100 greater. Inspection of the estimated first-order autocorrelation parameter for model four indicates that this is not "statistically significant," which is supported by the difference of more than 10 between the DIC values for this model and model six in which this term is absent. The importance of these quadratic terms for Day is not surprising in light of the residual plot observed after the initial analysis in Case Study 5.4 (see Figure 5.15).

It is interesting to note that in each of the first three models where no quadratic Day terms are included, the parameter Phi(AR1) is estimated to be quite large with relatively high precision. An advantage of the Bayesian approach is the ease with which more refined inferences can be made on parameters such as these, which is particularly difficult using classical methods, for example, Figure 9.7 shows the posterior distribution with calculated percentiles. This serial correlation term is clearly trying to model the lack of fit due to the curvature in the true growth curve over time. In this case, it should have been obvious from Figure 5.15 that the lack of fit needed to be

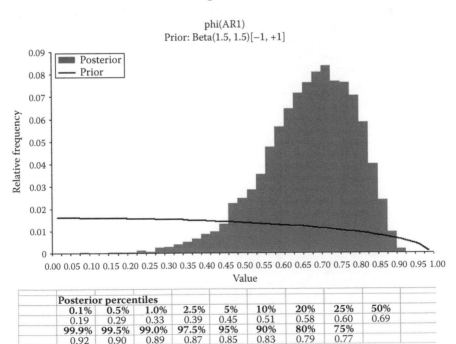

Posterior percentiles								
0.1%	0.5%	1.0%	2.5%	5%	10%	20%	25%	50%
0.19	0.29	0.33	0.39	0.45	0.51	0.58	0.60	0.69
99.9%	99.5%	99.0%	97.5%	95%	90%	80%	75%	
0.92	0.90	0.89	0.87	0.85	0.83	0.79	0.77	

FIGURE 9.7
Posterior with prior overlaid for the first-order autocorrelation parameter in model three.

addressed by modifying the assumption of a linear relationship between weight and time before considering the need to model serial correlation in the residuals. An obvious potential improvement to model six is to treat the quadratic terms for Day as random across rats like the linear terms have been, giving the intercepts, linear and quadratic terms a joint trivariate Normal prior distribution. Currently, this is not possible using BugsXLA.

Case Study 9.3 Modeling Serial Correlation in Epileptic Seizure Counts
Case Study 6.2 discusses a study in which the efficacy of a treatment for epilepsy in 59 patients was assessed. We reanalyze these data using a model that takes account of any serial correlation that might exist between seizure counts observed on the same patient. Initially we specify a model that includes all the fixed effect terms fitted before, but with no random effects and all the within patient correlation being explained by serial correlation. Figure 9.8 shows some of the data and the model specification form.

All the variables are as defined before. A model analogous to a first-order autoregressive process for the within patient residuals is specified by first selecting the "Longitudinal Model" check box. In the text box that appears, the name of the factor, patient, that defines the experimental unit is entered between curly brackets, "{ }," after specifying the covariance structure "AR1," as shown in Figure 9.8. This specifies a Poisson transition model introduced by Zeger and Qaqish (1988) that can be represented algebraically:

$$\eta_{ijkm} = \eta'_{ijkm} + \varphi(\log(y^*_{ijk(m-1)}) - \eta'_{ijk(m-1)}) \quad m = 2,3,4$$

$$\eta_{ijk1} = \eta'_{ijk1}$$

$$\eta'_{ijkm} = \mu + \delta_i + \alpha_j + \beta x_k + (\delta\beta)_i x_k + \gamma z_k$$

$$y^*_{ijk(m-1)} = \max(y_{ijk(m-1)}, d) \quad 0 < d < 1$$

$$\log(\lambda_{ijkm}) = \eta_{ijkm}$$

$$y_{ijkm} \sim \text{Pois}(\lambda_{ijkm})$$

$$i = 1,2 \quad j = 1,2 \quad k = 1,2,\ldots,59 \quad m = 1,2,3,4$$

where
  $y, x, z, \delta, \alpha, \beta, (\delta\beta)$ and γ are essentially as before
  φ determines the degree of autocorrelation between the counts on the same patient
  d is a constant needed to avoid a log(0) error; see also discussion below

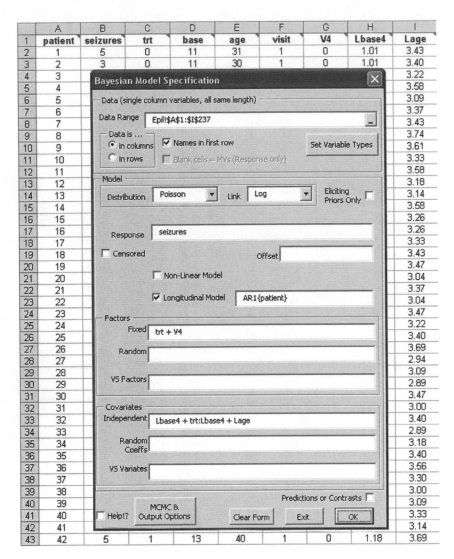

FIGURE 9.8
Data and initial model specification for serial correlation in epileptic seizure counts.

When there is no autocorrelation then φ equals zero and the model reduces to the simple Poisson log-linear model. When φ is less than zero, implying negative serial correlation, a prior response greater than its expectation decreases the expectation of the current response. When positive serial correlation exists, φ is greater than zero. One justification for this type of transition model is when the response is believed to have been generated from a size-dependent branching process. In such a process, the response at each generation can be thought of as the number of individuals in a population. The number of offspring for each individual is an independent Poisson random variable with expectation inversely proportional to the total number of individuals in the current generation raised to the power $(1 - \varphi)$. This leads to the response at the next generation being the sum of these Poisson variables. If the number of individuals drops to zero in any generation, then the population is restarted with d individuals. It should be evident that this process incorporates a crowding effect, in which each individual tends to adjust their number of offspring depending upon the current population size, reducing their numbers when the population is large. It is this feature that leads to a stationary process.

The prior for φ can be altered in the same way as the serial correlation parameter discussed in Case Study 9.2. The value of the constant d can also be altered. After checking that there were no convergence issues, the simple model MCMC settings were chosen (a burn-in of 5,000 followed by 10,000 samples), and the default priors were unchanged (see cells A22:B28 in Figure 9.9). Figure 9.9 shows the parameters in the model and a summary of their marginal posterior distributions as imported back into Excel.

Comparison of the fixed effects estimates here with those obtained in Case Study 6.2 (see Figure 6.10) shows many differences. Most notable is the much reduced posterior standard deviations in the model fitted here. This should not be taken as evidence of a better model fit, since this can occur simply by removing random effects from such a model. The serial correlation parameter is clearly greater than zero, indicating that there is additional correlation between counts from the same patient. However, just as in Case Study 9.2,

	A	B	C	D	E	F	G	H	I	J	K
1			Label	Mean	St.Dev.	2.5%	Median	97.5%		WinBUGS Name	
2			CONSTANT	1.9520	0.0553	1.8430	1.9520	2.0610		Beta0	
3			Intercept at 0	-2.8730	0.5407	-3.9450	-2.8700	-1.8170		alpha	
4		trt	1	-0.3731	0.0810	-0.5305	-0.3737	-0.2137		X.Eff[1,2]	
5		V4	1	-0.0988	0.0525	-0.2045	-0.0983	-2.144E-4		X.Eff[2,2]	
6			Lbase4	0.9259	0.0578	0.8124	0.9255	1.0390		V.Coeff[1,1]	
7			Lage	0.9596	0.1567	0.6525	0.9567	1.2700		V.Coeff[2,1]	
8		trt x Lbase4	1	0.6009	0.0861	0.4325	0.6006	0.7696		V.Coeff[3,2]	
9			phi(AR1)	0.3714	0.0372	0.2997	0.3710	0.4445		phi.AR1	
10	Note: CONSTANT & Factor effects are determined at the mean of the covariate(s).										
11	Interpret these cautiously when Factor x Covariate terms have been fitted.										
12											
13	Model	[Epil!$A$1:$I$237]									
14	Distribution	Poisson									
15	Link	Log									
16	Response	seizures									
17	Longi model	AR1(patient)									
18	Fixed	trt + V4									
19	Covariates	Lbase4 + trt:Lbase4 + Lage									
20											
21	Priors										
22	CONSTANT	N(mu=1.59, sigma=111)									
23	trt	N(mu=0, sigma=111)									
24	V4	N(mu=0, sigma=111)									
25	Lbase4	N(mu=0, sigma=149)									
26	Lage	N(mu=0, sigma=498)									
27	trt x Lbase4	N(mu=0, sigma=149)									
28	phi(AR1)	Beta(1.5,1.5) [-1,+1] ; 'Zero' d = 0.3									
29											
30	WinBUGS MCMC Settings										
31	Burn-In: 5000 Samples: 10000 (Thin:1; Chains:1)										
32	Run took 4.1 minutes										

FIGURE 9.9
Summaries of posterior distributions, model, priors, and MCMC settings for epilepsy trial.

TABLE 9.4

Summaries Used to Compare Different Longitudinal Models for Analysis of Epileptic Seizure Counts

Model No.	1	2	3	4
Longitudinal model	AR1{patient}	AR1{patient}	AR1{patient}	
Random effects		patient	patient/visit	patient/visit
Run time (min)	4.1	4.9	6.2	4.3
trt	−0.37 (0.08)	−0.34 (0.16)	−0.33 (0.16)	−0.33 (0.16)
trt x Lbase4	0.60 (0.09)	0.35 (0.24)	0.39 (0.23)	0.36 (0.22)
Phi(AR1)	0.37 (0.04)	−0.13 (0.05)	0.02 (0.10)	
pD	6.9	53	123	121
DIC	1541	1254	1158	1157

Most are posterior means with posterior standard deviations in brackets.

we should assess whether the simpler model in which this correlation is modeled by random effects provides a better fit. Table 9.4 provides summary statistics that can be used to compare four competing models in which only the random effects and serial correlation parameters were altered. It is clear that once a random effect term for patient is included in the model, then there is little evidence for any serial correlation in the counts. Using the DIC as guidance, models three and four are clearly superior to the first two models, with model four being preferred due to it being more parsimonious.

In both Case Studies 9.2 and 9.3, additional within experimental unit serial correlation was not found to improve the model fit. In general, if variables have been obtained with the purpose to explain some of the within unit variation, then it is recommended that random effects and random coefficient terms be considered first. It is often possible to simplify the explicit correlation structure needed for the errors by including such random effects. Unless there is good reason to believe in an autoregressive correlation structure, the φ parameter should not be considered unless it provides a clearly better model fit than these simpler random effect terms. For much more informed advice on how to identify an appropriate model to fit to longitudinal data refer to Fitzmaurice et al. (2004) or Diggle et al. (2002).

Case Study 9.4 Time Series of Presidential Approval Ratings

Jackman (2009) provides a Bayesian analysis of 96 monthly presidential job approval ratings for Ronald Reagan. We repeat here his analysis that involves fitting a linear regression model containing two covariates, inflation rate and unemployment percentage, with a correction for first-order serial correlation between the residual errors. Figure 9.10 shows some of the data and the model specification form.

Since this is a single time series, this is equivalent to there being only one experimental unit. Autoregressive residual errors are specified by using the notation "AR1{1}" in these cases. This model can be represented algebraically:

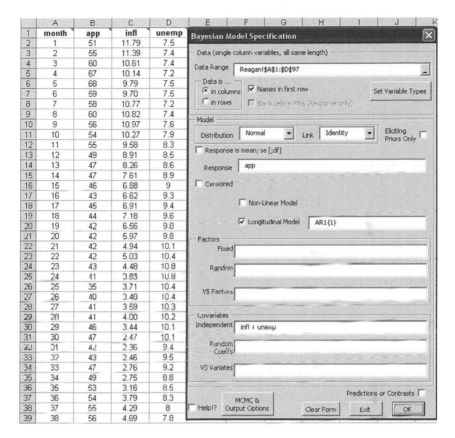

FIGURE 9.10
Data and model specification for presidential approval ratings.

$$y_i = \mu + \beta x_i + \gamma z_i + e_i \quad i = 1, 2, \ldots, 96$$

$$e_i \sim N(\varphi e_{i-1}, (1 - \varphi^2)\sigma^2) \quad i = 2, 3, \ldots, 96$$

$$e_1 \sim N(0, \sigma^2)$$

$$-1 \leq \varphi \leq 1$$

where
  y is the response variable app
  x is the covariate infl with associated regression coefficient β
  z is the covariate unemp with associated regression coefficient γ
  φ determines the degree of autocorrelation between the residual errors
  σ^2 is the unconditional variance of the residual errors

	A	B	C	D	E	F	G	H	I	J	K
1			Label	Mean	St.Dev.	2.5%	Median	97.5%		WinBUGS Name	
2			CONSTANT	53.6900	4.8380	45.0400	53.5400	63.5300		Beta0	
3			Intercept at 0	60.4500	14.1600	29.2600	61.7000	84.7500		alpha	
4			infl	0.0574	0.7275	-1.3500	0.0393	1.5530		V.Coeff[1,1]	
5			unemp	-0.9316	1.6300	-3.8060	-1.0720	2.6170		V.Coeff[2,1]	
6			SD(residual)	8.6930	3.1540	5.5090	7.8770	17.4900		sigma	
7			phi(AR1)	0.8871	0.0523	0.7765	0.8905	0.9783		phi.AR1	
8	Note: CONSTANT & Factor effects are determined at the mean of the covariate(s).										
9		Interpret these cautiously when Factor x Covariate terms have been fitted.									
10											
11	Model	[Reagan!$A$1:$D$97]									
12	Distribution	Normal									
13	Link	Identity									
14	Response	app									
15	Longi model	AR1{1}									
16	Covariates	infl + unemp									
17											
18	Priors										
19	CONSTANT	N(mu=53.1, sigma=773)									
20	infl	N(mu=0, sigma=302)									
21	unemp	N(mu=0, sigma=531)									
22	V(residual)	Inv-Gamma(0.001, 0.001)									
23	phi(AR1)	Beta(1.5,1.5) [-1,+1]									
24											
25	WinBUGS MCMC Settings										
26	Burn-In: 5000 Samples: 10000 (Thin:1; Chains:1)										
27	Run took 40 seconds										

FIGURE 9.11
Summaries of posterior distributions, model, priors, and MCMC settings for presidential approval ratings.

The simple model MCMC settings were chosen (a burn-in of 5,000 followed by 10,000 samples), and the default priors were unchanged (see cells A19:B23 in Figure 9.11). Checking functions for individual observations were requested as explained in Case Study 3.1 (see Figure 3.18). Figure 9.11 shows the parameters in the model and a summary of their marginal posterior distributions as imported back into Excel.

It is clear that there is very strong serial correlation in the monthly approval ratings and little evidence of any effects due either to the inflation or unemployment rates in the associated month. Figure 9.12 shows four of the plots on the model checks sheet created. The most noticeable feature on the simple time series plot of the approval ratings (top left graph) is the sudden jump that occurs between months 71 and 72. Inspection of the other plots suggests that some other months' ratings are also a little extreme relative to the assumed model.

Given the very marked shift in approval ratings that occurred after the 71st month, this should prompt further investigation into potential causes of this effect. Here we simply assume that the ratings before and after this point need to be considered as having different underlying mean levels, with the cause still to be determined. To model this, we defined a factor, split, which takes the level "0" for the first 71 months and "1" for the remaining months. This factor was added as a fixed effect term to the previous model and the analysis rerun using the same MCMC settings and default priors as before. Figure 9.13 shows the usual summary statistics as imported back into Excel. By including a term that nominally accounts for the shift that occurred,

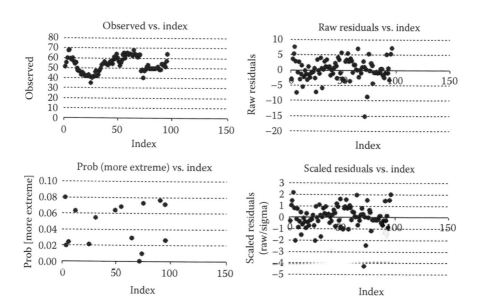

FIGURE 9.12
Model checking plots (versus index only) for presidential approval ratings.

	A	B	C	D	E	F	G	H	I	J	K
1			Label	Mean	St.Dev.	2.5%	Median	97.5%		WinBUGS Name	
2			CONSTANT	57.6100	1.2690	55.2700	57.6100	59.9800		Beta0	
3			Intercept at 0	101.3000	7.8570	82.6700	102.3000	113.5000		alpha	
4		split 1		-16.6700	2.3120	-20.9000	-16.7600	-11.7900		X.Eff[1,2]	
5			infl	-0.3236	0.3400	-1.0560	-0.3057	0.3142		V.Coeff[1,1]	
6			unemp	-5.5910	0.9888	-7.0760	-5.7380	-3.1030		V.Coeff[2,1]	
7			SD(residual)	4.1070	0.9792	3.1150	3.8680	6.7190		sigma	
8			phi(AR1)	0.6096	0.1225	0.3873	0.6021	0.8786		phi.AR1	
9	Note: CONSTANT & Factor effects are determined at the mean of the covariate(s).										
10	Interpret these cautiously when Factor x Covariate terms have been fitted.										
11											
12	**Model**	[Reagan!$A$1:$E$97]									
13	Distribution	Normal									
14	Link	Identity									
15	Response	app									
16	Longi model	AR1{1}									
17	Fixed	split									
18	Covariates	infl + unemp									
19											
20	**Priors**										
21	CONSTANT	N(mu=53.1, sigma=773)									
22	split	N(mu=0, sigma=773)									
23	infl	N(mu=0, sigma=302)									
24	unemp	N(mu=0, sigma=531)									
25	V(residual)	Inv-Gamma(0.001, 0.001)									
26	phi(AR1)	Beta(1.5,1.5) [-1,+1]									
27											
28	**WinBUGS MCMC Settings**										
29	Burn-In: 5000 Samples: 10000 (Thin:1; Chains:1)										
30	Run took 47 seconds										

FIGURE 9.13
Summaries of posterior distributions, model, priors, and MCMC settings for presidential approval ratings after adding factor split to the model.

it is now seen that the approval ratings are negatively correlated with the unemployment rate, although there is still no evidence of a simple linear relationship with inflation. It should not be a surprise that the estimate of residual variation has reduced. Note also that the degree of serial correlation is smaller due to a significant amount of the variation in the response now being explained by the unemployment rate, which is itself serially correlated.

10

Robust Models

All models used to analyze data rely on assumptions to varying degrees. In the Bayesian approach to statistics these modeling assumptions are based on prior beliefs, although in practice, like in classical statistics, they are often informally updated in light of seeing the data; see the discussion in Chapter 8 on the dangers of over-fitting and under-representing uncertainty due to a reliance on a single chosen model following data-driven model selection. Clearly when the data come from designed experiments, the same prior beliefs should inform both the experimental design and the model with which the data will be analyzed. In these cases, it typically leads to the same models being applied as would be by classical statisticians, even though the theoretical justification can be quite different. Given that it is unlikely that we can be totally certain about all aspects of a model, it is important that we assess the sensitivity of the inferences to the assumptions made. The most obvious way to do this is to reanalyze the data with a range of different models reflecting alternative credible assumptions. In this chapter, we discuss some features of BugsXLA that allow models to be specified that are inherently more robust to certain types of assumptions.

A common concern when modeling data is the influence that extreme observations, or groups of observations, may have on the estimated quantities of interest. The Normal distribution, although the most commonly used to model residual errors, is notoriously sensitive to outliers. This problem is magnified as models of increasing complexity are fitted, since the influence of an outlying observation on the parameters in the model becomes increasingly more difficult to predict. The other common error distributions discussed in Chapters 4 and 6 also suffer from a lack of robustness to outliers. With the introduction of random effects, with their own assumed "population" distribution, there is another level at which outliers can be identified; a particular factor level mean might be extreme relative to the others. In the Bayesian approach, it is relatively easy to specify alternative error distributions that are much more robust than those used in standard models. This is done by adopting distributions with heavier or longer tails, representing the belief that extreme observations might occur. By doing this, these observations, which would have been considered outliers using a standard model, no longer strongly influence the parameters that are of primary interest. When outliers are anticipated in data that would otherwise be adequately modeled using a Normal Distribution, a common approach is to use a t-distribution in its place. BugsXLA provides the t-distribution as one of the error distributions that can be specified on the

"Model Specification" form. As will be discussed in Case Study 10.1, this introduces another parameter in the model that requires a prior distribution: the degrees of freedom associated with this t-distribution. BugsXLA also offers a t-distribution as an alternative to the Normal as the hierarchical prior for random effects, and when random coefficients are fitted, the multivariate t-distribution can be specified. In both these latter cases, the degrees of freedom cannot be estimated but is fixed at four. This simplification is made mainly because it is very rare that there are enough factor levels to estimate the degrees of freedom parameter very precisely. A fixed value of four is sufficiently small to provide a markedly heavier tailed distribution than the Normal.

Another way to define a heavy-tailed distribution is via a mixture of two Normals; essentially one distribution models the majority of the data, while the tails of the other much more dispersed distribution model the extreme observations. In this approach, the outliers are explicitly being modeled, and has the advantage over using a t-distribution of identifying those observations that are discordant. This approach is also easily generalized to other error distributions as discussed at the end of Case Study 10.1. More details on the Bayesian approach to specifying models for robust inference can be found in Chapter 17 of Gelman et al. (2004).

Case Study 10.1 Parallel Groups Clinical Study

Case Study 3.1 discussed a parallel groups clinical trial in which the efficacy of a new anti-inflammatory treatment for chronic obstructive pulmonary disease was assessed. Although the model checking undertaken did not raise any serious concerns, there were perhaps slightly more extreme residuals than one should expect by chance. Here we reanalyze these data illustrating the two approaches that BugsXLA provides to make the model more robust. In the first instance we replace the Normal with the t-distribution. The only change to the "Model Specification" form shown in Figure 3.1 is to select "t-dist" from the drop down "Distribution" list box instead of "Normal." The simple model MCMC settings (burn-in of 5,000 followed by 10,000 samples) were chosen. Figure 10.1 shows the form where the priors for the t-distribution parameters can be altered.

By default the scale parameter, sigma, is treated the same as in the Normal case. The degrees of freedom parameter is given an inverse-uniform distribution on the range [0, 0.5]. This range constrains the degrees of freedom to be larger than two, which is advisable since below two the t-distribution has infinite variance with unrealistically long tails. The inverse-uniform is used, rather than the uniform directly, to avoid putting too much prior weight on relatively large values for the degrees of freedom, which would effectively be representing a strong prior belief that the errors were approximately Normal. This belief would be counter to that which has motivated us to use a robust model in the first place. It is only recommended that robust models be used when there are reasons to believe that the Normal assumption will be inadequate. Only in cases where there are very many replicates is there sufficient

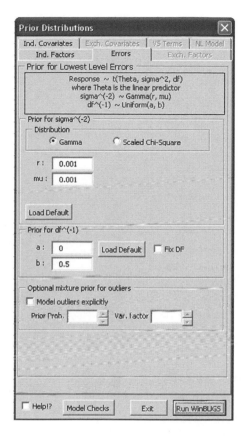

FIGURE 10.1

BugsXLA form for changing priors for t-distribution parameters.

information in the data to precisely estimate the degrees of freedom parameter. Hence, it may be better to fix this parameter at a small value rather than trying to estimate it. This can be done by setting the upper and lower bounds of the prior distribution to the same value, which should equal the reciprocal of the degrees of freedom required. BugsXLA offers a short cut via the "Fix DF" check box. On clicking this box, the user can specify the fixed value of the degrees of freedom required; it is necessary to then click on the "Update Prior" button that appears in order to set this prior distribution. Since four is a frequently used value, BugsXLA offers this as the default. In this particular analysis the default priors were left unchanged (see cells A23:B27 in Figure 10.2). Figure 10.2 shows the output as imported back into Excel.

Inspection of the summarized posterior distribution for the degrees of freedom parameter shows that it is very imprecisely estimated. This illustrates that even when a simple model such as this one is being fitted, it is probably necessary to have much more than 100 residual degrees of freedom in order to be able to reliably estimate this parameter. The analysis was rerun fixing a t_4-distribution giving essentially the same parameter estimates as shown here. Since these results are very similar to those shown in Figure 3.7,

	A	B	C	D	E	F	G	H	I	J	K
1			Label	Mean	St.Dev.	2.5%	Median	97.5%		WinBUGS Name	
2			d.f.	11.8200	23.3100	2.2810	5.4690	71.2700		df	
3			CONSTANT	-0.0317	0.0420	-0.1149	-0.0317	0.0537		Beta0	
4			Intercept at 0	1.938E-3	0.0636	-0.1210	2.913E-4	0.1315		alpha	
5		TRT	B	0.1074	0.0612	-0.0141	0.1077	0.2239		X.Eff[1,2]	
6		TRT	C	0.1402	0.0553	0.0314	0.1401	0.2507		X.Eff[1,3]	
7		TRT	D	0.0169	0.0600	-0.1016	0.0167	0.1350		X.Eff[1,4]	
8		TRT	E	0.0527	0.0576	-0.0624	0.0525	0.1652		X.Eff[1,5]	
9			FEV1_BASE	-0.0263	0.0468	-0.1211	-0.0250	0.0616		V.Coeff[1,1]	
10			SDparm(residual)	0.1518	0.0213	0.1134	0.1515	0.1943		sigma	
11			SD(residual)	0.2077	0.0533	0.1598	0.1945	0.3612		SD.res	
12	**Note: CONSTANT & Factor effects are determined at the mean of the covariate(s).**										
13	**Interpret these cautiously when Factor x Covariate terms have been fitted.**										
14											
15	**Model**		['clin study'!$A$1:$D$96]								
16	Distribution		t-dist								
17	Link		Identity								
18	Response		FEV1_CFB								
19	Fixed		TRT								
20	Covariates		FEV1_BASE								
21											
22	**Priors**										
23	CONSTANT		N(mu=0.0344, sigma=18.7)								
24	TRT		N(mu=0, sigma=18.7)								
25	FEV1_BASE		N(mu=0, sigma=39.8)								
26	Vparm(residual)		Inv-Gamma(0.001, 0.001)								
27	d.f.		Inv-Uniform(0,0.5)								
28											
29	**WinBUGS MCMC Settings**										
30	Burn-In: 5000 Samples: 10000 (Thin:1; Chains:1)										
31	Run took 22 seconds										

FIGURE 10.2
Summaries of posterior distributions, model, priors, and MCMC settings for clinical study using t-distribution.

this indicates that there are no extreme observations strongly influencing the estimated quantities of interest. Note that BugsXLA summarizes both the "SDparm(residual)" parameter, that is, the sigma parameter defined in Figure 10.1, as well as the "SD(residual)," which is the standard deviation of the t-distribution as defined in Appendix A. The latter quantity is the better analog to the residual standard deviation in a Normal model.

We now reanalyze these data using an approach first discussed by Box and Tiao (1968) in which the outliers are modeled explicitly. The error distribution is returned to the Normal, so that the model specification is exactly as shown in Figure 3.1. Figure 10.3 shows the form where the mixture prior can be specified. After selecting the "Model outliers explicitly" check box, the two settings appear as shown. These settings determine the mixture prior that will be used to model the residual errors. This model can be represented algebraically:

$$y_{ij} = \mu + \alpha_i + \beta x_{ij} + e_{ij} \quad i = 1, 2, \ldots, 5 \quad j = 1, 2, \ldots, n_i$$

$$e_{ij} \sim (1 - \pi) \, N(0, \sigma^2) + \pi \, N(0, k\sigma^2)$$

where
y, x, α, and β are as before
π is the prior probability that an observation is an outlier
k is the variance inflation factor

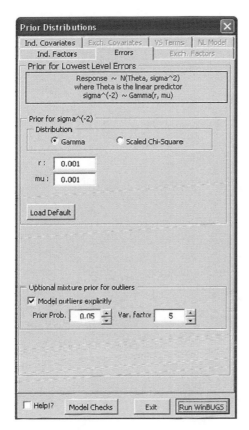

FIGURE 10.3
BugsXLA form for changing priors when modeling outliers explicitly.

Note that the loose notation for the residual errors is meant to represent a distribution that is a mixture of two Normals. BugsXLA uses the labels "Prior Prob." for π and "Variance factor" for k. The default settings, shown in Figure 10.3, are based on suggestions made by Box and Tiao (1968) and have been used here. Before running WinBUGS, the DIC and individual model checking functions were requested. Figure 10.4 shows the results of this analysis as imported back into Excel.

Once again the parameter estimates are very similar to those obtained previously, informing us that there are no extreme observations having undue influence. Note the message where the DIC statistics are usually displayed. This model is one for which WinBUGS is unable to compute the DIC. More generally, when DIC has failed to be imported, after ensuring that the burn-in is at least 5000, one should inspect the WinBUGS' log-file for more information. This can be done by turning off the "Auto Quit" option in the MCMC settings, or when a bugfolio has been created, viewing the file "BGX save.txt" (see Section 11.1 for more details).

When outliers are modeled explicitly, BugsXLA displays another message, shown in cell C13 in Figure 10.4. The WinBUGS model provides an estimate

	A	B	C	D	E	F	G	H	I	J	K
1			Label	Mean	St.Dev.	2.5%	Median	97.5%		WinBUGS Name	
2			CONSTANT	-0.0339	0.0441	-0.1214	-0.0340	0.0521		Beta0	
3			intercept at 0	0.0238	0.0661	-0.1091	0.0239	0.1522		alpha	
4		TRT	B	0.0974	0.0635	-0.0283	0.0964	0.2228		X.Eff[1,2]	
5		TRT	C	0.1407	0.0567	0.0305	0.1408	0.2547		X.Eff[1,3]	
6		TRT	D	0.0174	0.0643	-0.1051	0.0177	0.1442		X.Eff[1,4]	
7		TRT	E	0.0513	0.0615	-0.0708	0.0511	0.1738		X.Eff[1,5]	
8			FEV1_BASE	-0.0451	0.0465	-0.1343	-0.0453	0.0463		V.Coeff[1,1]	
9			SD(residual)	0.1718	0.0149	0.1442	0.1713	0.2030		sigma	
10	Note: CONSTANT & Factor effects are determined at the mean of the covariate(s).										
11	Interpret these cautiously when Factor x Covariate terms have been fitted.										
12											
13			Outliers explicitly modelled: 0 warning(s), see 'Mdl Chks(1)' for details.								
14											
15			No DIC statistics found; they may not be available for this model.								
16			View 'BGX save.txt' in Bugfolio, if created, for more information.								
17			Alternatively, check 'Response Node' name correct or try increasing Burn-In to at least 5000.								
18											
19											
20	Model		['clin study'!$A$1:$D$96]								
21	Distribution		Normal (outlier robust)								
22	Link		Identity								
23	Response		FEV1_CFB								
24	Fixed		TRT								
25	Covariates		FEV1_BASE								
26											
27	Priors										
28	CONSTANT		N(mu=0.0344, sigma=18.7)								
29	TRT		N(mu=0, sigma=18.7)								
30	FEV1_BASE		N(mu=0, sigma=39.8)								
31	V(residual)		Inv-Gamma(0.001, 0.001)								
32	Outliers		Prior Prob. = 0.05; Var. Factor = 5								
33											
34	WinBUGS MCMC Settings										
35	Burn-In: 5000 Samples: 10000 (Thin:1; Chains:1)										
36	Run took 27 seconds										

FIGURE 10.4
Summaries of posterior distributions, model, priors, and MCMC settings for clinical study when modeling outliers explicitly.

of the posterior probability that each observation is an outlier; this is simply the mean number of times that its error is treated as coming from the variance inflated component of the mixture distribution (see Appendix C for details). By default, BugsXLA counts any observation with a posterior probability of being an outlier greater than 50% as a warning, and displays the number found in the message displayed. This warning limit can be altered on the "BugsXLA Options" form as explained in Section 2.12. In this case, no warning cases were found, which is consistent with the lack of evidence of any influence on the estimates. These "outlier probabilities" are displayed and plotted on the model checks sheet created, referred to as "Mdl Chks(1)" in the message. Figure 10.5 shows these probability plots.

The explicit modeling of outliers can be included when using most of the other types of error distributions provided by BugsXLA. The mixture distribution is interpreted as defined in Table 10.1.

Case Study 10.2 Respiratory Tract Infections
Case Studies 4.1 and 6.1 discussed meta-analyses of a binomial response obtained from a series of clinical studies. We use these data to illustrate the explicit modeling of outliers in a GLMM, as well as the use of a t-distribution

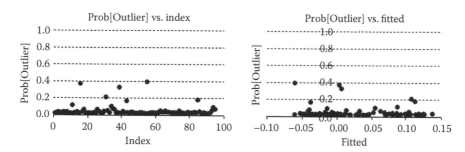

FIGURE 10.5
Plots of outlier probabilities for clinical study.

TABLE 10.1

Mixture Distributions Used by BugsXLA When Modeling Outliers Explicitly

Error Distribution	Distribution without Outliers	Outlier Component of Mixture
Normal	$N(\mu, \sigma^2)$	$N(\mu, k\sigma^2)$
t-distribution	$t(\mu, \sigma^2, v)$	$t(\mu, k\sigma^2, v)$
Log-normal	$Log\ N(\mu, \sigma^2)$	$Log\ N(\mu, k\sigma^2)$
Gamma	$Ga(r, \mu)$	$Ga(r/k, \mu/k)$
Poisson	$Pois(\lambda)$	$Pois(\lambda^*)$ with $\lambda^* \sim N(\lambda, k\lambda)$
Binomial	$Bin(\pi, n)$	$Bin(\pi^*, n)$ with $\pi^* \sim U(0, 1)$

to improve the robustness of the random effects assumption to extremes at this level of the model hierarchy. In the first instance, we fit a model similar to that in Case Study 6.1, but exclude the interaction between study and treatment. The simple model MCMC settings were chosen, with the default priors apart from the addition of a mixture distribution to model the outliers as shown in Table 10.1. Figure 10.6 shows the output as imported back into Excel.

BugsXLA reports that it has identified two extreme observations. Note that due to the nature of the robust model, the influence of these observations will have been automatically reduced. Figure 10.7 shows the plot of the outlier probabilities. The two observations identified, due to their outlier probabilities exceeding 0.5, are in rows 21 and 43 of Figures 4.1 and 6.1; they are both from the control group with very high infection rates (44/47 and 23/23 respectively). By down weighting the influence of these observations, as well as to a lesser extent a few others, this has led to a smaller treatment effect estimate. When back-transformed onto the odds-ratio scale, the treatment to control odds-ratio has posterior mean (standard deviation) of 0.37 (0.06), which is closer to one than those shown in Table 6.1.

It should be recognized that the model being fitted here, like the one in Case Study 4.1, assumes that the treatment effect is exactly the same for all studies. Since we think it is more reasonable to assume that at least some of the variation is due to heterogeneity of the treatment effect between studies,

	A	B	C	D	E	F	G	H	I	J	K
1			Label	Mean	St.Dev.	2.5%	Median	97.5%		WinBUGS Name	
2			CONSTANT	-0.8624	0.2795	-1.4020	-0.8716	-0.3143		Beta0	
3		trt	T	-1.0060	0.1763	-1.4370	-0.9775	-0.7315		X.Eff[1,2]	
4			SD(study)	0.9871	0.2204	0.6220	0.9637	1.4790		sigma.Z[1]	
5											
6			Outliers explicitly modelled: 2 warning(s), see 'Mdl Chks(1)' for details.								
7											
8	Model		[data!$A$1:$D$45]								
9	Distribution		Binomial (outlier robust)								
10	Link		Logit								
11	Response		infected/total								
12	Fixed		trt								
13	Random		study								
14											
15	Priors										
16	CONSTANT		N(mu=0, sigma=100)								
17	trt		N(mu=0, sigma=100)								
18	study		Norm(0,tau^2); tau ~ Half-N(sigma=5)								
19	Outliers		Prior Prob. = 0.05								
20											
21	WinBUGS MCMC Settings										
22	Burn-In: 5000 Samples: 10000 (Thin:1; Chains:1)										
23	Run took 22 seconds										

FIGURE 10.6
Summaries of posterior distributions, model, priors, and MCMC settings for respiratory tract meta-analysis when modeling outliers explicitly.

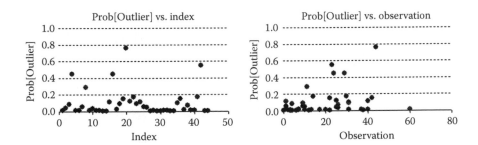

FIGURE 10.7
Plots of outlier probabilities for respiratory tract meta-analysis.

the study-by-treatment interaction term should be included as a random effect. When this is done and outliers explicitly modeled, no extreme observations are identified. One might still be concerned that the assumption of a Normal Distribution for the random effects is unduly influencing the estimates. We now relax this assumption by assuming a t-distribution with four degrees of freedom. The model specification is as shown in Figure 6.1. Figure 10.8 shows how the random effects distribution is changed. The "t(4)" distribution is selected from the drop down list and the "Save Changes" button alters the specified distribution in the list box at the top. The "study × trt" term was also changed to have a t-distribution; it is first necessary to click on the row labeled "study × 'trt'" to denote which random effect's prior is being altered. It is important to note that changes are only made after clicking on the "Save Changes" button.

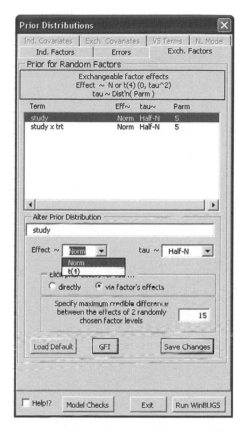

FIGURE 10.8
BugsXLA form for changing prior distribution of random effects.

Figure 10.9 shows the output as imported back into Excel. The two parameters representing the random effects are given the prefix "SDparm" to distinguish them from the standard deviation parameter associated with the Normal Distribution. BugsXLA adds a comment to these labels (cells C4 and C5 in Figure 10.9), stating that each represents the scale parameter of a t_4-distribution, which has standard deviation equal to root-two times its scale parameter. The treatment effect has hardly changed from that obtained with Normal distributed random effects; the posterior mean (standard deviation) of the treatment to control odds-ratio is 0.27 (0.06), compared to 0.25 (0.06) in Table 6.1.

Case Study 10.3 Random Coefficients Modeling of Rat Growth Curves
Case Study 5.4 showed how random regression coefficients could be used to model the growth curves of the weights of 30 young rats. Here we illustrate how the assumption of a multivariate Normal Distribution for the random effects of intercepts and slopes can be altered to a multivariate t_4-distribution. The model specification is unchanged from that shown in Figure 5.12. Figure 10.10 shows how the random coefficients prior distribution is changed.

	A	B	C	D	E	F	G	H	I	J	K
1			Label	Mean	St.Dev.	2.5%	Median	97.5%		WinBUGS Name	
2			CONSTANT	-0.6914	0.2551	-1.2140	-0.6902	-0.1638		Beta0	
3		trt	T	-1.3480	0.2192	-1.8160	-1.3410	-0.9358		X.Eff[1,2]	
4			SDparm(study)	0.7879	0.2151	0.4381	0.7658	1.2760		sigma.Z[1]	
5			SDparm(study x trt)	0.4742	0.1284	0.2658	0.4599	0.7598		sigma.Z[2]	
6											
7	Model		[data!$A$1:$D$45]								
8	Distribution		Binomial								
9	Link		Logit								
10	Response		infected/total								
11	Fixed		trt								
12	Random		study + study:trt								
13											
14	Priors										
15	CONSTANT		N(mu=0, sigma=100)								
16	trt		N(mu=0, sigma=100)								
17	study		t(4)(0,tau^2); tau ~ Half-N(sigma=3.75)								
18	study x trt		t(4)(0,tau^2); tau ~ Half-N(sigma=3.75)								
19											
20	WinBUGS MCMC Settings										
21	Burn-In: 5000 Samples: 10000 (Thin:1; Chains:1)										
22	Run took 32 seconds										

FIGURE 10.9
Summaries of posterior distributions, model, priors, and MCMC settings for respiratory tract meta-analysis with t-distributed random effects.

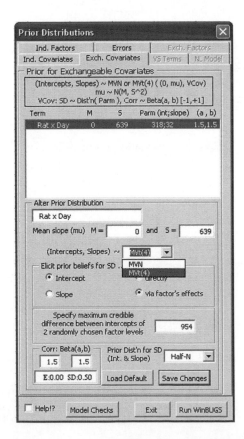

FIGURE 10.10
BugsXLA form for changing prior distribution of random coefficients.

	A	B	C	D	E	F	G	H	I	J	K
1			Label	Mean	St.Dev.	2.5%	Median	97.5%		WinBUGS Name	
2			CONSTANT	242.5000	2.8170	236.9000	242.5000	248.3000		Beta0	
3			Intercept at 0	106.9000	2.5680	101.9000	107.0000	112.0000		alpha	
4		(Slope mean)	Rat x Day	6.1610	0.1074	5.9500	6.1600	6.3740		W.MuCoeff[1]	
5		(Corr)	Rat x Day	0.5880	0.1661	0.2105	0.6096	0.8506		Cmvn.Beta[1]	
6		(Intercept)	SDparm(Rat x Day)	12.1700	2.1210	8.5210	11.9900	16.9500		sigma.WZ[1]	
7		(Slope)	SDparm(Rat x Day)	0.4285	0.0903	0.2733	0.4212	0.6320		sigma.Wcoeff[1]	
8			SD(residual)	6.0970	0.4630	5.2730	6.0650	7.0900		sigma	
9	Note: CONSTANT & Factor effects are determined at the mean of the covariate(s).										
10	Var-Cov(slope-intercept) terms are also for centred data.										
11											
12	**Model**	[Rats!$A$1:$C$151]									
13	Distribution	Normal									
14	Link	Identity									
15	Response	Weight									
16	Random Coeffs	Rat/Day									
17											
18	**Priors**										
19	CONSTANT	N(mu=243, sigma=6000)									
20	Rat x Day	MVt((0,mu), V, 4); mu ~ N(mean=0, sigma=639)									
21		V: sd(int) ~ Half-N(sigma=318)									
22		V: sd(slope) ~ Half-N(sigma=32)									
23		V: Correlation ~ Beta(1.5,1.5) [-1,+1]									
24	V(residual)	Inv-Gamma(0.001, 0.001)									
25											
26	**WinBUGS MCMC Settings**										
27	Burn-In: 5000 Samples: 10000 (Thin:1 Chains:1)										
28	Run took 75 seconds										

FIGURE 10.11

Summaries of posterior distributions, model, priors, and MCMC settings for rat growth curves with multivariate t-distributed random coefficients.

In a similar manner to how the random effects prior distribution was changed in Case Study 10.2, it is simply a matter of selecting "MVt(4)" from the drop down list box. If more than one set of random coefficients are in the model, they either all have to be given a multivariate Normal or all given a multivariate t_4 prior distribution. Unlike when choosing the distribution for the random effects, it is not possible to select a different option for each term. The simple model MCMC settings and default priors were chosen as reported in Figure 10.11, which shows the imported results. Comparison of the estimates with those shown in Figure 5.14 reveals that they have not changed much. This indicates that the standard model with Normal distributed random effects is probably adequate in this instance.

11

Beyond BugsXLA: Extending the WinBUGS Code

Clearly, BugsXLA can only offer a finite number of analysis options. This chapter discusses how an experienced WinBUGS user can extend an analysis initially specified using BugsXLA. This might be to make inferences on additional quantities of interest, or extend the model to one not currently offered by BugsXLA. In order to do this, it is first necessary to be aware of the files created by BugsXLA each time an analysis is specified. By default, these files are deleted after importing the results back into Excel, but they can be saved by requesting that a bugfolio be created on the 'MCMC Output & Options' form; this is done by selecting the 'Create Bugfolio' check box (see Figure 2.7). Alternatively, BugsXLA can be made to create a bugfolio every time by changing the default setting on the 'Options' form (see Section 2.12), or if the decision to save all the files is only made at the point when the results are being imported, the 'Keep IO Dir (BugF#)' check box can be selected on the 'Import Results' form (see Section 2.10). Whichever method is used, all the files are saved in a sub-folder of the one in which the Excel workbook is stored. The name of this sub-folder is 'BugF' by default, or with integer suffix, but can be changed by the user if specified on the 'MCMC Output & Options' form. This sub-folder, the bugfolio, contains the following plain text files that can be read using any standard text editor such as Microsoft's Notepad:

BGX DataRect.txt

This contains the data being analyzed in the rectangular format readable by WinBUGS. Any factors will have had their levels converted into integers. It may also contain some indicator variables used to specify the model in WinBUGS.

BGX DataList.txt

This contains all the other data that are of a different length to that contained in 'BGX DataRect.txt'. Examples include the number of observations, number of levels for each factor, and some values used to define the prior distributions.

BGX Inits1.txt

This contains the initial values used by WinBUGS to initiate the MCMC process. If multiple chains are requested, then there will be a separate file for each chain. Refer to Appendix B for more information on how BugsXLA sets initial values.

BGX Script.txt

This is the WinBUGS script that BugsXLA instructs WinBUGS to run.

BGX WinBUGS Code.txt

This is the WinBUGS code that defines the analysis. Appendix C provides a general explanation of how BugsXLA codes each component of the model.

BGX save.txt

This is the saved log file created by WinBUGS. The results of the analysis are read from this file and imported back into Excel.

If posterior samples are requested, then two additional files will exist:

BGX CODA1.txt

This contains all the individual sample values. If multiple chains are requested, then there will be a separate file for each chain.

BGX CODAIndex.txt

This contains the index needed to interpret the values contained in 'BGX CODA1.txt'.

It is not the intention of this chapter to explain every aspect of every possible analysis that BugsXLA can produce. Rather it is to illustrate the general approach to extending the code using a few examples. Appendix C provides more details on the generic code elements that BugsXLA produces, which together with the information in this chapter should be sufficient for an experienced WinBUGS user wishing to extend some BugsXLA-generated code.

In the rest of this chapter, it is assumed that the reader is very familiar with WinBUGS. Since no detailed instructions on how to carry out the WinBUGS operations are provided, most of this chapter will not be useful to the novice.

11.1 Using BugsXLA's WinBUGS Utilities

Once BugsXLA has created all the files, it is possible to rerun the analysis by opening up the script file 'BGX Script.txt' in WinBUGS and running it. The user can then work with the model and results as they would for any other WinBUGS analysis. For those preferring to work inside Excel as much as possible, BugsXLA offers some additional utilities to facilitate this (Section 2.13

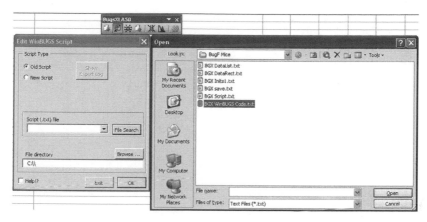

FIGURE 11.1
Locating and loading the WinBUGS program for editing using Notepad.

provides details). Here we illustrate how they can be used to make a minor alteration to a WinBUGS program, rerun the analysis, and import the results back into Excel. We use the initial analysis described in Case Study 4.5 where a Weibull model was fitted to censored mice lifetime data. The analysis is rerun requesting that the bugfolio, 'BugF Mice', be created. On completion of the standard BugsXLA analysis, we use the 'Edit WinBUGS Script' utility to edit the WinBUGS code. Although, as the title suggests, the main purpose is to edit script files, since all it does is open a text file using Notepad, it can also be used to edit the code. Figure 11.1 illustrates the process by which the program file is located and loaded into Notepad. Clicking on the Notepad icon on the main BugsXLA toolbar (highlighted in Figure 11.1) brings up the 'Edit WinBUGS Script' form. By default it looks for files with 'script' in their name, so we need to ignore any suggestions it has made and click on the 'File Search' button to locate the folder 'BugF Mice' containing the WinBUGS program. On finding the folder, we click on 'BGX WinBUGS Code.txt' and this enters this filename into the appropriate field on the 'Edit WinBUGS Script' form. Clicking on the 'OK' button opens the file in Notepad.

We will not try to explain all the code here (refer to Appendix C), but want to focus on the following part where the likelihood is defined:

```
log(mu[j]) <- Beta0 + sum(X.row[, j])
### Weibull Distribution: mu[] is the median ###
mu.weib[j] <- log(2)*pow(mu[j], -r.weib)
#   ### Weibull Distribution: mu[] is the mean ###
#   log(mu.weib[j]) <- r.weib*(loggam(1+(1/r.weib))-log(mu[j]))
Y[j] ~ dweib(r.weib, mu.weib[j])I(cen.lo[j], cen.hi[j])
```

By default, BugsXLA defines the median, not the mean, of the Weibull Distribution as being log-linked to the linear predictor. However, it also provides the code that can be used to change it to the mean; this is remarked out

FIGURE 11.2
Running a WinBUGS script.

in the code snippet above. We edit this code by removing the '#' symbol at the beginning of the line defining the mean, and adding a '#' symbol to remark out the line defining the median. This file is saved and Notepad closed. Since this very simple change to the code does not introduce any new nodes, or dramatically change the model, it is not necessary to edit any of the other files in the bugfolio. This analysis can be run by using the 'Run WinBUGS (Script)' utility as illustrated in Figure 11.2. Clicking on the crude WinBUGS icon on the main BugsXLA toolbar (highlighted in Figure 11.2) brings up the 'Run WinBUGS (Script)' form. Although it may have identified the correct script to run, it is recommended that the 'File Search' button be used to make sure it is working with the file in the current bugfolio. Once this has been loaded into the form, clicking on the 'OK' button opens up WinBUGS and runs the script.

Unlike when specifying a Bayesian analysis using the main BugsXLA 'Model Specification' form, once WinBUGS has finished running, the results are not automatically imported back into Excel. This is because these WinBUGS utilities have been designed to be used independently. One might want to create code and script files in WinBUGS itself, and only use these tools to export data from Excel into WinBUGS format (using the 'Export to WinBUGS' utility as discussed in Section 2.13) and import the results back into Excel. Figure 11.3 illustrates how to import results from the log file saved by WinBUGS. Clicking the icon showing an arrow pointing into a document (highlighted in Figure 11.3) brings up the 'Import Results from WinBUGS' form. As before, it is recommended that the 'File Search' button be used to make sure the correct log file is imported. Once this has been loaded into the form, clicking on the 'OK' button displays the familiar 'Import Results' form.

Figure 11.4 shows the results as imported back into Excel using this tool. Clearly BugsXLA is unable to provide meaningful labels when importing results directly from the saved log file. These can be obtained by referring

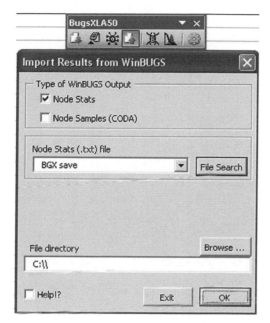

FIGURE 11.3
Importing the results from WinBUGS.

	A	B	C	D	E	F
1	**Name**	**Mean**	**St.Dev.**	**2.5%**	**Median**	**97.5%**
2	r.weib	3.1450	0.3271	2.5170	3.1370	3.8040
3	Beta0	3.1960	0.0759	3.0550	3.1920	3.3470
4	X.Eff[1,1]	3.807E-8	1.001E-5	-1.976E-5	2.372E-7	2.008E-5
5	X.Eff[1,2]	0.3642	0.1189	0.1385	0.3630	0.5998
6	X.Eff[1,3]	0.1060	0.1069	-0.1033	0.1069	0.3178
7	X.Eff[1,4]	-0.1223	0.1120	-0.3411	-0.1228	0.1004
8	Pred.Ave[1]	24.5000	1.8720	21.2100	24.3300	28.4300
9	Pred.Ave[2]	35.3100	3.2870	29.6400	34.9800	42.8500
10	Pred.Ave[3]	27.2400	2.1730	23.3800	27.0700	31.9600
11	Pred.Ave[4]	21.6900	1.7700	18.6000	21.5400	25.5400
12	Pred.Ave[5]	1.1180	0.1198	0.9019	1.1130	1.3740
13	Pred.Ave[6]	0.7782	0.0940	0.6017	0.7745	0.9673
14	Pred.Ave[7]	1.2640	0.1422	1.0020	1.2620	1.5530
15						
16		**Y**				
17	**Dbar**	494.6760				
18	**Dhat**	489.7490				
19	**pD**	4.9260				
20	**DIC**	499.6020				

FIGURE 11.4
Summaries of posterior distributions and DIC statistics as imported using BugsXLA's WinBUGS importing tool.

back to the output displayed after the original analysis. Column J in Figure 4.20 and column U in Figure 4.21 show the same names as are in column A of Figure 11.4, with their more informative labels in columns B:C and N, respectively. Comparison of the estimates obtained using log(mean) as the linear predictor with those obtained using log(median) indicates that in this case the differences are insignificant.

In Case Study, 10.1 BugsXLA returned the message,

> View 'BGX Save.txt' in Bugfolio, if created, for more information,

when it could not return the DIC statistics requested. This file can also be read using the 'Edit WinBUGS Script' utility. After following the advice given above on finding and loading a text file into Notepad, the following line is seen in the WinBUGS log file:

```
command #Bugs:dic.set cannot be executed (is greyed out)
```

Although this does not explain why WinBUGS could not set the DIC, it at least rules out the possibility that the DIC was calculated and BugsXLA simply failed to read it from the file.

11.2 Editing the Initial MCMC Values

Case Study 6.4 describes the analysis of a complex experiment involving both mixture and process variables with the aim of improving the quality of a film-manufacturing process. In this example, it was not possible to use BugsXLA's novel model MCMC settings, in which three chains are run, due to a problem caused by the automatically generated initial starting values. Since BugsXLA v5.0 does not offer any facility to alter these initial values, the only option is to edit the 'BGX Inits#.txt' files manually and rerun the script file. To do this, it is necessary to request a bugfolio, rerun the analysis using the novel model MCMC settings and quit WinBUGS after it completes its failed analysis. It is then possible to locate and open the files containing the three sets of initial values. Figure 11.5 shows these three files.

The first set of initial values, contained in 'BGX Inits1.txt', is the same as used when a single chain is requested. Since we know that WinBUGS runs fine when these are used they do not need to be edited. One or both of the second and third sets of values are too extreme. It is recommended that copies of the original files are made so that the edits can be easily compared to the original values. This could be useful since it might take more than one iteration to obtain initial values that allow WinBUGS to complete the analysis. The variable names used in these files are the generic names used by BugsXLA

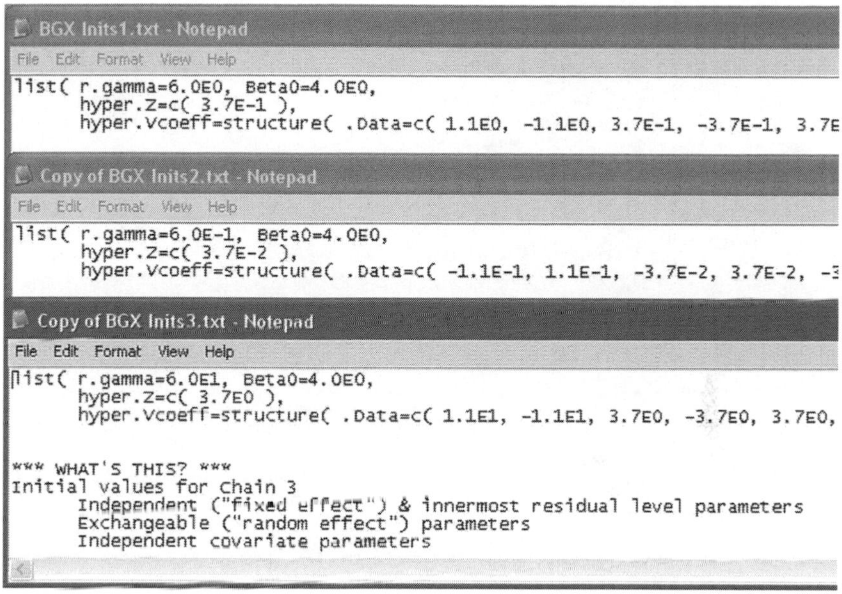

FIGURE 11.5
Automatically generated initial values when novel model settings used to analyze Case Study 6.4.

when creating WinBUGS code. Details are provided in Appendix C that can be used to decipher their meaning. The files themselves provide a very brief description of what they contain, shown in Figure 11.5 under the line,

```
"*** WHAT'S THIS? ***".
```

Rather than try to determine better initial values through a deep understanding of the parameters in the model, we suggest that a more pragmatic approach be taken in the first instance. Appendix B describes how BugsXLA determines initial values. It attempts to choose values for the first chain that should not be too extreme relative to values supported by the likelihood. Most of the initial values for the second and third chains are set an order of magnitude smaller and larger, respectively, than the first set. This can be seen in Figure 11.5, for example, the initial values for r.gamma are 6, 0.6, and 60 for the first, second, and third chains, respectively. Since this model has a log-link, an order of magnitude change is huge. Reducing the multiplication factor from ten to three, say, should still produce dispersed initial values on the log scale. Figure 11.6 shows the initial values after they have been edited.

The approach explained in Section 11.1 can then be used to rerun the script, which will now initialize the second and third chains with the new values. In this instance these first set of changes solves the problem and WinBUGS successfully completes the analysis. This enables convergence to be assessed

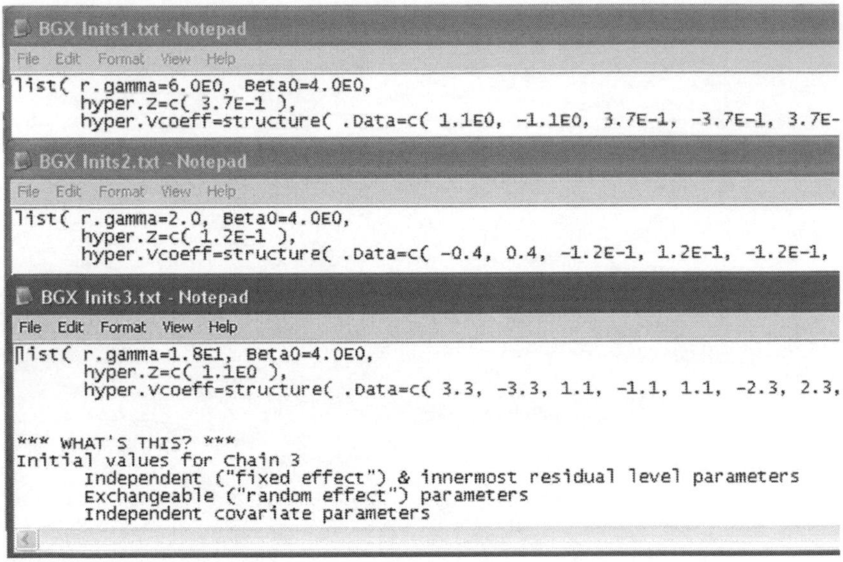

```
BGX Inits1.txt - Notepad
File  Edit  Format  View  Help
list( r.gamma=6.0E0, Beta0=4.0E0,
      hyper.z=c( 3.7E-1 ),
      hyper.vcoeff=structure( .Data=c( 1.1E0, -1.1E0, 3.7E-1, -3.7E-1, 3.7E-
```

```
BGX Inits2.txt - Notepad
File  Edit  Format  View  Help
list( r.gamma=2.0, Beta0=4.0E0,
      hyper.z=c( 1.2E-1 ),
      hyper.vcoeff=structure( .Data=c( -0.4, 0.4, -1.2E-1, 1.2E-1, -1.2E-1,
```

```
BGX Inits3.txt - Notepad
File  Edit  Format  View  Help
list( r.gamma=1.8E1, Beta0=4.0E0,
      hyper.z=c( 1.1E0 ),
      hyper.vcoeff=structure( .Data=c( 3.3, -3.3, 1.1, -1.1, 1.1, -2.3, 2.3,


*** WHAT'S THIS? ***
Initial values for chain 3
      Independent ("fixed effect") & innermost residual level parameters
      Exchangeable ("random effect") parameters
      Independent covariate parameters
```

FIGURE 11.6
Edited initial values used to analyze Case Study 6.4 with three chains.

using the techniques discussed in Section 1.2. More generally it might be necessary to edit the initial values again, in which case a better understanding of the meaning of the variables could be useful (refer to Appendix C).

11.3 Estimating Additional Quantities of Interest

Case Study 6.1 describes a random effects meta-analysis of a series of clinical studies. A common objective of such a meta-analysis is to derive evidenced based prior distributions for a future study. Spiegelhalter et al. (2004, Chapter 5) discuss the various ways one could specify a relationship between the treatment effects in these historical studies and the treatment effect in a future study. Here we assume complete exchangeability between all of the treatment effects, so that

$$\theta_h, \theta_f \sim N(\mu_\theta, \sigma_\theta^2) \quad h = 1, 2, \dots, 22$$

where
  θ_h are the historical treatment effects from each of the 22 studies
  θ_f is the future treatment effect for which we wish to derive a prior distribution

As Spiegelhalter et al. (2004) point out, it is important to realize that the prior distribution for θ_f is the predictive distribution of the treatment effect in a future study, not the posterior distribution of the 'population mean' effect μ_θ. Hence, the posterior summary provided for 'trt T' (=μ_θ) in Figure 6.4 could not be used to derive the prior distribution for θ_f. One could be forgiven for thinking that the predictive distribution for a future study's treatment effect could be obtained by specifying the BugsXLA prediction shown in Table 11.1.

BugsXLA interprets this as the contrast between the mean of treated patients in a future study with the mean of control patients in a different future study. This is almost certainly not a prediction of any interest, and leads to an odds-ratio estimate dramatically more imprecise than for a within study contrast. The reason BugsXLA does not interpret this statement as we would wish, is that it is not able to recognize the need to cancel out study effects but not study-by-treatment interaction effects when evaluating the contrast, and so it includes both. Using the algebraic notation of Case Study 6.1:

$$\theta_f = \gamma_2 - \gamma_1 + (s\gamma)_{f2} - (s\gamma)_{f1}$$

whereas the predictions defined in Table 11.1 are

$$\gamma_2 - \gamma_1 + s_f - s_{f*} + (s\gamma)_{f2} - (s\gamma)_{f*1}$$

where f and f* are two different future studies.

We now show how the WinBUGS code created by BugsXLA can be edited and rerun to obtain the correct prior (predictive) distribution. As before, we start by requesting that a bugfolio is created. The analysis is rerun with the erroneous predictions defined in Table 11.1 specified. The WinBUGS program is located in the bugfolio and can be opened using Notepad. Figure 11.7 shows the last few lines of this file, 'BGX WinBUGS Code.txt'.

We only explain enough of the code here to justify the changes that will be made (refer to Appendix C for more details). The WinBUGS stochastic nodes are related to the parameters of the model as follows:

$$\texttt{X.Eff[1,2]} \equiv \gamma_2$$

TABLE 11.1

Erroneous Contrast Statement for Predictive Distribution of Future Study's Treatment Effect

F	G
Predictions for...	
trt	study
c(T,C)	~

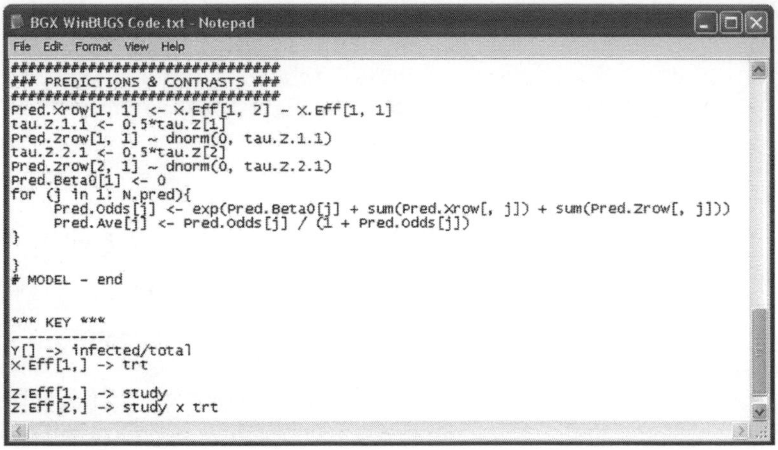

```
BGX WinBUGS Code.txt - Notepad
File  Edit  Format  View  Help
###############################
### PREDICTIONS & CONTRASTS ###
###############################
Pred.Xrow[1, 1] <- X.Eff[1, 2] - X.Eff[1, 1]
tau.Z.1.1 <- 0.5*tau.Z[1]
Pred.Zrow[1, 1] ~ dnorm(0, tau.Z.1.1)
tau.Z.2.1 <- 0.5*tau.Z[2]
Pred.Zrow[2, 1] ~ dnorm(0, tau.Z.2.1)
Pred.Beta0[1] <- 0
for (j in 1: N.pred){
    Pred.Odds[j] <- exp(Pred.Beta0[j] + sum(Pred.Xrow[, j]) + sum(Pred.Zrow[, j]))
    Pred.Ave[j] <- Pred.Odds[j] / (1 + Pred.Odds[j])
}

}
# MODEL - end

*** KEY ***
-----------
Y[] -> infected/total
X.Eff[1,] -> trt

Z.Eff[1,] -> study
Z.Eff[2,] -> study x trt
```

FIGURE 11.7
Lines of WinBUGS code where predictions and contrasts are defined.

$$X.Eff[1,1] \equiv \gamma_1$$

$$tau.Z[1] \equiv 1/\sigma_1^2$$

$$tau.Z[2] \equiv 1/\sigma_2^2$$

The WinBUGS deterministic nodes are

Pred.Xrow[1, 1] : 'population mean' effect, $\mu_\theta = \gamma_2 - \gamma_1$
tau.Z.1.1   : precision of the difference between two study effects, $0.5/\sigma_1^2$

with Pred.Zrow[1, 1] being a draw from the predictive distribution of this difference

tau.Z.2.1   : precision of the difference between two interaction effects, $0.5/\sigma_2^2$

with Pred.Zrow[2, 1] being a draw from the predictive distribution of this difference

N.pred   : equals the number of predictions specified, which is one
Pred.Odds[1] : predicted odds-ratio
Pred.Ave[1]   : meaningless; would be a probability if prediction was not a contrast

The predicted log(odds-ratio) is

Pred.Beta0[1] + sum(Pred.Xrow[,1]) + sum(Pred.Zrow[,1])

Comparing the formula for the predicted log(odds-ratio) with the earlier definitions of the WinBUGS nodes, it superficially appears that each MCMC simulation provides a draw from the Normal distribution:

$$N(\gamma_2 - \gamma_1, 2(\sigma_1^2 + \sigma_2^2))$$

However, it should be remembered that all four of the parameters defining this Normal Distribution are themselves being simultaneously estimated. Hence, the draws are from a distribution derived by integrating this Normal Distribution with respect to the four-dimensional posterior distribution for γ_1, γ_2, σ_1, and σ_2. This is automatically computed as part of the MCMC run.

In order to obtain the correct prior (predictive) distribution for the treatment effect in a future study, we need to remove the between study variation from the prediction. Also, since the parameter will enter the model on the log-scale, we need to add a new deterministic node that defines the predicted log(odds-ratio). The study variation is removed by editing out the node tau.Z.1.1 and setting Pred.Zrow[1, 1] to zero as shown in Figure 11.8. It is recommended that remark characters (#) are used, rather than straight deletion, so that it is clear how the code has been edited, and that comments are added to the top of the file (not shown) as a reminder of why the changes have been made. Note that both of the edited lines cannot be simply remarked out, since WinBUGS would otherwise crash when it tried to sum over an empty element of Pred.Zrow. A new line has also been added defining the predicted log(odds-ratio). Although the code could be simplified somewhat from the generic code automatically generated by BugsXLA, this has not been done to make it clearer how the original code has been edited.

```
BGX WinBUGS Code.txt - Notepad
File  Edit  Format  View  Help
###############################
### PREDICTIONS & CONTRASTS ###
###############################
Pred.Xrow[1, 1] <- X.Eff[1, 2] - X.Eff[1, 1]
#####tau.Z.1.1 <- 0.5*tau.Z[1]
Pred.Zrow[1, 1] <- 0 #####~ dnorm(0, tau.Z.1.1)
tau.Z.2.1 <- 0.5*tau.Z[2]
Pred.Zrow[2, 1] ~ dnorm(0, tau.Z.2.1)
Pred.Beta0[1] <- 0
for (j in 1: N.pred){
    Pred.Odds[j] <- exp(Pred.Beta0[j] + sum(Pred.Xrow[, j]) + sum(Pred.Zrow[, j]))
    Pred.Ave[j] <- Pred.Odds[j] / (1 + Pred.Odds[j])
    ##### 1 new line #####
    Log.OddsRatio[j] <- Pred.Beta0[j] + sum(Pred.Xrow[, j]) + sum(Pred.Zrow[, j])
}

}
# MODEL - end

*** KEY ***
-----------
Y[] -> infected/total
X.Eff[1,] -> trt

Z.Eff[1,] -> study
Z.Eff[2,] -> study x trt
```

FIGURE 11.8
WinBUGS code edited to provide predictive distribution of future study's treatment effect.

	A	B	C	D	E	F
1	**Name**	**Mean**	**St.Dev.**	**2.5%**	**Median**	**97.5%**
2	Beta0	-0.6202	0.2760	-1.1940	-0.6192	-0.0926
3	X.Eff[1,1]	-1.335E-7	9.999E-6	-1.954E-5	-6.734E-8	1.931E-5
4	X.Eff[1,2]	-1.4200	0.2342	-1.9180	-1.4090	-0.9957
5	sigma.Z[1]	1.0630	0.2312	0.6632	1.0450	1.5780
6	sigma.Z[2]	0.5880	0.1519	0.3368	0.5712	0.9344
7	Pred.Ave[1]	0.2245	0.1412	0.0380	0.1953	0.5869
8	Pred.Odds[1]	0.3669	0.7383	0.0395	0.2427	1.4200
9	Log.OddsRatio[1]	-1.4300	0.8950	-3.2310	-1.4160	0.3510

FIGURE 11.9
Imported summary of prior (predictive) distribution of future study's treatment effect, Log. OddsRatio.

Once again, the approach explained in Section 11.1 can then be used to rerun the script and import the results. But before doing this, it is necessary to edit the script file, 'BGX Script.txt', so that WinBUGS is instructed to compute summary statistics for the new node Log.OddsRatio. This can be done using BugsXLA's 'Edit WinBUGS Script' utility. It is simply necessary to add the line:

```
set(Log.OddsRatio)
```

immediately after the final set command that already exists: set(Pred. Odds). Figure 11.9 shows the results as imported back into Excel using BugsXLA's 'Import from WinBUGS' utility. The prior distribution required is labeled Log.OddsRatio. Given the close match between the percentiles of this quantity's distribution, and those inferred by a Normal Distribution with the same mean and standard deviation, it is probably adequate to approximate this prior by the Normal Distribution:

$$N(-1.43, 0.895)$$

As should be expected, this is more dispersed than the posterior distribution for the 'population mean' effect, which, referring back to Figure 6.4, has a mean (standard deviation) of −1.40 (0.23). The difference between the means of these two distributions is simply due to MCMC error.

Appendix A: Distributions Referenced in BugsXLA

Distribution	Notation in BugsXLA	Density	Domain of x
Bernoulli	$x \sim \text{Bern}(\pi)$	$\pi^x(1-\pi)^{1-x}$	0 or 1
Beta	$x \sim \text{Beta}(\alpha, \beta)$	$\Gamma(\alpha+\beta)/(\Gamma(\alpha)\Gamma(\beta))\, x^{\alpha-1}(1-x)^{\beta-1}$	[0, 1]
Binomial	$x \sim \text{Bin}(\pi, n)$	$n!/(x!(n-x)!)\, \pi^x(1-\pi)^{n-x}$	0, 1, ..., n
Categorical	$x \sim \text{Cat}(\pi_{(1:c)})$	π_x	1, 2, ..., c
Exponential	$x \sim \text{Exp}(\lambda)$	$= \text{Ga}(1, \lambda)$	$[0, \infty)$
Gamma	$x \sim \text{Ga}(r, \mu)$	$\mu^r \exp(-\mu x)\, x^{r-1}/\Gamma[r]$	$[0, \infty)$
Half-Cauchy	$x \sim \text{Half-C}(\tau^2)$	$(0.5\pi\tau(1+(x/\tau)^2))^{-1}$	$[0, \infty)$
Half-Normal	$x \sim \text{Half-N}(\tau^2)$	$(0.5\pi\tau^2)^{-0.5} \exp(-x^2/2\tau^2)$	$[0, \infty)$
Log-Normal	$x \sim \text{LogN}(\mu, \sigma^2)$	$(2\pi\sigma^2 x^2)^{-0.5} \exp(-(\log(x)-\mu)^2/2\sigma^2)$	$[0, \infty)$
Normal	$x \sim \text{N}(\mu, \sigma^2)$	$(2\pi\sigma^2)^{-0.5} \exp(-(x-\mu)^2/2\sigma^2)$	$(-\infty, \infty)$
Poisson	$x \sim \text{Pois}(\lambda)$	$\lambda^x \exp(-\lambda)/x!$	0, 1, ..., ∞
Scaled Chi-square	$x \sim \text{ScChiSq}(s^2, \nu)$	$= \text{Ga}(\nu/2, \nu s^2/2)$	$[0, \infty)$
t-distribution	$x \sim \text{t}(\mu, \sigma^2, \nu)$	$\Gamma((\nu+1)/2)/(\Gamma(\nu/2)\,(\nu\pi\sigma^2)^{0.5})$ $(1+\nu^{-1}(x-\mu)^2/\sigma^2)^{-(\nu+1)/2}$	$(-\infty, \infty)$
Uniform	$x \sim \text{U}(a, b)$	$1/(b-a)$	[a, b]
Weibull	$x \sim \text{Weib}(r, \theta)$	$r\theta x^{r-1} \exp(-\theta x^r)$	$[0, \infty)$

The first and second parameters of the Gamma Distribution, r and μ, are sometimes referred to as scale and shape, respectively.

The first parameter of the Weibull Distribution, r, is sometimes referred to as the shape.

Distribution	Notation in BugsXLA	Mean	Median	Mode	Variance	CV
Bernoulli	$x \sim \text{Bern}(\pi)$	π			$\pi(1 - \pi)$	
Beta	$x \sim \text{Beta}(\alpha, \beta)$	$\alpha/(\alpha + \beta)$		$(\alpha - 1)/(\alpha + \beta - 2)$	$\alpha\beta/((\alpha + \beta)^2(\alpha + \beta + 1))$	$(\beta/(\alpha(\alpha + \beta + 1)))^{\frac{1}{2}}$
Binomial	$x \sim \text{Bin}(\pi, n)$	$n\pi$	$n\pi$*¹	$(n + 1)\pi$*¹	$n\pi(1 - \pi)$	$((1 - \pi)/n\pi)^{\frac{1}{2}}$
Categorical	$x \sim \text{Cat}(\pi_{(1:c)})$			Not meaningful		
Exponential	$x \sim \text{Exp}(\lambda)$	λ^{-1}	$\log(2)/\lambda$	0	λ^{-2}	1
Gamma	$x \sim \text{Ga}(r, \mu)$	r/μ		$(r - 1)/\mu$	r/μ^2	$r^{-\frac{1}{2}}$
Half-Cauchy	$x \sim \text{Half-C}(\tau^2)$		τ	0		
Half-Normal	$x \sim \text{Half-N}(\tau^2)$	$\tau(2/\pi)^{\frac{1}{2}}$	$0.67\,\tau$	0	$\tau^2(1 - 2/\pi)$	0.76
Log-Normal	$x \sim \text{LogN}(\mu, \sigma^2)$	$\exp(\mu + 0.5\sigma^2)$	$\exp(\mu)$	$\exp(\mu - \sigma^2)$	$(\exp(\sigma^2) - 1)\exp(2\mu + \sigma^2)$	$(\exp(\sigma^2) - 1)^{\frac{1}{2}}$
Normal	$x \sim \text{N}(\mu, \sigma^2)$	μ	μ	μ	σ^2	σ/μ
Poisson	$x \sim \text{Pois}(\lambda)$	λ	$\approx \lambda + \frac{1}{3} + 0.02\lambda^{-1}$*¹	λ*¹	λ	$\lambda^{-\frac{1}{2}}$
Scaled Chi-square	$x \sim \text{ScChiSq}(s^2, \nu)$	s^{-2}		$(\nu - 2)/\nu s^2$	$2\nu^{-1}\,s^{-4}$	$(2/\nu)^{\frac{1}{2}}$
t-distribution	$x \sim \text{t}(\mu, \sigma^2, \nu)$	μ	μ	μ	$\nu\sigma^2/(\nu - 2)$	$\sigma\nu^{\frac{1}{2}}(\nu - 2)^{-\frac{1}{2}}/\mu$
Uniform	$x \sim \text{U}(a, b)$	$(a + b)/2$	$(a + b)/2$	$(a + b)/2$	$(b - a)^2/12$	
Weibull	$x \sim \text{Weib}(r, \theta)$	$\Gamma(1 + r^{-1})\theta^{1/r}$	$(\log(2)/\theta)^{1/r}$	$((r - 1)/r\theta)^{1/r}$		$(\Gamma(1 + 2r^{-1})\,\Gamma^{-2}(1 + r^{-1}) - 1)^{\frac{1}{2}}$

CV: coefficient of variation = standard deviation/mean

*¹: nearest integer

Empty cells imply that the value cannot be expressed in a simple form or that it is not useful.

Also, the following multivariate distributions are referenced, see Table A1 of Gelman et al. (2004) for more details:

Multivariate Normal: $\mathbf{x} \sim \text{MVN}(\boldsymbol{\mu}, \boldsymbol{\Sigma})$

Multivariate t: $\mathbf{x} \sim \text{MVt}(\boldsymbol{\mu}, \boldsymbol{\Sigma}, \nu)$

Multinomial: $\mathbf{x} \sim \text{Multi}(\boldsymbol{\pi}, n)$

BugsXLA uses the multinomial-Poisson transformation when modeling multinomial data in order to obtain faster MCMC sampling. The idea is to pretend that the counts in the categories are actually independent Poisson variables, with additional parameters added to account for the finite sample sizes; see Baker (1995) for details.

Appendix B: BugsXLA's Automatically Generated Initial Values

BugsXLA attempts to specify initial values that are not so extreme relative to values supported by the likelihood and prior that WinBUGS is unable to start the MCMC process. Values for the first chain are supposed to be more "central" than those specified for any additional chains requested. This is achieved by using the following factor in the formulae used to define the initial values:

$$C_i = 1, 0.1, 10, 0.3, 3 \quad i = 1, 2, 3, 4, 5 \text{ (number of chains)}$$

Note that some parameters are given the same initial values for each chain.

In order to define initial values that are not too discordant with the likelihood, the formulae for some of them involve the sample mean and variance of the response variable. These are denoted M_Y and V_Y below, with S_Y being the standard deviation. When a link function other than the identity is used, these statistics are of the link transformed response with adjustments to avoid infinite values. When the response is binomial or categorical, V_Y is set equal to 1, and M_Y is set equal to 0 unless the complementary log-log link is used, in which case M_Y is set equal to -0.367; the values for M_Y represent a probability of 0.5. Similarly, M_X and S_X, where X denotes a covariate, are used in the formulae for some of the regression coefficients. In order to define initial values that are not too discordant with the prior, the formulae for some of them involve the prior mean and standard deviation of the parameter.

The WinBUGS nodes for which the initial values are being set are shown in square brackets. The reader may need to refer to Appendix C to be certain of the meaning of each WinBUGS node. When printed to the file 'BGX Inits#.txt', most values are rounded to two significant figures.

The whole process by which BugsXLA determines initial values, including adding some degree of user control over their specification, will be reviewed during development of the next version.

B.1 Scale and Shape Parameters

$N(\cdot, \sigma^2)$, $\text{LogN}(\cdot, \sigma^2)$, $t(\cdot, \sigma^2, \cdot)$ error distribution

σ^{-2} [tau] $\sim \text{Ga}(a, b)$

inits = $10 \, C_i / V_Y$

unless $(a/b) \pm 4 \, a^{\frac{1}{2}}/b$ closer to prior mean (a/b)

$t(\cdot, \cdot, \ \nu)$ error distribution
  ν^{-1} [inv.df] $\sim U(a, b)$
  inits $= (a + b)/2 + 0.9 \log_{10}(C_i) (b - a)/2$

Weib$(r, \cdot)$
  r [r.weib] $\sim Ga(a, b)$
  inits $= 1 + C_i$
  unless $(a/b) \pm 4a^{\frac{1}{2}}/b$ closer to prior mean (a/b)

Ga$(r, \cdot)$
  r [r.gamma] $\sim Ga(a, b)$
  inits $= 10 \ C_i/V_Y$
  unless $(a/b) \pm 4a^{\frac{1}{2}}/b$ closer to prior mean (a/b)

Generalized logit link
  [m.gnlgt] $\sim Ga(a, b)$
  inits $= C_i$
  unless $(a/b) \pm 4a^{\frac{1}{2}}/b$ closer to prior mean (a/b)

Summary Stats Response (Normal) with degrees of freedom specified
  A vector of length number of observations [N.obs]
  [inv.se2] $\sim Ga(a, b)$
  inits $= 10 \ C_i/V_Y$
  unless $(a/b) \pm 4a^{\frac{1}{2}}/b$ closer to prior mean (a/b)

B.2 Model Constants

Multinomial (additional parameters used in multinomial-Poisson transformation)
  A vector of length number of observations [N.obs]
  [MN.lambda] $\sim N(\mu, \sigma^2)$
  inits $= M_Y$
  unless $\mu \pm 4\sigma$ closer to prior mean μ

Unordered categorical
  A vector of length number of categories [MVdims]
  [CF.Eff] $\sim N(\mu, \sigma^2)$
  inits $= M_Y$
  unless $\mu \pm 4\sigma$ closer to prior mean μ

Ordered categorical
> A vector of length number of categories − 1
> [CF.a] ~ N(·,·)
>
> inits = 0.1, 0.2, ..., (N.cats-1)/10

All models bar the categoricals
> [Beta0] ~ N(μ, σ²)
>
> inits = M_Y
>
> unless μ ± 4σ closer to prior mean μ

B.3 Factors

Fixed effects
> [X.Eff] ~ N(·,σ²) or t(·,σ²,·)
>
> inits = ± 0.3 C_i S* ± allocated "pseudo-randomly"
>
> S* = min(S_Y, σ)
>
> (inits − 0 when constrained to zero)

Random effects
> [hyper.Z] ~ hyper-prior
>
> inits = 0.3 C_i S*
>
> S* = min(S_Y, PriorScale) for Half-N(·), Half-C(·) and U(·,·)
>
> S* = max(3/V_Y, PriorScale) for Ga(·,·) and ScChiSq(·,·)

hyper.Z is standard deviation for Half-N(·), Half-C(·) and U(·,·), but precision for Ga(·,·) and ScChiSq(·,·)

Half-N(τ) hyper-prior
> PriorScale = τ

Half-C(τ) hyper-prior
> PriorScale = 3τ

U(0, b) hyper-prior
> PriorScale = b/12

Ga(r, μ) hyper-prior
> PriorScale = r/μ

ScChiSq(s², ν) hyper-prior
> PriorScale = s^{-2}

B.4 Covariates

Independent

[hyper.Vcoeff] ~ N(μ, τ^2) or Half-N(τ)

inits = $\mu \pm 0.3\,C_i\,S^*$ or $0.3\,C_i\,S^* \pm$ allocated "pseudo-randomly"

S* = min(S_Y/S_X, τ)

Exchangeable

[W.MuCoeff] ~ N(μ, σ^2) or t($\mu, \sigma^2, \cdot$)

inits = $\mu \pm 0.3\,C_i\,S^* \pm$ allocated "pseudo-randomly"

S* = min(S_Y/S_X, σ)

U(0, b) hyper-prior

[sigma.WZ] ~ hyper-prior

inits = $C_i\,S^*/11$

$S^* = \min(S_Y, b)$

[sigma.Wcoeff] ~ hyper-prior

inits = $C_i\,S^*/12$

$S^* = \min(S_Y/S_X, b)$

Half-N(τ) or Half-C(τ) hyper-prior

[hyper.WZ] ~ hyper-prior

inits = $0.3\,C_i\,S^*$

$S^* = \min(S_Y, \tau)$

[hyper.Wcoeff] ~ hyper-prior

inits = $0.3\,C_i\,S^*$

$S^* = \min(S_Y/S_X, \tau)$

[CmvnStar.Beta] ~ Beta(α, β)

inits = $\alpha/(\alpha + \beta) \pm C_i\,(1 - |\alpha/(\alpha + \beta)|)\,/12 \pm$ allocated "pseudo-randomly"

B.5 Variable Selection

[VSpi] ~ Beta(α, β)

inits = $\alpha/(\alpha + \beta) + 0.9\,\log_{10}(C_i)\,S^*$

$S^* = \min\,(\alpha/(\alpha + \beta),\ 1 - \alpha/(\alpha + \beta))$

[VSgamma] ~ U(a, b)

inits = $M_g + 0.9\,\log_{10}(C_i)\,(b - M_g)$

$M_g = (a + b)/2$

Factors
   [VSfac.In] ~ Bern(·)
   inits = 0 or 1 "pseudo-randomly"
   [VSfac.Beta.star] ~ N(·,·)
   inits = 0

Covariates
   [VScov.In] ~ Bern(·)
   inits = 0 or 1 "pseudo-randomly"
   [VScov.Beta.star] ~ N(·,·)
   inits = M_g S*
   N(·,·), Log-N(·,·) or t(·,·,·) error distribution
      $S^* = (V_Y/(10\ C_j))^{1/2}$
   Else
      $S^* = C_i$ "sigma constant"

B.6 Censored Observations and Outliers

[Y] ~ error distribution
   LE X
   inits = 0.999999 X
   GE X
   inits = 1.000001 X
   IN A B
   inits = (A + B)/2
[out] ~ Bern(·)
   inits = 0 or 1 "pseudo-randomly"

B.7 Non-Linear Model

[NL.Emax] ~ N(μ, σ²)
   inits = μ ± 0.3 C_i S* ± allocated "pseudo-randomly"
   $S^* = \min(S_Y, \sigma)$

[NL.EmaxGrp] ~ N(μ, σ^2)
   inits = $\mu \pm 0.3$ C_i S^* $\pm$ allocated "pseudo-randomly"
   $S^* = \min(S_Y, \sigma)$

[NL.EmaxGrp.SD] ~ Half-N(τ)
   inits = 0.3 C_i τ

[NL.logED50] ~ N(μ, σ^2)
   inits = $\mu \pm 0.3$ C_i S^* $\pm$ allocated "pseudo-randomly"
   $S^* = \min(R_x, \sigma)$

$R_x = (\log(X_{max}) - \log(X_{min}))/4$
   X_{max} is largest and X_{min} is smallest non-zero value of covariate

[NL. logED50Grp] ~ N(μ, σ^2)
   inits = $\mu \pm 0.3$ C_i S^* $\pm$ allocated "pseudo-randomly"
   $S^* = \min(R_x, \sigma)$

[NL. logED50Grp.SD] ~ Half-N(τ)
   inits = 0.3 C_i τ

[NL.logHill] ~ N(μ, σ^2)
   inits = $\mu \pm 0.3$ C_i σ $\pm$ allocated "pseudo-randomly"

[NL. logHillGrp] ~ N(μ, σ^2)
   inits = $\mu \pm 0.3$ C_i σ $\pm$ allocated "pseudo-randomly"

[NL. logHillGrp.SD] ~ Half-N(τ)
   inits = 0.3 C_i τ

B.8 Longitudinal Model

[phi.AR1] or [phiStar.AR1] ~ Beta(α, β)
   inits = $\alpha/(\alpha + \beta) \pm C_i$ $(1 - |\alpha/(\alpha + \beta)|)/12 \pm$ allocated "pseudo-randomly"

[tau.meas] ~ Ga(r, μ)
   inits = r/μ

Appendix C: Explanation of WinBUGS Code Created by BugsXLA

It is assumed that the reader of this appendix is familiar with the WinBUGS programming language. The purpose is to explain the generic code elements that BugsXLA creates, but only in sufficient detail that an experienced WinBUGS user could edit the code if necessary. It is not intended that every aspect of every possible analysis will be covered here. We first define how the different types of variables are included in the code.

C.1 Factor Effects

Fixed effects are denoted by the node name X.Eff[·,]. These are matrices with the first dimension denoting which factor or interaction term it represents, and the second dimension denoting the level for the factor. Instead of increasing the number of dimensions to accommodate interaction terms, the multidimensional aspect of their levels is collapsed into a single dimension, for example, an interaction between two factors having three and four levels is collapsed into a single dummy factor with 12 levels. The levels for each factor associated with each observation is stored in another matrix, X[·,·], which is read from the rectangular data file 'BGX DataRect.txt'. The first and second dimensions of X[·,·] denote the observation number and factor, respectively. These are combined to become part of the linear predictor using the code:

```
for (j in 1: N.obs){
    for (i in 1: N.X){
        X.row[i, j] <- X.Eff[i, X[j, i]]
    }
        mu[j] <- Beta0 + sum(X.row[, j])
}
```

where N.obs and N.X are scalars defining the number of observations and fixed effect factors, respectively, with both being read from the list formatted data file 'BGX DataList.txt'. Beta0 is the model constant.

Random and Variable Selection effects are defined in the same way. Their node names are Z.Eff[·,·] and VSfac.Eff[·,·], with their levels being stored in Z[·,·] and VSfac[·,·], respectively. It should be clear how all these different types of effects can be combined in the linear predictor simply by including additional sum(·) terms to the line defining mu[j].

When the response is multinomial or categorical, each factor's effect is implicitly an interaction with the response factor. This is implemented in the code by expanding the second dimension of the matrix representing each factor's effects, as illustrated here:

```
for (j in 1: N.obs){
    for (k in 1: MV.dims){
        for (i in 1: N.X){
            X.row[i, k + MV.dims*(j - 1)] <- X.Eff[i, k +
                MV.dims*(X[j, i] - 1)]
        }
    }
}
```

The scalar `MV.dims` equals the number of categories in the response, and is read from the list formatted data file 'BGX DataList.txt'.

C.2 Covariate Coefficients

Independent coefficients are denoted by the node name `V.Coeff[·,·]`. In order to accommodate covariate-by-factor interactions, all coefficients are treated as factor-by-covariate interaction terms. Standard coefficient parameters are "paired up" with a dummy factor that is always set to its first level. The value of the covariate and its associated factor are stored in the matrices `V[·,·]` and `V.X[·,·]`, respectively, which are read from the rectangular data file 'BGX DataRect.txt'. All covariates are automatically centered and scaled in order to improve MCMC performance. This is done using code:

```
### Centre & scale covariates (independent coefficients) ###
### mean.V[] & sd.V[] read from 'BGX DataList.txt'
for (i in 1: N.V){
    for (j in 1: N.obs){
        sc.V[j, i] <- (V[j, i] - mean.V[i]) / sd.V[i]
    }
    for (j in 1: levs.V[i]){
        sc.V.Coeff[i, j] <- V.Coeff[i, j] * sd.V[i]
    }
    ### Calculate ind. coeff. part of intercept at 0 ###
    mean.VCoeff0[i] <- V.Coeff[i, 1]*mean.V[i]
}
alpha <- Beta0 - sum(mean.VCoeff0[])
```

The mean and standard deviation of each covariate, `mean.V[·]` and `sd.V[·]`, are calculated by BugsXLA and written to the list formatted data file 'BGX

DataList.txt' along with the number of covariates, N.V, and the number of levels associated with each covariate's factor, levs.V[·]. Note that V.Coeff[·,·] is defined so that it represents the covariate's coefficient in the original metric, not the unit variance scaled metric. Also, the node alpha is defined to equal the intercept when all covariates are set to zero, leaving Beta0 as the usual model constant when all covariates are at their mean values. Note that due to rounding when calculating the means of the covariates, slight discrepancies might exist when comparing parameters that are affected by this value. This does not, of course, fundamentally affect the interpretation of the model. These coefficients are combined to become part of the linear predictor using the code:

```
for (j in 1: N.obs){
    for (i in 1: N.V){
        V.row[i, j] <- sc.V[j, i] * sc.V.Coeff[i, V.X[j, i]]
    }
    mu[j] <- Beta0 + sum(V.row[, j])
}
```

Although the code could be written more succinctly using WinBUGS' inprod(·,·) function, the approach adopted by BugsXLA is faster to execute. Random and Variable Selection coefficients are defined in essentially the same way. The node names for random coefficients are W.Coeff[·,·] with their covariate values and associated factor levels stored in the matrices W[·,·] and W.Z[·,·], respectively. Variable Selection coefficients are VScov.Coeff[·,·] with their covariate values and associated factor levels stored in the matrices VScov[·,·] and VScov.X[·,·], respectively. Note that when random coefficients are fitted, the intercepts are given node names WZ.Eff[·,·] and are included as factor effects.

C.3 Error Distributions and Link Functions

The generalized linear model approach to specifying the relationship between a response and a set of explanatory variables makes it possible to build a model in a very simple way. The code described above for defining the linear predictor is simply related to the error distribution via the link function. For example, in an NLM with both factors and covariates, the following code would be used:

```
for (j in 1: N.obs){
    mu[j] <- Beta0 + sum(X.row[, j]) + sum(V.row[, j])
    Y[j] ~ dnorm(mu[j], tau)
}
```

where Y[·] is the vector of response values, which are read from the rectangular data file 'BGX DataRect.txt'. If instead the response was a binomial or Poisson variate, the code inside the 'for loop' would be replaced by either:

```
logit(mu[j]) <- Beta0 + sum(X.row[, j]) + sum(V.row[, j])
Y[j] ~ dbin(mu[j], N[i])
```

or

```
log(mu[j]) <- Beta0 + sum(X.row[, j]) + sum(V.row[, j])
Y[j] ~ dpois(mu[j])
```

Other error distributions and link functions are coded using obvious variants of this approach. Ordinal and ordered categorical data require extra code to represent the proportional odds model as illustrated here:

```
for (j in 1: N.obs){
    ### Relate probability response = k to cumulative
        probability of > k ###
    p[j, 1] <- 1 - Q[j, 1]
    for (k in 2: N.cats - 1){
        p[j, k] <- Q[j, k - 1] - Q[j, k]
    }
    p[j, N.cats] <- Q[j, N.cats - 1]
    for (k in 1: N.cats - 1){
        logit(Q[j, k]) <- mu.Q[j, k]
        mu.Q[j, k] <- -CF.a[k] + sum(X.row[, j]) +
            sum(V. row[, j])
    }
    Y[j, 1:N.cats] ~ dmulti(p[j,], N[j])
    N[j] <- sum(Y[j,])
}
```

where N.cats equals the number of categories in the response, and is read from the list formatted data file 'BGX DataList.txt'. The matrix Q[·,·] represents the cumulative probability of being in a category higher than the index of its second dimension. CF.a[·] are the cut points of the latent variable.

When the data consists of the summary statistics mean, standard error, and associated degrees of freedom (see Case Study 5.3) additional code is needed as illustrated here:

```
for (j in 1: N.obs){
    mu[j] <- Beta0 + sum(X.row[, j]) + sum(V.row[, j])
    Y[j] ~ dnorm(mu[j], inv.se2[j])
    Y.var[j] <- pow(Y.se[j], 2)
    Y.var[j] ~ dgamma(Y.varG1[j], Y.varG2[j])
    Y.varG1[j] <- Y.df[j] / 2
    Y.varG2[j] <- Y.varG1[j] * inv.se2[j]
}
```

The response triplicate is in the three vectors Y[·], Y.se[·], and Y.df[·], all of which are read from the rectangular data file 'BGX DataRect.txt'. Note that the sampling distribution of each estimate of variance is the usual scaled Chi-Square ($\equiv$ Gamma) implied by the assumption of Normal errors.

C.4 Prior Distributions

The code defining the prior distribution for a scalar quantity should be easily understood. The priors for the fixed effects of factors are coded in a reasonably straightforward manner:

```
### X Effects: independent ("Fixed"), corner constraints via
    tiny priorSD.X[] ###
### priorSD.X[] read from 'BGX DataList.txt'
for (i in 1: N.X){
    for (j in 1: levs.X[i]){
        priorInvV.X[i, j] <- pow(priorSD.X[j, i], -2)
        X.Eff[i, j] ~ dnorm(0, priorInvV.X[i, j])
    }
}
```

As stated in the code comment, the constrained to zero levels of these effects is implemented by specifying a mean of zero and a very small standard deviation for their prior distribution.

The prior for the random effects of a factor requires code that is a little more complicated, mainly due to the fact that BugsXLA provides the option to choose between five different distributions. The code when the default Half-Normal Distribution is used is shown here:

```
### Z Effects: assumed exchangeable ('Random') ###
### Half-N prior: hyper.Z is SD of prior for Z.Eff[] ###
for (j in 1: 5){
    Z.Eff[1, j] ~ dnorm(0, tau.Z[1])
}
tau.Z[1] <- pow(sigma.Z[1], -2)
sigma.Z[1] <- hyper.Z[1] + 1.E-20
hyper.Z[1] ~ dnorm(0, 1.15)I(0,)
```

Since it is possible to specify a different prior distribution for each random effect in the model, it is not possible to include all the random effects in an outer loop as is done for fixed effects. For this reason, the index of the first dimension of Z.Eff[·, ·] is stated explicitly (1 above), and the number of levels of the factor is also stated explicitly in the 'for loop' (5 above). The addition of 10^{-20} to the node sigma.Z[·] is to avoid a zero error that occasionally occurred causing WinBUGS to crash. Note that the value of the Half-Normal scale parameter entered by the user of BugsXLA is also explicitly stated, 1.15 here. When the Uniform Distribution is used the only change to the code is an obvious one to the final line.

When the Half-Cauchy Distribution is used, the code becomes:

```
### Half-C prior: hyper.Z is xi in Gelman's parameterization ###
for (j in 1: 5){
    Z.Eff[1, j] <- hyper.Z[1]*Z.eta[1,j]
    Z.eta[1,j] ~ dnorm(0, tauZ.eta[1])
}
tau.Z[1] <- pow(sigma.Z[1], -2)
hyper.Z[1] ~ dnorm(0, 25)
tauZ.eta[1] ~ dgamma(0.5, 0.5)
sigma.Z[1] <- abs(hyper.Z[1])/sqrt(tauZ.eta[1])
```

Although the Cauchy Distribution is the same as a t-distribution with one degree of freedom, it is not possible to code this up directly in WinBUGS because it only provides a t-distribution with two or more degrees of freedom. This code is based on the parameter expansion technique discussed in Gelman et al. (2004). More generally, they suggest that this approach makes the computations for the t-distribution more efficient, although BugsXLA only uses it for the Half-Cauchy Distribution.

When either the Gamma or scaled Chi-Square Distribution is used the code is

```
### Gamma prior: hyper.Z is Precision of prior for Z.Eff[] ###
for (j in 1: 5){
    Z.Eff[1, j] ~ dnorm(0, hyper.Z[1])
}
sigma.Z[1] <- 1/sqrt(hyper.Z[1])
hyper.Z[1] ~ dgamma(0.001,0.001)
```

utilizing the fact that the Chi-Square is a special case of the Gamma (see Appendix A). Note that when either of these two distributions is used, the stochastic node hyper.Z[1] is the precision of the random effects' hyper-distribution, while with the other three distributions it is the standard deviation.

The prior for independent covariate coefficients is coded as shown here:

```
### Covariate Coefficients: assumed independent ("Fixed") ###
### priorMN.Vcoeff[] & priorSD.Vcoeff[] read from 'BGX
    DataList.txt'
for (i in 1: N.V){
```

```
    for (j in 1: levs.V[i]){
        priorInvV.Vcoeff[i, j] <- pow(priorSD.Vcoeff[i, j], -2)
        hyper.Vcoeff[i, j] ~ dnorm(priorMN.Vcoeff[i, j],
          priorInvV.Vcoeff[i, j])
    }
}
for (j in 1: levs.V[1]){
    V.Coeff[1, j] <- hyper.Vcoeff[1, j]
}
```

The reason for the second loop round the number of levels is because BugsXLA provides the option to constrain the sign of the coefficient by specifying a Half-Normal prior distribution. If this is done, the code inside this second loop is altered to

```
V.Coeff[1, j] <- abs(hyper.Vcoeff[1, j])
```

when the coefficient is constrained to be positive, and the same with '−' in front of the abs(·) function when it is constrained to be negative.

When exchangeable intercepts and covariate coefficients are fitted, the code defining the prior distribution is quite complex. It is shown below in parts to aid explanation.

```
for (i in 1: N.W){
    sigma2.WZ[i] <- pow(sigma.WZ[i], 2)
    sigma2.Wcoeff[i] <- pow(sigma.Wcoeff[i], 2)
    ### Hierarchical SD parameters given Half-Normal priors
    ### hyper.parmWZ[] & hyper.parmWcoeff[] read from 'BGX
        DataList.txt'
    sigma.WZ[i] <- abs(hyper.WZ[i]) + 1.E-20
    hyper.WZ[i] ~ dnorm(0, hyper.tauWZ[i])
    hyper.tauWZ[i] <- pow(hyper.parmWZ[i], -2)
    sigma.Wcoeff[i] <- abs(hyper.Wcoeff[i]) + 1.E-20
    hyper.Wcoeff[i] ~ dnorm(0, hyper.tauWcoeff[i])
    hyper.tauWcoeff[i] <- pow(hyper.parmWcoeff[i], -2)
```

This first part defines Half-Normal prior distributions for the standard deviation parameters of the random intercepts and coefficients. It also defines the variances that are used in the Bivariate Normal Distribution later.

```
### priorMN.Wcoeff[] & priorSD.Wcoeff[] read from 'BGX
    DataList.txt'
W.MuCoeff[i] ~ dnorm(priorMN.Wcoeff[i], priorInvV.Wcoeff[i])
priorInvV.Wcoeff[i] <- pow(priorSD.Wcoeff[i], -2)
```

This second part defines the Normal prior distribution for the population mean covariate coefficient.

```
for (j in 1: levs.W[i]){
    WZ.Eff[i, j] <- Wmvn.Beta[i, j, 1]
    W.Coeff[i, j] <- Wmvn.Beta[i, j, 2]
    Wmvn.Beta[i, j, 1:2] ~ dmnorm(Mmvn.Beta[i, 1:2],
                                    Omvn.Beta[i, 1:2, 1:2])
}
```

This third part defines the Bivariate Normal prior distribution for the intercepts and the coefficients, with the final part below equating the components of the bivariate mean and precision matrix to the appropriate parameters.

```
### CORRmvn.A[] & CORRmvn.B[] read from 'BGX DataList.txt'
Mmvn.Beta [i, 1] <- 0
Mmvn.Beta[i, 2] <- W.MuCoeff[i]
Omvn.Beta[i, 1:2, 1:2] <- inverse(Vmvn.Beta[i, 1:2, 1:2])
Vmvn.Beta[i, 1,1] <- sigma2.WZ[i]
Vmvn.Beta[i, 2,2] <- sigma2.Wcoeff[i]
CmvnStar.Beta[i] ~ dbeta(CORRmvn.A[i], CORRmvn.B[i])
Cmvn.Beta[i] <- 2*(CmvnStar.Beta[i] - 0.5)
Vmvn.Beta[i, 1,2] <- Cmvn.Beta[i]*sqrt(Vmvn.Beta[i, 1,1]*
                                        Vmvn.Beta[i, 2,2])
Vmvn.Beta[i, 2,1] <- Vmvn.Beta[i, 1,2]
}
```

This final part also defines the Beta prior distribution for the correlation between the intercepts and the coefficients.

The code defining the prior distributions for the Variable Selection parameters is very complex and will not be explained in this book. The complexity when VS factors are included in the model is such that an additional data file is created containing the large arrays used in the code. This file is called 'BGX DataVSfac.txt'. Refer to Woodward and Walley (2009) for details of the prior distribution used for Variable Selection factors.

C.5 Predictions and Contrasts

When predictions or contrasts are specified, additional code is added to the bottom of the program to calculate these. The code will be explained by considering a model containing one fixed effects factor (X), one random effects factor (Z) and one covariate (V). Factor X has levels 'A' and 'B'. If the predictions and contrasts defined in Table C.1 were specified then the code below it would be created.

TABLE C.1

Predictions and Contrasts Associated
with Following Code

Predictions for ...		
X	Z	V
		9
c(A, B)		
	~	

```
Pred.Xrow[1, 1] <- X.Eff[1, 1]
Pred.Xrow[1, 2] <- X.Eff[1, 1] - X.Eff[1, 2]
Pred.Xrow[1, 3] <- X.Eff[1, 1]
Pred.Zrow[1, 1] <- 0
Pred.Zrow[1, 2] <- 0
Pred.Zrow[1, 3] ~ dnorm(0, tau.Z[1])
Pred.Vrow[1, 1] <- (9 - mean.V[1])*V.Coeff[1, 1]
Pred.Vrow[1, 2] <- 0
Pred.Vrow[1, 3] <- 0
Pred.Beta0[1] <- Beta0
Pred.Beta0[2] <- 0
Pred.Beta0[3] <- Beta0
for (j in 1: N.pred){
    Pred.Ave[j] <- Pred.Beta0[j] + sum(Pred.Xrow[, j]) +
    sum(Pred.Zrow[, j]) + sum(Pred.Vrow[, j])
}
```

The matrix `Pred.Xrow[·,·]` defines the contribution of the fixed factor effects to the predictions and contrasts. `Pred.Zrow[·,·]` and `Pred.Vrow[·,·]` do the same for the random factor effects and independent covariates, respectively. The first dimension denotes which term it represents, and the second dimension denotes the prediction or contrast being estimated. It can be seen how empty cells in Table C.1 have been interpreted as the zero constrained level for the fixed effect, the population mean, that is, zero, for the random effect, and the mean of the covariate, that is, zero because the covariates have been centered in the model fitted. `Pred.Beta0[·]` is a vector that determines whether the model constant is included in the prediction; when a contrast is specified the model constant cancels out. Note how the '~' notation for random effects is interpreted, producing a random draw from the parameter's random effects distribution. `N.pred`, which equals three in this case, is read from the list formatted text file 'BGX DataList.txt'.

C.6 Model Checking

When individual observation model checking functions are requested for an NLM, the following code is added:

```
### Model Checking Functions
fit[j] <- mu[j]
res[j] <- Y[j] - mu[j]
sc.res[j] <- (Y[j] - mu[j])/sigma
Y.rep[j] ~ dnorm(mu[j], tau)
p.hi[j] <- step(Y.rep[j] - Y[j])
p.lo[j] <- 1 - p.hi[j]
```

The fitted values, fit[·], residuals, res[·], and scaled residuals, sc.res[·], are defined in an obvious way. The Bayesian p-value defining the probability of a more extreme observation is obtained by producing a replicated vector of observations, Y.rep[·]. The vectors p.hi[·] and p.lo[·] will be Bernoulli variables whose means are used to estimate this Bayesian p-value; it is set equal to the smaller of these two means. When other error distributions are being modeled, code very similar to that above is created.

C.7 Non-Linear Models

The covariate in an Emax model is scaled to improve MCMC performance.

```
### Scale non-linear covariate ###
NLx.median <- ranked(NLx[], Mid.N.obs); Mid.N.obs <- round
    (N.obs/2)
for (j in 1: N.obs){
    sc.NLx[j] <- NLx[j]/NLx.median
}
```

where NLx[·] is the covariate, which is read from the rectangular data file 'BGX DataRect.txt'. Scaling the covariate only affects the ED_{50} parameter(s), which is scaled in the code later to ensure the estimates returned are in the original metric. The code shown below is for the case when grouped Emax curves are fit; when a single curve is fit, the code is similar and easier to follow. We illustrate here the three types of priors that can be fitted: fixed effects, random effects, and constant. The fixed effects prior is shown for the ED_{50} parameters:

```
NL.ED50.InvV <- pow(4.89, -2)
for (i in 1:5){
```

```
log(NL.ED50Grp[i]) <- NL.logED50Grp[i]
NL.logED50Grp[i] ~ dnorm(3.61, NL.ED50.InvV)
sc.NL.ED50Grp[i] <- NL.ED50Grp[i]/NLx.median
}
```

The parameters of the Log-Normal prior distribution set by the user of BugsXLA are explicitly coded into the program. Note the line where the ED_{50} parameters are scaled to match the scaled covariate. The random effects prior is shown for the Emax parameters:

```
NL.Emax ~ dnorm(0, NL.Emax.InvV)
NL.Emax.InvV <- pow(12, -2)
for (i in 1:5){
    NL.EmaxGrp[i] ~ dnorm(NL.Emax, NL.EmaxGrp.InvV)
}
NL.EmaxGrp.InvV <- pow(NL.EmaxGrp.SD, -2)
NL.EmaxGrp.SD ~ dnorm(0, NL.EmaxSD.InvV)I(0,)
NL.EmaxSD.InvV <- pow(12, -2)
```

The Half-Normal prior distribution for the standard deviation parameter of the random effects distribution is defined in a similar way to that done for random effects factors. The prior that defines the parameter as being the same for all levels of the grouping factor, 'constant', is shown for the Hill parameters:

```
log(NL.Hill) <- NL.logHill
NL.logHill ~ dnorm(0, NL.Hill.InvV)
NL.Hill.InvV <- pow(0.354, -2)
for (i in 1:5){
    NL.HillGrp[i] <- NL.Hill
}
```

This code needs no additional explanation.

C.8 Longitudinal Models

When the residuals are modeled with autoregressive serial correlation, the code defining the likelihood is modified as shown here:

```
for (j in 1: N.obs){
    mu[j] <- mu.AR1[j] + pow(phi.AR1, Time.d[j]) *
        e[Time.f[j], Unit[j]]
    mu.AR1[j] <- Beta0 + sum(X.row[, j])
    e[Time.f[j]+1, Unit[j]] <- Y[j] - mu.AR1[j]
```

```
    Y[j] ~ dnorm(mu[j], tau.AR1[j])
    tau.AR1[j] <- tau/(1 - pow(phi.AR1,
      2*Time.d[j])*(1- equals(Time.f[j], 1)))
}
for (u in 1:30){
    e[1, u] <- 0
}
```

A first-order autoregressive correlation structure in the residuals is induced by defining a matrix e[·,·] that represents the deviation of the observed response from its expectation marginal to the multivariate distribution of the residuals, that is, the expectation ignoring the influence of all the previous residuals. The second dimension of e[·,·] denotes the experimental unit on which the deviation occurred, and the first dimension denotes the within experimental unit time order of the deviation. The vectors Unit[·] and Time.f[·] containing the indices for e[·,·] are read from the rectangular formatted data file 'BGX DataRect.txt'. When a variate defining the time of each observation is specified in the model, another vector of values, Time.d[·], is also read from 'BGX DataRect. txt'. Time.d[·] contains the time differences between successive observations on each unit. If a time variate is not defined in the model then the code above is slightly simpler, since the difference is taken to be one unit of time in every case. Note how the variance of the residual errors is set via the node tau.AR1[·] such that it takes a different value for the first observation, in time order, on each experimental unit. When the jth observation is the first on an experimental unit

```
    tau.AR1[j] = tau
```

otherwise

```
    tau.AR1[j] = tau/(1 - pow(phi.AR1, 2*Time.d[j]).
```

Note that when a time variate is specified, the observations do not have to be in time order in the data set. The vector Time.f[·] ensures that the serial correlation occurs between the residuals of successive observations. This code is based on the formulation detailed in Chapter 5 of Diggle et al. (2002).

When measurement error is added to the serial correlation, the code shown above is amended as below:

```
    e.meas[j] ~ dnorm(0, tau.meas)
    mu[j] <- mu.AR1[j] + pow(phi.AR1, Time.d[j]) *
      e[Time.f[j], Unit[j]] + e.meas[j]
    mu.AR1[j] <- Beta0 + sum(X.row[, j])
    e[Time.f[j]+1, Unit[j]] <- Y[j] - mu.AR1[j] - e.meas[j]
```

A measurement error stochastic node, e.meas[·], is explicitly defined and added to the deterministic node defining the mean response, mu[·]. Since the measurement error should not affect the serial correlation, e.meas[·] is subtracted from the response when calculating e[;,].

When the AR1 Poisson transition model is specified, the likelihood is modified as shown here:

```
for (j in 1: N.obs){
    log(mu[j]) <- mu.AR1[j] + phi.AR1 * e[Time.f[j], Unit[j]]
    mu.AR1[j] <- Beta0 + sum(X.row[, j])
    e[Time.f[j]+1, Unit[j]] <- log(max(Y[j], 0.3)) -
        mu.AR1[j]
    Y[j] ~ dpois(mu[j])
}
for (u in 1:59){
    e[1, u] <- 0
}
```

The code is similar to that for the NLM case earlier, although here we show the case when no time variate has been specified.

C.9 Robust Models

When the t-distribution or multivariate t-distribution is used to make the model more robust, the only change to the code shown above is to replace the Normal or multivariate Normal with the robust alternative, that is, replace 'dnorm' and 'dmnorm' with 'dt' and 'dmt', respectively, noting the need to specify an additional parameter for the degrees of freedom. When outliers are modeled explicitly, the likelihood is modified as shown here:

```
Y[j] ~ dnorm(mu[j], tau.mix[j, outP1[j]])
tau.mix[j, 1] <- tau
tau.mix[j, 2] <- tau/5
sigma.mix[j] <- pow(tau.mix[j, outP1[j]], -0.5)
outP1[j] <- out[j] + 1
out[j] ~ dbern(0.05)
```

The residual precision, which is usually a scalar stochastic node named tau, is replaced by a matrix tau.mix[·, ·]; its first dimension indexes the observation number, and the second dimension denotes whether the observation is being treated as an outlier. The node tau still exists in the program, but it is now defined as the precision, conditional on the model parameters, of those observations not being treated as outliers. A vector outP1[j] takes the value two if the jth observation is being treated as an outlier, and the value one otherwise. This vector is defined via a Bernoulli stochastic node, out[·], whose probability parameter is that set by the user of BugsXLA. The value of tau.mix[j, 2], the conditional precision of outliers, is set equal to tau divided by the factor set by the user of BugsXLA. Analogous approaches to that described here are used when alternative error distributions are specified and outliers are explicitly modeled.

Appendix D: Explanation of R Scripts Created by BugsXLA

When R scripts are requested additional files are saved to the bugfolio that facilitate further analysis using the R computing environment. R is a free software package available for download from http://www.r-project.org. It is not the intention here to provide any explanation of how to use R, so the interested user should refer to one of the many documents or books that does provide such an introduction. Some of the best introductory documentation can be obtained for free online, for example, Venables et al. (2010). The reference card created by Short (2004) is also recommended. Numerous texts exist explaining how to fit the types of models discussed in this book, albeit from a classical perspective. Verzani (2005) provides an entry level introduction to exploring the data and fitting simple Normal Linear Models. The two books of Faraway (2004, 2006) explain in more detail how to fit NLMs and GLMs as well as introducing NLMMs and GLMMs.

BugsXLA creates four additional files when R scripts are requested.

BGX DataCSV.csv

This contains all the data defined in the 'Data Range' on the model specification form. The data are stored in comma separated format, which is convenient for reading into R.

BGX R BRugs.R

This is an R script that can be used to rerun the analysis specified using the BRugs package. BRugs is a package that provides the functionality of the WinBUGS MCMC computing engine within R. This R script mimics the WinBUGS script, 'BGX Script.txt', by instructing WinBUGS to read in the data and initial values, and run the code created by BugsXLA. The BRugs package is dynamically linked to WinBUGS and so the results of the analysis are automatically stored within R. The R script also includes code to create more meaningful names than the generic node names created by BugsXLA. It includes a call to an R function that is defined in the file 'BGX R Functions.R'.

BGX R Functions.R

This is an R script that contains a function called from within 'BGX R BRugs.R'. It facilitates the extraction of a vector or matrix of the posterior samples, with a meaningful name added, from the BRugs created internal store of results.

The R scripts created to work with the BRugs package will not be discussed in any more detail. These may be useful for experienced users of R and BRugs, who can utilize the additional functionality of BRugs to explore the WinBUGS analysis in more detail. Note that the results obtained from running the identical analysis using BRugs or WinBUGS directly will not be the same, although the differences should be due to MCMC error. However, I have on occasions found that BRugs has failed to converge satisfactorily in cases where the initial analysis using WinBUGS via BugsXLA has worked fine.

BGX R EDA.R

This is an R script that provides the starting point for an exploratory data analysis using the R environment. When the simpler models are being fitted, it also provides analogous code to undertake a classical analysis of the data. Be warned, there is no guarantee that the automatically generated statements are correct, and so this should only be used after checking these personally.

We illustrate the type of R commands provided by discussing the R script 'BGX R EDA.R' created for Case Study 3.1. The first part of the script reads in the data and defines any factors that are included. The data are stored in an R data frame named BGX.df. Unless the data file is huge, it is advisable to check the data are as expected; the script displays the data to the screen to facilitate this check.

```
### Read in the data and define the factors ###
setwd('C:/BRugF/')
BGX.df <- read.csv('BGX DataCSV.csv')
BGX.df
BGX.df$TRT <- factor(BGX.df$TRT)
```

The second part suggests some exploratory plots utilizing the lattice package. The first plot is a simple all pair wise scatter plot of the data (see Figure D.1). A more sophisticated scatter plot, stratified by the factor is also produced (see Figure D.2). When there are many variables in the model, many of the plots suggested here will not be useful. The purpose is purely to provide a reminder of the type of code that can be used to produce potentially useful plots.

```
### Exploratory plot(s) of the data ###
### (not all are guaranteed to be useful) ###
library(lattice)
plot(BGX.df)
xyplot(FEV1_CFB ~ FEV1_BASE | TRT, data = BGX.df)
```

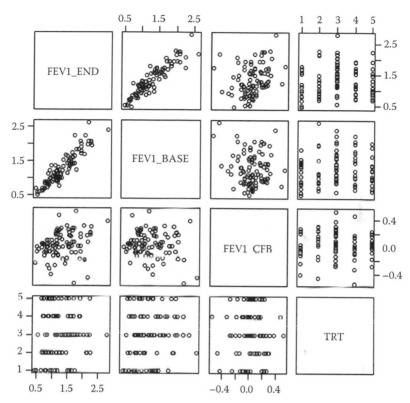

FIGURE D.1
Plot produce by R command 'plot(BGX.df)'.

The final part of the script is where an attempt is made to fit an equivalent classical model to the data. Note the strong warning given. In this particular case, the simple ANCOVA model is appropriate. The summary of the fitted model displayed is shown and discussed in Case Study 3.1. The plot command for the fitted model produces a series of plots that are useful for model checking.

```
#######################################################
### Non-Bayesian likelihood based analyses.     ###
### NB It is likely the code will NOT give      ###
###    an equivalent model to the one fitted    ###
###    via BugsXLA. Currently there is not      ###
###    even a guarantee that the code will      ###
###    be understood by R!                      ###
###    CHECK FOLLOWING CODE BEFORE RUNNING!!!   ###
#######################################################
BGX.fm <- lm(FEV1_CFB ~ TRT + FEV1_BASE, data = BGX.df)
summary(BGX.fm)
plot(BGX.fm)
```

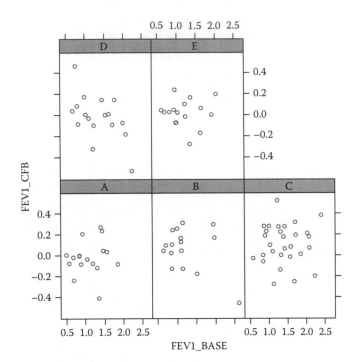

FIGURE D.2
Plot produce by R command 'xyplot(FEV1_CFB ~ FEV1_BASE | TRT, data = BGX.df)'.

Appendix E: Troubleshooting

BugsXLA will only function if the English language version of Excel is set; this may be a Microsoft Office or even Windows level setting (see Figure E.1 for an example of a Norwegian user's solution). If problems persist after making this change, click on the 'Clear Old Analysis' button found on the 'BugsXLA Options' form. As a last resort, undertake a clean reinstall as discussed at the end of this appendix.

The pathnames in WinBUGS' scripts have a maximum length of about 120 characters. Although BugsXLA attempts to maintain a check on the length, on occasions pathnames that are too long are created. Unfortunately, WinBUGS does not usually give an error message that clearly indicates the problem. Sometimes it just fails to do anything. If you suspect this is an issue, the only solution is to move your work to a folder closer to the root directory.

A similar problem to that caused by excessively long pathnames has also been observed when the pathname contains a character that WinBUGS does not like. The only example that has been brought to my attention was when the use of a hyphen '-' caused a problem. If WinBUGS fails to run and your pathname contains non-alphanumeric characters, other than spaces, it is advisable to try again after renaming the folders and files. If you do discover characters that cause a problem, please email me so that I can make others aware of the known issues.

Chapter 11 discusses ways to interrogate the WinBUGS log file to help diagnose run time problems. However, WinBUGS is notoriously bad at providing error messages that are decipherable by the layman, so please set your expectations accordingly.

If you discover any other general issues and identify the solution, please email so that I can share this information with other users. My email address can be found at my website: www.bugsxla.philwoodward.co.uk. Refer to the "Troubleshooting" page on this website for the latest information.

E.1 Undertaking a Clean Reinstall of BugsXLA

If BugsXLA becomes corrupted for some reason, your only option is to undertake a clean reinstall. This is done by first removing BugsXLA from Excel:

1. Click on the 'Unload BugsXLA' button found on BugsXLA's Options form. If BugsXLA will not even load up its Options form then ignore this step.

(a)

(b)

(c)

FIGURE E.1
(a and b) Changes to language settings needed in order for BugsXLA to work. (c) Note that the local language can still be used as the 'Text Services and Input Languages' setting.

2. Remove the BugsXLA toolbar from Excel's memory:

    EXCEL 2000–2003

        Tools: Customize

        Select 'BugsXLA' from the list of Toolbars

        Delete (confirm OK)

EXCEL 2007

> Right mouse click on any BugsXLA icon and
> select 'delete custom toolbar'.

3. Remove the BugsXLA add-in:

EXCEL 2000–2003

> Tools: Add-Ins

EXCEL 2007

> Select MS Office Button: Excel Options: Add-Ins (see left hand side)
> At the bottom of the form ensure Manage: Excel Add-Ins is selected
> Click "Go…"

Uncheck 'BugsXLA' from the list of Add-Ins

OK

4. Exit Excel.

Delete all the downloaded BugsXLA files from your hard drive. Sometimes Windows has the annoying habit of making copies of files such as Excel add-ins, and so in extreme cases it might be necessary to undertake a complete search of your hard drive in order to locate and delete these copies. Finally, download BugsXLA again from my website and follow the instructions for adding it to Excel as an add-in. If BugsXLA still fails to work, then I suspect you may have an issue with your installation of Microsoft Office/Excel itself. At least one previous user found that reinstalling Microsoft Office fixed his problem.

References

Agresti, A. (2007). *An Introduction to Categorical Data Analysis*. 2nd Edn. John Wiley & Sons, Inc., New York.

Baker, S.G. (1995). The multinomial-Poisson transformation. *The Statistician* 43, 495–504.

Baker, W.L., Baker, E.L., and Coleman, C.I. (2009). Pharmacologic treatments for chronic obstructive pulmonary disease: A mixed-treatment comparison meta-analysis. *Pharmacotherapy* 29(8), 891–905.

Bartlett, M.S. (1936). The square-root transformation in analysis of variance. *Supplement to the Journal of the Royal Statistical Society* 3(1), 68–78.

Benjamini, Y. and Hochberg, Y. (1995). Controlling the false discovery rate: A practical and powerful approach to multiple testing. *Journal of the Royal Statistical Society (B)* 57, 289–300.

Berger, J. (2006). The case for objective Bayesian analysis. *Bayesian Analysis* 1(3), 385–402.

Bliss, C.I. (1935). The calculation of the dosage-mortality curve. *Annals of Applied Biology* 22, 134–167.

Bolstad, W.M. (2007). *Introduction to Bayesian Statistics*. 2nd Edn. John Wiley & Sons, Inc., New York.

Box, G.E.P. (1983). An apology for ecumenism in statistics, in *Scientific Inference, Data Analysis, and Robustness*. Academic Press, New York, pp. 51–84.

Box, G.E.P. (2000). *Box on Quality and Discovery: With Design, Control and Robustness*. John Wiley & Sons, Inc., New York.

Box, G.E.P. and Bisgaard, S. (1987). The scientific context of quality improvement. *Quality Progress* 4(2), 54–62.

Box, G.E.P., Hunter, W.G., and Hunter, J.S. (1978). *Statistics for Experimenters*. John Wiley & Sons, Inc., New York.

Box, G.E.P. and Meyer, R.D. (1993). Finding the active factors in fractioned screening experiments. *Journal of Quality Technology* 25, 94–105.

Box, G.E.P. and Tiao, G.C. (1968). A Bayesian approach to some outlier problems. *Biometrika* 55(1), 119–129.

Box, G.E.P. and Tiao, G.C. (1973). *Bayesian Inference in Statistical Analysis*. Wiley Classics, New York.

Brooks, S.P. and Gelman, A. (1998). Alternative methods for monitoring convergence of iterative simulations. *Journal of Computational and Graphical Statistics* 7, 434–455.

Brown, H. and Prescott, R. (2006). *Applied Mixed Models in Medicine*. 2nd Edn. John Wiley & Sons Ltd., New York.

Chipman, H., Hamada, M., and Wu, C.F.J. (1997). A Bayesian variable-selection approach for analyzing designed experiments with complex aliasing. *Technometrics* 39(4), 372–381.

Clarke, G.M. and Cooke, D. (2004). *A Basic Course in Statistics*. 5th Edn. Hodder Arnold, London, U.K.

Cochran, W.G. and Cox, G.M. (1992). *Experimental Designs*. 2nd Edn. Wiley Classics Library. John Wiley & Sons, Inc., New York.

Collaborative Group*. (1993). Meta-analysis of randomised controlled trials of selective decontamination of the digestive tract. *British Medical Journal* 307, 525–532. (*Selective Decontamination of the Digestive Tract Trialists' Contamination Group.)

Collett, D. (2003). *Modelling Survival Data in Medical Research*. Chapman & Hall/CRC, Boca Raton, FL.

Cornell, J.A. (2002). *Experiments with Mixtures: Designs, Models and the Analysis of Mixture Data*. 3rd Edn. John Wiley & Sons, Inc., New York.

Cox, D.R. (1958). *Planning of Experiments*. John Wiley & Sons, Inc., New York.

Davies, O. (1954). *The Design and Analysis of Industrial Experiments*. John Wiley & Sons, Inc., New York.

Davis, T.P. (1995). Analysis of an experiment aimed at improving the reliability of transmission centre shafts. *Lifetime Data Analysis* 1, 275–306.

Diggle, P.J., Heagerty, P., Liang, K.Y., and Zeger, S.L. (2002). *Analysis of Longitudinal Data*. 2nd Edn. Oxford University Press, New York.

Dobson, A.J. (1988). *An Introduction to Statistical Modelling*. Chapman & Hall, New York.

Dobson, A.J. (2002). *An Introduction to Generalized Linear Models*. 2nd Edn. Chapman & Hall/CRC, Boca Raton, FL.

Draper, D. (1995). Assessment and propagation of model uncertainty (with discussion). *Journal of the Royal Statistical Society, B* 57(1), 45–97.

Draper, N. and Smith, H. (1998). *Applied Regression Analysis*. 3rd Edn. John Wiley & Sons, Inc., New York.

Edwards, A.W.F. (1992). *Likelihood*. Expanded Edition. The John Hopkins University Press, Baltimore, MD.

Faraway, J.J. (2004). *Linear Models with R*. Chapman & Hall/CRC, Boca Raton, FL.

Faraway, J.J. (2006). *Extending the Linear Model with R*. Chapman & Hall/CRC, Boca Raton, FL.

Fisher, R.A. (1925). *Statistical Methods for Research Workers*. Oliver and Boyd, Edinburgh, Scotland.

Fitzmaurice, G.M., Laird, N.M., and Ware, J.H. (2004). *Applied Longitudinal Analysis*. John Wiley & Sons, Inc., Hoboken, NJ.

Fleiss, J.L. (1986). *The Design and Analysis of Clinical Experiments*. John Wiley & Sons, Inc., New York.

Gelfand, A.E., Dey, D.K., and Chang, H. (1992). Model determination using predictive distributions with implementation via sampling-based methods, in *Bayesian Statistics 4* (eds. J.M. Bernardo, J.O. Berger, A.P. Dawid, and A.F.M. Smith). Oxford University Press, Oxford, U.K., pp. 147–168.

Gelfand, A.E., Hills, S.E., Racine-Poon, A., and Smith, A.F.M. (1990). Illustration of Bayesian inference in normal data models using Gibbs sampling. *Journal of the American Statistical Society* 85, 972–985.

Gelfand, A.E. and Smith, A.F.M. (1990). Sampling based approaches to calculating marginal densities. *Journal of the American Statistical Association* 85, 398–409.

Gelman, A. (2005). Analysis of variance—Why it is more important than ever (with discussion). *Annals of Statistics* 33, 1–53.

Gelman, A. (2006). Prior distributions for variance parameters in hierarchical models. *Bayesian Analysis* 1(3), 515–553.

Gelman, A., Carlin, J., Stern, H., and Rubin, D. (2004). *Bayesian Data Analysis*. 2nd Edn. CRC Press/Chapman & Hall, Boca Raton, FL.

Gelman, A. and Rubin, D. (1992). Inference from iterative simulation using multiple sequences. *Statistical Science* 7, 457–511.

Grieve, A.P. (1987). Applications of Bayesian software: Two examples. *Statistician* 36, 283–288.

Hand, D. (1981). *Discrimination and Classification*. Wiley, Chichester, U.K.

Henkin, E. (1986). The reduction of variability of blood glucose levels, in *Fourth Supplier Symposium on Taguchi Methods*. American Supplier Institute, Dearborn, MI, pp. 758–785.

Higgins, J.P.T., Thompson, S.G., and Spiegelhalter, D.J. (2009). A re-evaluation of random-effects meta-analysis. *Journal of the Royal Statistical Society (A)* 172, 137–159.

Hill, A.V. (1910). The possible effect of the aggregation of molecules of haemoglobin on its dissociation curves. *Proceedings of the Physiological Society* 40, iv–vii.

Jackman, S. (2009). Reagan dataset. http://jackman.stanford.edu/mcmc/index.php (last updated December 12, 2010; accessed May 15, 2011).

Jones, B. and Kenward, M.G. (2003). *Design and Analysis of Cross-Over Trials*. 2nd Edn. Chapman & Hall/CRC, Boca Raton, FL.

Kuo, L. and Mallick, B. (1998). Variable selection for regression models. *Sankyha B* 60, 65–81.

Lawson, J. (2008) Bayesian interval estimates of variance components used in quality improvement studies. *Quality Engineering* 20, 334–345.

Lee, P.M. (2004). *Bayesian Statistics: An Introduction*. 3rd Edn. Hodder Arnold, London, U.K.

Lindley, D.V. (1985). *Making Decisions*. 2nd Edn. John Wiley & Sons, Inc., New York.

Lindley, D.V. and Smith, A.F.M. (1972). Bayes estimates for the linear model (with discussion). *Journal of the Royal Statistical Society (B)* 34, 1–42.

Lunn, D.J., Thomas, A., Best, N., and Spiegelhalter, D. (2000) WinBUGS—A Bayesian modelling framework: Concepts, structure, and extensibility. *Statistics and Computing* 10, 325–337.

Margolese, M.S. (1970). Homosexuality: A new endocrine correlate. *Hormones and Behavior* 1, 151–155.

Marshall, E.C. and Spiegelhalter, D.J. (2007). Identifying outliers in Bayesian hierarchical models: A simulation-based approach. *Bayesian Analysis* 2(2), 409–444.

McCullagh, P. and Nelder, J. (1989). *Generalized Linear Models*. 2nd Edn. Chapman & Hall, London, U.K.

McCulloch, C.E. and Searle, S.R. (2001). *Generalized, Linear and Mixed Models*. John Wiley & Sons, Inc., New York.

McDonald, G. and Schwing, R.C. (1973). Instabilities of regression estimates relating to air pollution and mortality. *Technometrics* 15, 463–481.

Meyer, R.D. and Wilkinson, R.G. (1998). Bayesian variable assessment. *Communications in Statistics—Theory and Methods* 27(11), 2675–2705.

Miller, A.E., Thompson, W.L., Mortenson, D.C., and Moore, C. (2010). Protocol for ground-based monitoring of vegetation in the Southwest Alaska Network. Natural Resource Report NPS/SWAN/NRR–2010/205. National Park Service, Fort Collins, CO. Available from the Natural Resource Publications Management website: http://www.nature.nps.gov/publications/NRPM

Mood, A.M., Graybill, F.A., and Boes, D.C. (1974). *Introduction to the Theory of Statistics*. 3rd Edn. McGraw-Hill, Inc., New York.

Myers, R., Montgomery, D., and Vining, G. (2002). *Generalized Linear Models: With Applications in Engineering and Sciences*. John Wiley & Sons, Inc., New York.

Neal, R. (1998). Suppressing random walks in Markov chain Monte Carlo using ordered over-relaxation, in *Learning in Graphical Models* (ed. M.I. Jordan). Kluwer Academic Publishers, Dordrecht, the Netherlands, pp. 205–230.

Nelder, J.A. (1977). A reformulation of linear models (with discussion). *Journal of the Royal Statistical Society (A)* 140, 48–77.

Nelder, J. and Wedderburn, R. (1972). Generalized linear models. *Journal of the Royal Statistical Society, Series A* 132, 370–384.

Ntzoufras, I. (2009). *Bayesian Modeling Using WinBUGS*. John Wiley & Sons, Inc., Hoboken, NJ.

O'Hagan, A., Buck, C.E., Daneshkhah, A., Eiser, J.R., Garthwaite, P.H., Jenkinson, D.J., Oakley, J.E., and Rakow, T. (2006). *Uncertain Judgements. Eliciting Experts' Probabilities*. John Wiley & Sons Ltd., New York.

O'Hara, R.B. and Sillanpaa, M.J. (2009). A review of Bayesian variable selection methods: What, how and which. *Bayesian Analysis* 4(1), 85–118.

Plummer, M. (2010). *JAGS: Just Another Gibbs Sampler*. http://calvin.iarc.fr/~martyn/software/jags/ (accessed May 15, 2011).

Prentice, R.L. (1976). A generalization of the probit and logit methods for dose response curves. *Biometrics* 32, 761–768.

Raiffa, H. (1968). *Decision Analysis*. McGraw-Hill, Inc., New York.

Robinson, T.J., Anderson-Cook, C.M., and Hamada, M.S. (2009). Bayesian analysis of split-plot experiments with nonnormal responses for evaluating nonstandard performance criteria. *Technometrics* 51(1), 56–64.

Robinson, T.J., Myers, R.H., and Montgomery, D.C. (2004). Analysis considerations in industrial split-plot experiments with nonnormal responses. *Journal of Quality Technology* 36, 180–192.

Rosenbaum, P.R. (2009). *Design of Observational Studies*. Springer, New York.

Senn, S. (2002). *Cross-Over Trials in Clinical Research*. 2nd Edn. John Wiley & Sons, Ltd., New York.

Senn, S. (2007). *Statistical Issues in Drug Development*. 2nd Edn. John Wiley & Sons, Ltd., New York.

Shafer, J. (1982). Lindley's paradox (with discussion). *Journal of the American Statistical Association* 77, 325–334.

Short, T. (2004). R reference card. http://cran.r-project.org/doc/contrib/Short-refcard.pdf (accessed May 15, 2011).

Smith, A.F.M. (1986). Some Bayesian thoughts on modelling and model choice. *The Statistician* 35, 97–102.

Smith, T.C., Spiegelhalter, D.J., and Thomas, A. (1995). Bayesian approaches to random-effects meta-analysis: A comparative study. *Statistics in Medicine* 14, 2685–2699.

Spiegelhalter, D., Abrams, K., and Myles, J. (2004). *Bayesian Approaches to Clinical Trials and Health-Care Evaluation*. John Wiley & Sons, New York.

Spiegelhalter, D., Best, N., Carlin, B., and van der Linde, A. (2002). Bayesian measures of model complexity and fit (with discussion). *Journal of the Royal Statistical Society, Series B* 64, 583–640.

Spiegelhalter, D., Thomas, A., Best, N., and Gilks, W. (1996). *BUGS 0.5 Manual (Version ii)*. MRC Biostatistics Unit, Cambridge, U.K.

Spiegelhalter, D., Thomas, A., Best, N., and Lunn, D. (2003). *WinBUGS User Manual, Version 1.4*. MRC Biostatistics Unit, Cambridge, U.K.

Thall, P.F. and Vail, S.C. (1990). Some covariance models for longitudinal count data with overdispersion. *Biometrics* 46, 657–671.

Tierney, L. (1994). Markov chains for exploring posterior distributions. *Annals of Statistics* 22, 1701–1762.

Venables, W.N., Smith, D.M., and the R Development Core Team. (2010). An introduction to R (v2.11.1). http://cran.r-project.org/doc/manuals/R-intro.pdf (accessed May 15, 2011).

Verzani, J. (2005). *Using R for Introductory Statistics*. Chapman & Hall/CRC, Boca Raton, FL.

Walkenbach, J. (2007). *Excel 2007 Charts*. Wiley Publishing Inc., New York.

Wolstenholme, L.C. (1999). *Reliability Modelling: A Statistical Approach*. Chapman & Hall/CRC, Boca Raton, FL.

Woodward, P. (2005). BugsXLA. Bayes for the common man. *Journal of Statistical Software* 14, 5. http://www.jstatsoft.org/v14/i05

Woodward, P. and Walley, R. (2009). Bayesian variable selection for fractional factorial experiments with multilevel categorical factors. *Journal of Quality Technology* 41(3), 228 240.

Wu, C.F.J. and Hamada, M. (2000). *Experiments. Planning, Analysis and Parameter Design Optimization*. John Wiley & Sons, Inc., New York.

Zeger, S.L. and Qaqish, B. (1988). Markov regression models for time series: A quasi-likelihood approach. *Biometrics* 44, 1019–1031.

Index

Printed and bound by CPI Group (UK) Ltd, Croydon, CR0 4YY

24/10/2024

01778278-0016